Advances in Mycorrhizal Science and Technology

Damase Khasa
Centre d'étude de la forêt
Université Laval
Québec, QC G1V 0A6

Yves Piché
Centre d'étude de la forêt
Université Laval
Québec, QC G1V 0A6

Andrew P. Coughlan
Centre d'étude de la forêt
Université Laval
Québec, QC G1V 0A6

NRC·CNRC
NRC Research Press
Ottawa 2009

© 2009 National Research Council of Canada
All rights reserved. No part of this publication may be reproduced in a retrieval system, or transmitted by any means, electronic, mechanical, photocopying, recording or otherwise, without the prior written permission of the National Research Council of Canada, Ottawa, ON K1A 0R6, Canada. Printed on acid-free paper. ∞

Published exclusively in Canada and non-exclusively in other territories of the world (excluding Europe, Asia, the Middle East, Africa, and South America) by NRC Research Press as ISBN 978–0–660–19883–5. NRC No. 46861.

Published exclusively in Europe, Asia, the Middle East, Africa, and South America and non-exclusively in other territories of the world (excluding Canada) by CABI Publishing, a Division of CAB International, with the ISBN 978–1–84593–586–3.

CABI is a trading name of CAB International.
CABI Head Office, Nosworthy Way, Wallingford, Oxfordshire, 0X10 8DE, UK
Tel: +44 (0)1491 832111   Fax: +44 (0)1491 833508
E-mail: cabi@cabi.org   Web site: www.cabi.org

USA, the Caribbean, Central America (including Mexico), Australia and Oceania remain open for either party to supply with copies of the work.

Library and Archives Canada cataloguing in publication data

Khasa, Damase; Piché, Yves; Coughlan, Andrew P.

Advances in mycorrhizal science and technology

Includes bibliographical references and an index.
Issued by the National Research Council of Canada.
Co-published by CABI Publishing.
ISBN 978–0–660–19883–5

1. Mycorrhizas. 2. Mycorrhizal fungi. 3. Mycorrhizal plants.
4. Mycorrhizas in agriculture. 5. Symbiosis.
I. Khasa, Phambu, 1958– II. Piché, Yves, 1950– III. Coughlan, Andrew P., 1961– IV. National Research Council Canada V. CABI Publishing.

QK604.2.M92 A38 2009   579.5'1785   C2009-980042-X

NRC Monograph Publishing Program

Editor: P.B. Cavers (University of Western Ontario)

Editorial Board: W.G.E. Caldwell, OC, FRSC (University of Western Ontario); M.E. Cannon, FCAE, FRSC (University of Calgary); K.G. Davey, OC, FRSC (York University); M.M. Ferguson (University of Guelph); S. Gubins (Annual Reviews); B.K. Hall, FRSC (Dalhousie University); P. Hicklenton Ph.D, P.Ag.; W.H. Lewis (Washington University); A.W. May, OC (Memorial University of Newfoundland); B.P. Dancik, Editor-in-Chief, NRC Research Press (University of Alberta)

Inquiries: Monograph Publishing Program, NRC Research Press, National Research Council of Canada, Ottawa, ON K1A 0R6, Canada. Web site: pubs.nrc-cnrc.gc.ca

Dedication

This monograph is dedicated to Yolande Dalpé and J. André Fortin, both of whom have devoted their scientific careers to the study of mycorrhizal symbioses in Canada and abroad.

Dr. Yolande Dalpé

Yolande Dalpé was born in Waterloo (Québec, Canada) in 1948. She received her Doctorat-es-Sciences (D.Sc.; Physiology of Fungi) from Université Paul-Sabatier (Toulouse, France) in 1981. In the same year, she was appointed as a research scientist at Agriculture and Agri-Food Canada (Ottawa, Ontario), where she is currently program leader of the mycology section. Yolande has trained numerous scientists and graduate students in the identification and taxonomy of arbuscular mycorrhizal fungi and has published over 130 scientific papers. She is an associate editor of *Botany* and was a committee member of the Natural Sciences and Engineering Research Council of Canada (NSERC). Yolande is currently responsible for the development and management of the Glomeromycota in vitro collection (GINCO) and was a founding member of the International Mycorrhiza Society.

Professor J. André Fortin

J. André Fortin was born in Québec (Québec, Canada) in 1937. He received his Ph.D. in Forest Biology from Université Laval in 1966. In the same year he joined Université Laval's Faculté de foresterie et de géomatique, where he was appointed professor in 1976. André has trained nearly 60 graduate students, has published over 175 scientific papers and technical reports, and has co-edited two books on mycorrhizas. In 1983, he was elected Fellow of the Royal Society of Canada. André went on to found the Centre de recherche en biologie forestière (CRBF) at Université Laval in 1984 and the Institut de recherche en biologie végétale (IRBV) at the Université de Montréal in 1990. He retired from the latter in 1996.

During his career, he served as vice-president of the Canadian Botanical Association and was a member of the Board of Governors of the International Foundation for Science. He was also an advisor to the National Research Council of Canada and associate editor of *Botany*. In recent years, he founded the Association pour la commercialisation des champignons forestiers and, like Yolande, was a founding member of the International Mycorrhiza Society.

Contents

Dedication iii

Preface vii

List of Contributors ix

Overview xi

Chapter 1 1

Mycorrhizas in Canadian forest and agricultural ecosystems

Shannon M. Berch, Marcia A. Monreal, and Gavin Kernaghan

Chapter 2 15

From a germinating spore to an established arbuscular mycorrhiza: signalling and regulation

José M. Garcia-Garrido and Horst Vierheilig

Chapter 3 39

Growth and branching of asymbiotic, presymbiotic, and extraradical AM fungal hyphae: clarification of concepts and terminology

Christine Juge, Andrew P. Coughlan, J. André Fortin, and Yves Piché

Chapter 4 51

Interactions between arbuscular mycorrhizal fungi and soil microorganisms

Laëtitia Lioussanne, Marie-Soleil Beauregard, Chantal Hamel, Mario Jolicoeur, and Marc St-Arnaud

Chapter 5 71

Arbuscular mycorrhiza: where nature and industry meet

Tandra Fraser, Atul Nayyar, Walid Ellouze, Juan Carlos Perez, Keith Hanson, Jim Germida, Zadok Bouzid, and Chantal Hamel

Chapter 6 87

The relative field mycorrhizal dependency concept and its usefulness in agronomy

Christian Plenchette and J. André Fortin

Chapter 7 93

Extraction, propagation, and conservation of arbuscular mycorrhizal fungi

Yolande Dalpé

Chapter 8 105

Industrial perspective of applied mycorrhizal research in Canada

Susan Parent and Peter Moutoglis

Chapter 9 115

Mycorrhizal fungi in Canadian forest nurseries and field performance of inoculated seedlings

Ali M. Quoreshi, Gavin Kernaghan, and Gary A. Hunt

Chapter 10 129

Ectomycorrhizal inoculation for boreal forest ecosystem restoration following oil sand extraction: the need for an initial three-step screening process

Grégory Bois and Andrew P. Coughlan

Chapter 11 139

Technological transfer: the use of ectomycorrhizal fungi in conventional and modern forest tree nurseries in North Africa

Mohammed S. Lamhamedi, Mohammed Abourouh, and J. André Fortin

Chapter 12 153

Ectomycorrhiza in the neotropics with emphasis on lowland forests

Bradley R. Kropp

Chapter 13 161

Ecophysiology of sporocarp development of ectomycorrhizal basidiomycetes associated with boreal forest gymnosperms

J. André Fortin and Mohammed S. Lamhamedi

Chapter 14 175

Ecology and management of edible ectomycorrhizal mushrooms in eastern Canada

Marie-France Gévry and Normand Villeneuve

Index 193

Preface

Mycorrhizal symbioses are widespread and fundamental to terrestrial ecosystems; they ensure the completion of the life cycles of the plants and fungi involved. This factor alone justifies continued research to improve our understanding of their functional biology and ecology. Furthermore, mycorrhizal symbioses have shaped plant evolution, and today they play an important role in the succession and overall health and stability of plant communities. In short, these associations offer an almost unlimited range of research possibilities. This fact is reflected in the wealth of scientific literature that has been published on the subject in recent years. These new findings have improved our understanding of these complex symbiotic associations; they have also stimulated the development and application of new technologies. As a result, mycorrhizal fungi are being more commonly used in plant production systems and restoration programs. This volume considers some of these research fields and their applications and covers topics ranging from plant–fungal communication to commercial harvesting of edible mycorrhizal forest mushrooms.

Although this volume has an undeniably Canadian bias, reflecting the fact that the editors all currently live and conduct much of their scientific work in Canada, the topics covered are of global interest. This is enhanced by the facts that ecosystems in the Northern Nearctic share many affinities with those of the Northern Palearctic and that the functional role of a given mycorrhizal symbiosis is similar regardless of its geographical location.

Canadian scientists have long been at the forefront of mycorrhizal research; while many are worthy of special mention, two from eastern Canada, Yolande Dalpé and J. André Fortin, have been particularly active and have directly and indirectly influenced several generations of scientists working in this domain. It is, therefore, with great pleasure that we dedicate this volume to them. The 14 chapters, which provide either a general review or the results of original research, have, for the most part, been written by former collaborators or students of these two scientists or by their students' students or collaborators, many of whom are now important figures in international mycorrhizal research. The diversity of the subjects covered in the different chapters reflects the wide range in personal interests of the authors.

The success of a book like this is largely dependent on those who contribute. As editors, we sincerely thank all the authors who accepted our invitation to submit manuscripts and shared our excitement about assembling this volume. In particular, we thank them for complying with deadlines and responding politely to our editorial comments. They deserve the credit for the success of their chapters and the book as a whole. The manuscripts were edited by DK, YP, and APC and returned to the authors for their approval and updates if necessary. Each chapter was subsequently peer-reviewed by two independent referees chosen by NRC Research Press. We thank the reviewers for their valuable comments, which have greatly improved this volume. The production of this book would not have been possible without the collaboration of the team at NRC Research Press and we would particularly like to thank Suzanne Kettley, Laura Stewart, and Lisa Corda for their patience and care at all stages of production and help in bringing this project to fruition.

In a collaborative work such as this, it is impossible to cover all topics and compromises have been made. As a result, some may find that certain important topics have been ignored or only briefly touched on. For this and other failings of the book, we as editors take full responsibility. However, our hope is that this book, which is for students and scientists alike, will motivate others to undertake research in this field and to help solve some of the current and future questions surrounding these essential symbiotic associations.

Damase Khasa
Yves Piché
Andrew P. Coughlan

Québec, Québec
April 2008

List of Contributors

Mohammed Abourouh
Centre de Recherche Forestière, Boîte postale 763, Agdal, Rabat, Maroc

Email: abourouhmohamed@hotmail.com

Marie-Soleil Beauregard
Institut de recherche en biologie végétale, Jardin botanique de Montréal, 4001, Sherbrooke Est, Montréal, QC H1X 2B2, Canada

Shannon M. Berch
BC Ministry of Forests and Range, Research Branch Laboratory, P.O. Box 9536, Victoria, BC V8W 9C4, Canada

Email: Shannon.Berch@gov.bc.ca

Grégory Bois
PHYTOREM® S.A, 30, avenue Charles de Gaulle, 13140 Miramas, France

Email: boisgregory@gmail.com

Zadok Bouzid
Département de Biologie, Faculté des Sciences Mathématiques, Physiques et Naturelles, Université Tunis El Manar, Campus Universitaire, Tunis 1060, Tunisia

Email: sadok.bouzid@fst.rnu.tn

Andrew P. Coughlan
Centre d'étude de la forêt, Pavillon C.-E.-Marchand, Université Laval, Québec, QC G1V 0A6, Canada

Email: andrew.coughlan@rsvs.ulaval.ca

Yolande Dalpé
Eastern Cereal and Oilseed Research Centre, Agriculture and Agri-Food Canada, 960 Carling Avenue, Ottawa, ON K1A 0C6, Canada

Email: yolande.dalpe@agr.gc.ca

Walid Ellouze
Environmental Health / Water and Nutrients, Agriculture and Agri-Food Canada, 1 Airport Road, P.O. Box 1030, Swift Current, SK S9H 3X2, Canada

Email: walid.ellouz@umontreal.ca

J. André Fortin
Centre d'étude de la forêt, Pavillon C.-E.-Marchand, Université Laval, Québec, QC G1V 0A6, Canada

Email: j.andre.fortin@videotron.ca

Tandra Fraser
Environmental Health / Water and Nutrients, Agriculture and Agri-Food Canada, 1 Airport Road, P.O. Box 1030, Swift Current, SK S9H 3X2, Canada

Email: tdf241@yahoo.com

José M. Garcia-Garrido
Departamento de Microbiología de Suelos, Estación Experimental del Zaidín, CSIC, E-18008 Granada, Spain

Email: Josemanuel.garcia@eez.csic.es

Jim Germida
Department of Soil Science, University of Saskatchewan, 51 Campus Drive, Saskatoon, SK S7N 5A8, Canada

Email: jim.germida@usask.ca

Marie-France Gévry
Chaire de recherche sur la forêt habitée, Centre d'études nordiques, Université du Québec à Rimouski, 300, allée des Ursulines, Rimouski, QC G5L 3A1, Canada

Email: mf_gevry@hotmail.com

Chantal Hamel
Environmental Health / Water and Nutrients, Agriculture and Agri-Food Canada, 1 Airport Road, P.O. Box 1030, Swift Current, SK S9H 3X2, Canada

Email: hamelc@agr.gc.ca

Keith Hanson
Environmental Health / Water and Nutrients, Agriculture and Agri-Food Canada, 1 Airport Road, P.O. Box 1030, Swift Current, SK S9H 3X2, Canada

Email: hanson@agr.gc.ca

Gary A. Hunt
Department of Natural Resource Sciences, Thompson Rivers University, 900 McGill Road, Kamloops, BC V2C 5N3, Canada

Email: gahunt@tru.ca

Mario Jolicoeur
Unité de recherche Bio-P2, Département de génie chimique, École Polytechnique de Montréal, Boîte postale 6079, succursale centre-ville, Montréal, QC H3C 3A7, Canada

Email: Mario.jolicoeur@polymtl.ca

List of Contributors

Christine Juge
Centre d'étude de la forêt, Pavillon C.-E.-Marchand, Université Laval, Québec, QC G1V 0A6, Canada

Email: christine.juge@meidje.net

Gavin Kernaghan
Biology Department, Mount Saint Vincent University, 166 Bedford Highway, Halifax, NS B3M 2J6, Canada

Email: Gavin.Kernaghan@msvu.ca

Bradley R. Kropp
Biology Department, Utah State University, Logan, UT 84322, USA

Email: brkropp@biology.usu.edu

Mohammed S. Lamhamedi
Direction de la recherche forestière, ministère des Ressources naturelles et de la Faune, 2700, rue Einstein, Québec, QC G1P 3W8, Canada

Email: Mohammed.lamhamedi@mrnf.gouv.qc.ca

Laëtitia Lioussanne
Institut de recherche en biologie végétale, Jardin botanique de Montréal, 4001, Sherbrooke Est, Montréal, QC H1X 2B2, Canada

Email: lioussal@videotron.ca

Marcia A. Monreal
Agriculture and Agri-Food Canada, Research Branch, P.O. Box 1000A, Brandon, MB R7A 5Y3, Canada

Email: monreal@agr.gc.ca

Peter Moutoglis
BioSyneterra Solutions, Carrefour industriel et expérimental de Lanaudière, 801, route 344, Boîte postale 3158, L'Assomption, QC J5W 4M9, Canada

Email: pmoutoglis@yahoo.com

Atul Nayyar
Environmental Health / Water and Nutrients, Agriculture and Agri-Food Canada, 1 Airport Road, P.O. Box 1030, Swift Current, SK S9H 3X2, Canada

Email: nayyara@agr.gc.ca

Susan Parent
Premier Tech, 1, Avenue Premier, Rivière-du-Loup, QC G5R 6C1, Canada

Email: pars@premiertech.com

Juan Carlos Perez
Environmental Health / Water and Nutrients, Agriculture and Agri-Food Canada, 1 Airport Road, P.O. Box 1030, Swift Current, SK S9H 3X2, Canada

Email: perezjc@agr.gc.ca

Yves Piché
Centre d'étude de la forêt, Pavillon C.-E.-Marchand, Université Laval, Québec, QC G1V 0A6, Canada

Email: ypiche@rsvs.ulaval.ca

Christian Plenchette
INRA, UMR BGA, 17, rue Sully, 21065 Dijon, CEDEX, France

Email: plenchet@dijon.inra.fr

Ali M. Quoreshi
Symbiotech Research Inc. Unit 201, 509 11th Avenue, Nisku, AB T9E 7N5, Canada

Email: symbiotech.ali@2020seedlabs.ca

Marc St-Arnaud
Institut de recherche en biologie végétale, Jardin botanique de Montréal, 4001, Sherbrooke Est, Montréal, QC H1X 2B2, Canada

Email: Marc.st-arnaud@umontreal.ca

Horst Vierheilig
Estación Experimental del Zaidín, CSIC, Apdo. 419, E-18008 Granada, Spain

Email: horst.vierheilig@eez.csic.es

Normand Villeneuve
Ministère des Ressources naturelles et de la Faune, Direction de l'environnement et de la protection des forêts, 880, chemin Sainte-Foy, local 6.50, Québec, QC G1S 4X4, Canada

Email: Normand.Villeneuve@mrnf.gouv.qc.ca

Overview

More than 30 dedicated experts from different fields of mycorrhizal research have contributed to this volume. The topics covered represent only a small part of the investigations carried out by Canadian research groups and scientists in Canada and overseas, work done in Canadian laboratories by scientists from overseas, and some of the national and international applications of these studies. From the outset, our intention was that each chapter be a discrete and complete entity and that the different chapters be complementary. While we have strived to keep overlap to a strict minimum, a degree of similarity remains among contributions. We feel that this is important for the comprehension of the respective chapters and thus adds overall strength to the book. For ease of use, we have chosen to provide a reference list at the end of each chapter. The majority of the work on mycorrhizal fungi conducted in Canada has focused on arbuscular mycorrhizal (AM) and ectomycorrhizal associations. Moreover, many of these studies have been conducted in a strictly agricultural or forestry context. Both of these aspects are reflected in the present volume, which contains a balance of shorter technical chapters interspersed with longer review chapters.

In Chapter 1, Berch, Monreal, and Kernaghan set the stage for the reader by providing a general introduction to the different mycorrhizal associations, all seven of which occur in Canada. The taxonomic and functional diversity of these symbioses are outlined and, where pertinent, particular attention is given to their roles in resource-producing ecosystems. Moreover, the authors demonstrate how the contributions made by Canadian scientists, Canadian research groups, and international scientists working in Canada fit into the global mycorrhizal research picture.

All mycorrhizal associations originate from the successful colonization of a host root by a mycorrhizal fungal hypha issued from a germinating spore or an already existent mycelium within the soil matrix. Root colonization is the result of complex communication between the fungus and the host. In Chapter 2, Garcia-Garrido and Vierheilig provide a cutting-edge account of the signalling and regulation events involved in the establishment and maintenance of arbuscular mycorrhizas. Comparisons between AM fungal systems and N-fixing leguminous systems are highlighted. This chapter is essential to anyone interested in the mechanism underlying the formation of this ecologically widespread and agriculturally important symbiosis, especially in light of the rapid advances being made in the field of molecular biology.

Understanding and correctly describing the different hyphal growth patterns of AM fungi is extremely important, particularly when investigating the effects of different edaphic parameters on the fungal growth response. In Chapter 3, Juge, Coughlan, Fortin, and Piché consider the three extraradical phases of AM fungal growth: the asymbiotic germ tube, the presymbiotic hyphae, and the symbiotic extraradical mycelium. The authors provide a detailed historical analysis of the different terms that have been used to describe hyphal development in these different phases and propose the adoption of a coherent terminology. Additionally, they provide a detailed conceptual drawing that unites these phases. By clarifying and standardizing the terms used, the authors aim to facilitate the work of future researchers.

Mycorrhizal fungi often form a large proportion of the soil microbial biomass, but they are not alone within the soil matrix. Other inhabitants of this underground ecosystem include a wide range of beneficial and harmful microorganisms, which can affect mycorrhizal fungi and their plant hosts. In Chapter 4, Lioussanne, Beauregard, Hamel, Jolicoeur, and St-Arnaud provide a thorough, thoughtful, and up-to-date review of the interactions of AM fungi with other soil-borne microorganisms. This chapter will become more relevant, especially in agriculture, as more environmentally friendly methods of plant production are sought.

In Chapter 5, Fraser, Nayyar, Ellouze, Perez, Hanson, Germida, Bouzid, and Hamel provide a highly informative summary of some of the different roles of AM fungal associations in nature. Included among these are the effect of AM fungi on plant nutrient uptake, plant water status, and soil aggregation; interplant linkages and spatio-temporal variation in fungal species are also discussed. The authors then consider the AM fungal association in an agricultural context, examining how some agricultural practices (e.g., rotation, fertilizer, tillage, inoculants, plant breeding, and pesticides) influence the formation of arbuscular mycorrhizas.

Overview

It is widely accepted that plants with coarse root systems are not as efficient, in terms of nutrient uptake, as plants with finer root systems with root hairs. In theory, the former are more dependent on mycorrhizal fungal associates to carve out their niches. In Chapter 6, Plenchette and Fortin revisit this concept of mycorrhizal dependency (MD; i.e., how much a plant depends on mycorrhizal fungi to acquire nutrients and grow) and link it broadly to agricultural practices. The authors highlight the fact that agricultural breeding programs may inadvertently select for cultivars with reduced MD and increased nutrient demand and therefore move agriculture away from more sustainable practices. The authors explore these concepts and propose an improved MD index, the $RFMD^P$, a relative field measurement of MD at a given P level.

In order to conduct research on AM fungi and to develop inocula for use in plant production systems, we need to be able to isolate the mycobionts involved and grow them under controlled conditions. In Chapter 7, Dalpé describes the methods currently used to extract AM fungal propagules from soil and plant roots. The author then provides valuable information on the propagation and preservation of AM fungi in pot cultures and under in vitro conditions. The author considers laboratory- and field-based approaches, giving a thorough and thoughtful review of the pros and cons of the approaches that have been tested and (or) used worldwide. This chapter provides an essential overview for anyone interested in scientific research and technological development involving these critically important fungi.

Recent research has shown that plants in diverse production systems (e.g., agriculture and forestry) can benefit from inoculation with mycorrhizal fungi. The benefits, which include enhanced plant survival, growth, and development, have encouraged research into the development of commercial inocula. In Chapter 8, Parent and Moutoglis provide an insight into the historical development of mycorrhizal inoculum technology in Canada (with some references to international work). This chapter is written from an industrial perspective and provides information about the different mycorrhizal inocula currently available.

The successful establishment and subsequent growth of outplanted nursery-grown tree seedlings is inextricably linked to their ability to rapidly access nutrients and water. This may be enhanced by precolonization with efficient mycorrhizal fungi in the nursery. In Chapter 9, Quoreshi, Kernaghan, and Hunt provide a recent and informative overview of the use of ectomycorrhizal fungal inocula to colonize tree seedlings under nursery conditions and the subsequent performance of the seedlings following outplanting.

Mycorrhizal fungi can be critical in the successful restoration of certain disturbed and (or) polluted sites. In Chapter 10, Bois and Coughlan examine one particularly difficult case, the revegetation of highly saline tailing sands produced by the oil sands industry in northern Alberta. This is a relatively new field of research and the combined experience of the authors is used to outline a possible three-step approach to follow in order to attempt the revegetation of highly stressful substrates with ectomycorrhizal tree seedlings.

Because mycorrhizal fungi increase the active surface of the root system and improve plant water status, they can enhance tree growth in arid and semiarid environments. Therefore, they may have a vital role to play in the important quest to reduce global desertification. However, inoculation and plant production can be difficult in developing countries. In Chapter 11, Lamhamedi, Abourouh, and Fortin provide an interesting and up-to-date account of the challenges of producing conifer and hardwood seedlings of consistent "quality" in North Africa.

Tropical forests are predominantly composed of AM tree species; however, work done in Africa has shown the existence of distinct guilds of ectomycorrhizal tree species. On the other side of the Atlantic, information concerning ectomycorrhizal associations in neotropical forests is scanty at best. In Chapter 12, Kropp presents a review of a selection of interesting topics related to ectomycorrhizal fungal and plant species in this ecozone.

Ectomycorrhizal fungi typically produce sporocarps to ensure successful spore dispersal. In Chapter 13, Fortin and Lamhamedi provide a review of the investigations that have been conducted into the physiological mechanisms underlying initiation and development of sporocarps of ectomycorrhizal basidiomycetes when in association with members of the Pinaceae. Unpublished observations are combined with published ones, and the authors provide a critical analysis of what has been done and the direction that future research on this topic should take. The analysis and hypotheses put forward provide much food for thought.

Finally, the sporocarps of many ectomycorrhizal fungi are edible and those of a good number of species are highly sought after (e.g., chanterelles, cepes, matsutaki, and truffles). In Chapter 14, Gévry and Villeneuve provide a very useful integration and review on the topic of wild edible mushrooms in eastern Canada. This chapter provides a worldwide overview and a regional focus from an eastern Canadian (mostly Québec) perspective. Some of the information used to illustrate this chapter comes from recently conducted and as yet unpublished fieldwork, making it an important addition to the field of nontimber forest products. The impact of wild mushroom harvesting on the landscape is discussed.

Chapter 1

Mycorrhizas in Canadian forest and agricultural ecosystems

Shannon M. Berch, Marcia A. Monreal, and Gavin Kernaghan

Introduction

Canada is large (979.1 × 10^6 ha of land) and diverse (consisting of 15 major terrestrial ecozones: Arctic Cordillera, Northern Arctic, Southern Arctic, Taiga Cordillera, Taiga Plains, Taiga Shield, Hudson Plains, Boreal Cordillera, Boreal Plains, Boreal Shield, Prairie, Montane Cordillera, Pacific Maritime, Atlantic Maritime, and Mixedwood Plains; Canadian Council on Ecological Areas; www.ec.gc.ca/soer-ree/English/vignettes/default.cfm), and forestry and agriculture are two of the country's major economic forces. Based on Natural Resources Canada estimates (canadaforests.nrcan.gc.ca/statsprofile/forest/ca), Canada has 294.8 × 10^6 ha that are potentially available for commercial forest activities, 0.9 × 10^6 ha of which are harvested annually. In 2006, forest products contributed $33.2 billion to the Canadian gross domestic product (GDP) (Natural Resources Canada; canadaforests.nrcan.gc.ca/statsprofile/forest/ca). In 2006, there were 68 × 10^6 ha of farmland in use across Canada, 53% of which was in crops (Kittson et al. 2007). In 2006, agriculture contributed $14 billion to the GDP of Canada; this figure can be increased to $86 billion if value-added production (e.g., food processing, sales, and food services) is included (Kittson et al. 2007).

For many decades, the importance of mycorrhizal fungi to terrestrial ecosystems has been recognized, and their potential use in forestry and agriculture has been explored. Given the long history of Canadian involvement in mycorrhizal research and the importance of mycorrhizas in our diverse resource-producing ecosystems, we address the taxonomic and functional diversity of mycorrhizas in Canadian forest and agriculture ecosystems in this chapter and emphasize contributions made by Canadian researchers (citations given in boldface).

Arbuscular mycorrhizas

Arbuscular mycorrhizas are the most common "endotrophic" mycorrhiza. In these associations, hyphae of the fungal symbionts form appressoria and penetrate the root epidermal cells to reach the cortical cells of the roots. Within the cortex, the fungi produce intraradical hyphae and intracellular interfaces called arbuscules. In addition, some fungal species form enlarged lipid-filled hyphae, called vesicles. Extraradical hyphae and spores are produced at the root surface and in the rhizosphere.

Phylogenetic analysis using DNA sequences of genes that code for the small subunit of ribosomal RNA estimates that arbuscular mycorrhizal fungi originated between 353 and 462 million years ago in the Palaeozoic era (**Simon et al. 1993**). Benefits to plant hosts are attributed to improved water and nutrient uptake, in particular of those nutrients that are less mobile in soil (e.g., P, Zn, and Cu). Arbuscular mycorrhizal (AM) fungi can also serve as nutrient bridges between photosynthetic and certain nonphotosynthetic plants (Bidartondo et al. 2002) and influence plant biodiversity (van der Heijdden et al. 1998).

S.M. Berch.[1] BC Ministry of Forests and Range, Research Branch Laboratory, P.O. Box 9536, Victoria, BC V8W 9C4, Canada.
M.A. Monreal. Agriculture and Agri-Food Canada, Research Branch, P.O. Box 1000A, Brandon, MB R7A 5Y3, Canada.
G. Kernaghan. Biology Department, Mount Saint Vincent University, 166 Bedford Highway, Halifax, NS B3M 2J6, Canada.
[1]Corresponding author: (e-mail: Shannon.Berch@gov.bc.ca).

Taxonomic diversity

Approximately 150 AM fungal species have been described using spore morphology and characteristics associated with spore-bearing structures (Schüßler et al. 2001). **Simon et al. (1993)** published the first small subunit ribosomal DNA sequence-based phylogeny of the AM fungi, prompting a comprehensive phylogenetic reevaluation of these fungi. This has recently culminated in them being placed in the new phylum Glomeromycota, which consists of four orders (i.e., Archaeoporales, Paraglomerales, Diversiporales, and Glomerales) and nine clades that appear to resolve at the family level (Schüßler et al. 2001).

Arbuscular mycorrhizal fungi are ubiquitous and promiscuous: a single fungal species may form mycorrhizas with many plant hosts in many ecosystems. Approximately 80% of vascular plant species (representing approximately 1000 genera) of plants have been reported to form AM associations (Harley 1989). These include most agriculturally important crops (e.g., wheat, *Triticum aestivum* L.; flax, *Linum usitassimum* L.; barley, *Hordeum vulgare* L.; maize, *Zea mays* L.; sorghum, *Sorghum bicolor* (L.) Moench; sunflower, *Helianthus annuus* L.; soybeans, *Glycine max* (L.). Merr.; potato, *Solanum tuberosum* L.; grasses and forage crops, and fruit trees; see Chapter 5), herbs and tropical trees (**Kendrick 1992**). Horticultural plants such as roses, petunias, lilies, and tulips also form arbuscular mycorrhizas (**Peterson et al. 2004**), as do some forestry species such as western redcedar (*Thuja plicata* Donn ex D. Don; **Berch et al. 1991**) and sugar maple (*Acer saccharum* Marsh.; **Brundrett and Kendrick 1988a**).

There are two common types of AM associations formed in the root cortex: the *Arum-* and *Paris*-type (**Peterson et al. 2004**). In both types, intracellular structures are separated from the plant cell cytoplasm by a host-derived membrane and an interfacial matrix (Bonfante and Perotto 1995). This matrix is the main site for carbon transfer from the plant to the fungus and mineral nutrient transfer from the fungus to the plant (Bonfante and Perotto 1995).

Functional diversity

Soil properties, such as pH, can affect AM functional diversity by inducing changes in root colonization and species diversity. Sugar maple seedlings were studied under greenhouse conditions using forest soils from declining and healthy maple stands (**Coughlan et al. 2000**). Seedlings grown in the more acidic soil from the declining stand produced more AM fungal spores, but species diversity was lower. However, after liming the acidic soil, spores of several additional AM fungal taxa were found. Liming also increased root colonization in soil from both stand types, which may have improved host nutrient and water uptake.

Pesticides applied to soils can affect AM fungi. For example, spore and sporocarp abundance of certain species were reduced following treatments with simazine; however, the number of propagules of *Gigaspora calospora* (Nicol. & Gerd.) Gerd. & Trappe increased when simazine was applied at the rate of 3 kg·ha^{-1} (**Granger et al. 1995**).

Intraradical hyphae can persist in decaying roots. In natural and agricultural systems, these hyphae serve as an important source of inoculum for the colonization of roots of newly germinated seedlings. From an agricultural perspective, if a mycorrhizal crop follows a nonmycorrhizal one, the amount of soil inoculum can be reduced, delaying root colonization and reducing yield (see also Chapter 5). Studies have shown that mycorrhizal development of maize and flax were delayed when the previous crop was nonmycorrhizal canola (*Brassica napus* L.) (Gavito and Miller 1998). Furthermore, the capacity of AM fungal hyphae to withstand freezing and drying, surviving until soil conditions are appropriate for regrowth, is an important factor to consider when crops are direct seeded (zero or minimum tillage) after the soil has been frozen for several months during winter (**Addy et al. 1997; Miller et al. 1995**). Crops that form AM associations can also be positively affected by the increased inoculum potential provided by a previous mycorrhizal crop (**Dalpé and Monreal 2004**). Besides tillage operations, other agricultural practices, such as fertilizer applications, crop rotation, and liming, can affect AM potential and colonization levels. In a two-year study of turf grasses, creeping bentgrass (*Agrostis solonifera* L.) and Kentucky bluegrass (*Poa pratensis* L.) showed up to 60% higher AM colonization at a site with low P content (**Podeszfinski et al. 2002**).

In vitro culture of AM fungi on root-organ cultures has been used since the mid-1980s and allows certain aspects of AM fungal functional diversity to be studied (**Fortin et al. 2002**). Monoxenic cultures of several AM fungal species and strains are now commercially available (e.g., Glomeromycetes in vitro collection (GINCO); **Declerck and Dalpé 2001**), providing a tool for research into basic comparative analyses of root populations and strain potential, long-term propagation capabilities, and fungal adaptation (**Dalpé et al. 2005**).

Up to 20% of photosynthetically fixed carbon is transferred from the plant to AM fungi (Jakobsen and Rosendahl 1990; Leake et al. 2004). Use of the root-organ culture technique coupled with ^{13}C nuclear magnetic resonance spectroscopy allowed Bago et al. (2000) to determine that carbon was transferred from the plant to the fungus as glucose, glycogen, and trehalose. Nutrient uptake (e.g., P, Zn, S, and N) and translocation by AM fungi depends on uptake by extraradical hyphae, translocation from uptake location, and transfer to the plant cell by intraradical hyphae. Phosphorus transport studies using ^{32}P in a bicompartmental Petri dish showed that P translocation was also dependent on the fungal symbiont, being more efficient in runner hyphae (see Chapter 3 for definition) of *Glomus intraradices* Schenck & Smith than in *Glomus proliferum* Dalpé & Declerck (Rufyikiri et al. 2004).

Arbuscular mycorrhizas in Canadian forests

Brundrett and Kendrick (**1988***a*) found that 77% of the species examined in a sugar maple dominated hardwood forest in southern Ontario formed arbuscular mycorrhizas and compared this with related findings from other parts of Canada (e.g., **Malloch and Malloch 1981**; **Berch and Kendrick 1982**; **Girard 1985**) and elsewhere.

Taxonomic diversity

Although most forest plant species in Canada form arbuscular mycorrhizas (e.g., **Berch and Kendrick 1982**; **Brundrett and Kendrick 1988***b*), little is known concerning the diversity and distribution of AM fungi in Canadian forests. Nevertheless, *Acaulospora cavernata* Blaszk., *Glomus aggregatum* Schenck & Smith emend. Koske, *Glomus borealis* (Thaxt.) Trappe & Gerd., *Glomus clarum* Nicol. & Schenck, *Glomus constrictum* Trappe, *Glomus geosporum* (Nicol. & Gerd.) Walker, *Glomus hoi* Berch & Trappe, *Glomus macrocarpum* Tul. & Tul., *Glomus melanosporum* Gerd. & Trappe, *Glomus microaggregatum* Koske, Gemma & Olexia, *Glomus mosseae* (Nicol. & Gerd.) Gerd. & Trappe, *Glomus radiatum* (Thaxt.) Trappe & Gerd., *Glomus vesiculifer* (Thaxt.) Gerd. & Trappe, *Sclerocystis rubiformis* Gerd. & Trappe, and *Scutellospora calospora* (Nicol. & Gerd.) Walker & Sanders have been reported from forests in Ontario and Québec (**Thaxter 1922**; **Berch and Fortin 1984**; **Moutoglis and Widden 1996**; **Coughlan et al. 2000**), and a variety of largely unidentified *Glomus* spp. were collected from the rhizosphere of native ferns in southern Ontario (**Berch 1977**).

Functional diversity

In some British Columbia forests, western redcedar grows well in soils where P occurs mainly in organic forms, and it shows less response to P fertilization than other conifers. **Cade-Menun and Berch** (**1997**) determined that redcedar seedlings and their associate AM fungi grew well with high rates of glycerophosphate, ATP, and pyrophosphate as the P source. Low soil AM inoculum potential may play a role in excluding the arbuscular mycorrhiza-forming western redcedar from coniferous forest stands dominated by ectomycorrhizal-forming western hemlock (*Tsuga heterophylla* (Raf.) Sarg) on northern Vancouver Island (**Weber et al. 2005**).

Ericoid mycorrhizas

Plants in the Ericaceae, Epacridaceae, and Empetraceae (Ericales) form ericoid mycorrhizas (**Peterson et al. 2004**). Although the unique association between the fine "hair" roots of ericaceous plants and fungi was recognized early in the twentieth century (see Ternetz (1907) and Christoph (1921) in Bain (1937)), isolates remained unidentified until a perfect stage was formed and named (Pearson and Read 1973; Read 1974), anamorphs were recognized (**Couture et al. 1983**; **Dalpé 1989, 1991***a*; **Dalpé et al. 1989**; **Egger and Sigler 1993**), and DNA sequencing became routine (**Monreal et al. 1999**; **Berch et al. 2002**).

Taxonomic diversity

Molecular analysis of internal transcribed spacer (ITS2) sequences has revealed six or more species or species groups that form ericoid mycorrhizas worldwide (**Berch et al. 2002**), including *Rhizoscyphus ericae* (Read) Zhuang & Korf. (Helotiaceae and Leotiales) and *Oidiodendron* spp. (teleomorphs in *Myxotrichum, Byssoascus,* Myxotrichaceae, and Onygenales). *Rhizoscyphus ericae* was first described as *Pezizella ericae* (Read 1974), transferred to *Hymenoscyphus* (Kernan and Finocchio 1983) and then to the new genus *Rhizoscyphus* (Zhang and Zhuang 2004). *Rhizoscyphus ericae* isolated from ericoid mycorrhizas and grown in pure culture tends to express the arthroconidial anamorphic form, *Scytalidium* (**Egger and Sigler 1993**).

The comparison of ITS2 sequences of ericoid mycorrhizal fungal isolates from Australia (Chambers et al. 2000), the United Kingdom (Read 1974; Sharples et al. 2000), Italy (Bergero et al. 2000), the United States (**Egger and Sigler 1993**), and Canada (**Xiao and Berch 1996**; **Hambleton et al. 1998**; **Monreal et al. 1999**), showed that, regardless of geographic location, very few fungal species are involved in this association (**Berch et al. 2002**) and that their host range is broad.

Functional diversity

Ericaceous plants and their mycorrhizal fungi are now viewed as playing an important role in certain ecosystems. These fungi are capable of driving nutritional processes through their ability to efficiently access N from complex organic molecules present in the upper, organic-rich soil horizons (i.e., F and H), thereby affecting the accumulation and long-term storage of C (Read et al. 2004). *Rhizoscyphus* and *Oidiodendron* isolates produce extracellular enzymes and are able to break down organic material and N- and P-containing polymers (Read et al. 2004). Furthermore, mycorrhizal plants are able to grow on complex N sources, such as proteins (Bajwa et al. 1985; **Xiao and Berch 1999**), tannin-complexed protein (Bending and Read 1996), chitin (Leake and Read 1990), and *Vaccinium* litter (Kerley and Read 1998), whereas nonmycorrhizal plants are not.

Cultivated ericaceous crops such as cranberry and blueberry (Bain 1937; **Duclos and Fortin 1983**; **Litten et al. 1992**; **Stevens et al. 1996**; Scagel 2005) form ericoid mycorrhiza, as do wild-harvested ericaceous crops such as salal (*Gaultheria shallon* Pursh, harvested for floral greenery and berries) and lowbush blueberry (*Vaccinium angustifolium* Ait.) (**Couture et al. 1983**; **Dalpé 1986, 1991b**; **Xiao and Berch 1996, 1999**; **Hambleton and Currah 1997**).

Orchid mycorrhizas

The fungi that form orchid mycorrhizas play an important role in orchid seed germination and seedling establishment in nature (Rasmussen 1995; **Peterson et al. 1998, 2004**). Following germination, orchid mycorrhizal (OM) fungi supply C to an achlorophyllous orchid protocorm (**Peterson and Farquhar 1994**); there is no clear evidence that the fungus benefits from this relationship, and it has been suggested that orchid protocorms may be parasitic on these fungi (**Currah and Sherburne 1992**). Recently, it has been shown that photosynthate can pass from adult *Goodyera repens* Br. plants to the OM fungus and that nutrients pass from the fungus to the plant (Cameron et al. 2006, 2007).

Taxonomic diversity

There are over 17 000 orchid species divided between approximately 450 genera (**Peterson et al. 2004**). Although orchids occur throughout the world, there is a particularly high diversity in tropical ecosystems. Many species have extravagant floral forms and are important to the floricultural industry, and the seedpods of orchids in the genus *Vanilla* are important in the food industry (**Peterson et al. 2004**). All orchids form mycorrhizas; however, these can be divided into two very different types: mycoheterotrophic and autotrophic.

Orchids grow in diverse environments, with some species living belowground until flowering (Dixon et al. 2003) and never forming chlorophyll; these achlorophyllous species are termed mycoheterotrophic. Nevertheless, the majority of orchids are chlorophyllous and autotrophic once past the protocorm stage. Cultures of OM fungi isolated from cortical cells of colonized roots have been used to describe the fungal symbionts. Many of the fungal isolates are nonsporulating basidiomycetes; however, molecular techniques have allowed their identification. Autotrophic orchids form mycorrhizas with a variety of basidiomycetes including species of *Ceratorhiza* (sexual stages or teleomorphs in *Ceratobasidium*), *Epulorhiza* (teleomorphs in *Tulasnella* and *Sebacina*), and *Monoliopsis* (teleomorphs in *Thanatephorus* and *Waitea*) (**Currah et al. 1987, 1997**; **Currah and Sherburne 1992**; **Currah and Zelmer 1992**; **Zelmer and Currah 1995b**). Mycoheterotrophic orchids, such as *Corallorhiza* and *Cephalanthera*, form mycorrhizas with ectomycorrhizal-forming basidiomycetes including *Russula*, *Thelephora*, and *Tomentella* spp. (**Currah and Zelmer 1992**; Taylor and Bruns 1997). Host specificity appears to be quite high in orchid mycorrhizas. Molecular techniques show that the orchid *Cephalanthera austinae*

(Gray) Heller associates only with fungi in the Thelephoraceae, and *Corallorhiza maculata* (Raf.) Raf. occurs only with members of the Russulaceae (Taylor and Bruns 1997). Furthermore, two orchids within the genus *Corallorhiza* (*C. maculata* and *Corallorhiza mertensiana* (Bong.) Calder & Taylor) associate with different species in the Russulaceae (Taylor and Bruns 1999).

Functional diversity

There has been relatively little work on the function of orchid mycorrhizas; although there has been debate until quite recently as to whether the fungi in roots of autotrophic orchids are mycorrhizal, the simple morphology of these roots suggests that they are (**Peterson et al. 2004**). The hyphal coils (pelotons) formed in root cortical cells break down after a period of time and release nutrients (**Currah and Zelmer 1992**) and a carbon source (Alexander and Hadley 1985) to the plant. It has also been demonstrated that the minute seeds of orchids need to be colonized by OM fungi or require an exogenous carbon source to germinate and develop from an undifferentiated embryo to a protocorm and, subsequently, to a seedling (Arditti et al. 1990). During early symbiotic protocorm formation of the orchid *Platanthera hyperborea* (L.) Lindl. grown with the fungi *Rhizoctonia cerealis* van der Hoeven and *Ceratorhiza goodyerae-repentis* (Costantin & Dufour) Moore, the fungi enter through dead seed suspensor cells, trigger utilization of the seed's lipid and protein reserves, and allow protocorm development (**Richardson et al. 1992**). Tracer studies have shown that less sugar was transferred to mature *G. repens* plants than to seedlings (Alexander and Hadley 1985) and that P was transferred to the plant by the fungus (Alexander et al. 1984). It has now been demonstrated using adult *G. repens* plants in microcosms and radiotracers that *Ceratobasidium cornigerum* (Bourdot) Rogers can receive photosynthate from the plant in exchange for N (Cameron et al. 2006) and P (Cameron et al. 2007). Some OM fungi such as *Rhizoctonia solani* Kühn are known plant pathogens and may even become pathogenic for orchid protocorms in the presence of high concentrations of N (**Beyrle et al. 1995**).

The mycoheterotrophic orchids have an even more complex approach to nutrition. Two strains of symbiotic fungi isolated from rhizomes of the orchid *Corallorhiza trifida* Châtel. were found also to form ectomycorrhizas with lodgepole pine (*Pinus contorta* Dougl. ex Loud.) (**Zelmer and Currah 1995a**). This suggests that *C. trifida* is able to access photosynthates from neighbouring trees via an association with certain ectomycorrhizal (ECM) fungi. This tripartite symbiosis is an important survival strategy developed by achlorophyllous angiosperms (Leake 1994), and these mycoheterotrophs have been called the "cheating" orchids (Taylor and Bruns 1999).

Ectomycorrhizas

First described by Frank (1885), the ECM symbiosis involves the formation of a network of modified hyphae that surrounds, but does not enter, the root cells. The resulting structure, the Hartig net, is the location of exchange of water and nutrients for photosynthetically fixed carbon. In gymnosperms, the Hartig net typically surrounds both the root cortical and epidermal cells; in angiosperms, only the epidermal cells are colonized (**Peterson and Massicotte 2004**). A well-developed mantle of fungal tissue, somewhat analogous to the nonreproductive tissues of the associated sporocarp, is also characteristic of the ectomycorrhizas. Fungal mantles are anatomically highly variable and can often be used to identify the fungal symbiont to the genus or species level. Thorough descriptions of mantle anatomies can be found in Agerer (1987–2006), Ingleby et al. (1990) and **Goodman et al. (1996–1997)**. The fungal hyphae that emanate from the mantle into the soil environment may also be morphologically characteristic, sometimes possessing clamp connections, encrustations, or pigments (**Hutchison 1991**). In some species, hyphae may coalesce to form hyphal strands or rhizomorphs (Cairney et al. 1991). Ectomycorrhizal fungal colonization also transforms the overall morphology of the fine roots: root hairs are lost; cortical cells swell; and root tips may become arranged into pinnate, coralloid, or tuberculate systems (Agerer 1991).

Taxonomic diversity

Ectomycorrhizal fungi are mainly found in association with woody plants. In the northern hemisphere, these include members of the Betulaceae, Fagaceae, Pinaceae, Salicaceae, Rosaceae, Tiliaceae, Ulmaceae, and Cistaceae (Table 1.1) (Meyer 1973; Smith and Read 1997). Some species of herbaceous plants within the genera *Kobresia* and *Polygonum* also form ECM (**Massicotte et al. 1998**); however, these associations are not representative of the two genera. Some genera, such as *Salix* and *Populus*, may be

Table 1.1. Genera of dominantly ectomycorrhizal native Canadian plants and genera of fungi commonly encountered in Canadian surveys of ectomycorrhizas.

Host plant genus*	Fungal genus†
Trees	Agarics
Abies	*Amanita*
Alnus	*Boletus*
Betula	*Cortinarius*
Carpinus	*Dermocybe*
Carya	*Hebeloma*
Castanea	*Hygrophorus*
Fagus	*Inocybe*
Larix	*Laccaria*
Ostrya	*Lactarius*
Picea	*Leccinum*
Pinus	*Paxillus*
Populus	*Russula*
Pseudotsuga	*Suillus*
Quercus	*Tricholoma*
Salix	Hypogeous
Tilia	*Hysterangium*
Tsuga	*Rhizopogon*
Woody shrubs	
Alnus	Corticioid
Betula	*Amphinema*
Corylus	*Tomentella*
Dryas	*Piloderma*
Helianthemum	Ascomycetes
Salix	*Cenococcum geophilum*
	Phialocephala (and other dark septate fungi)
	Tuber
	Wilcoxina
	Thelephoroid
	Hydnellum
	Thelephora

Note: Plant genera listed are those that form ectomycorrhizas at most stages and under most conditions, although some groups may also form arbuscular mycorrhizas and ect-endomycorrhizas.

*Host plants are based on Malloch and Malloch (1981), Currah and Van Dyk (1986), Harley and Harley (1987), Brundrett and Kendrick (1990a, 1990b), and Smith and Read (1997).

†Fungal genera are based on Danielson (1984), Danielson et al. (1984), Visser (1995), Goodman et al. (1996–1997), Bradbury et al. (1998), Goodman and Trofymow (1998), Kranabetter and Wylie (1998), Durall et al. (1999), Hagerman et al. (1999), Kranabetter et al. (1999), Kernaghan and Harper (2001), Mah et al. (2001), Kranabetter and Friesen (2002), Sakakibara et al. (2002), Kernaghan et al. (2003), and DeBellis et al. (2006).

colonized by ECM and (or) AM fungi depending on plant age and soil conditions (Lodge and Wentworth 1990). The host range of a given ECM fungus may vary from extremely specific, species–species relationships to very broad relationships, with little or no specificity (**Molina et al. 1992**).

Several thousand species of fungi (mainly Basidiomycetes but also some Ascomycetes, e.g., *Tuber* spp.) form ECM associations worldwide (**Molina et al. 1992**), and high ECM fungal diversity occurs in the coastal, montane, and eastern mixed forests of Canada (**Villeneuve et al. 1991; Redhead 1993; Kernaghan and Currah 1998**). Moreover, the boreal forest, the largest forest region in Canada, remains poorly studied and may be a site of undiscovered fungal diversity. Ectomycorrhizal fungal genera commonly encountered during surveys in Canadian forests are given in Table 1.1.

Functional diversity

Species of ECM fungi can differ greatly in their effects on host plant growth and nutrient status (**Gagnon et al. 1987**; Burgess et al. 1993). This is due to variation in their abilities to access and transport available resources relative to their C demand. For example, ECM fungi vary in their ability to access different forms of N (including nitrate, ammonium, and organic forms such as amino acids and proteins) (Finlay et al. 1992; Chalot and Brun 1998), as well as in their ability to transport phosphate (Dighton et al. 1990). Much of this variability is likely related to the production of a range of extracellular enzymes, the levels of which vary considerably among ECM fungi (Courty et al. 2005).

Differences in patterns of resource acquisition are also thought to be involved in the changes in ECM fungal species composition that occur as forests age and soil nutrients become increasingly bound in organic complexes. Therefore, ECM fungi classified as "early" and "late" stage (Last et al. 1987) may differ in their enzymatic capabilities, with late-stage fungi possessing the systems needed to access the recalcitrant nutrient sources available in late-successional forests (Abuzinadah and Read 1986). Species of ECM fungi also vary widely in their response to other environmental factors including temperature (**Samson and Fortin 1986; Hutchison 1990**), moisture (Erland and Taylor 2002), salinity (**Kernaghan et al. 2002**; see also Chapter 10), pH (Hung and Trappe 1983), and anthropogenic N inputs (Lilleskov et al. 2002).

Arbutoid mycorrhizas

Arbutoid mycorrhizas are morphologically and anatomically similar to ectomycorrhizas, in that they possess fungal mantles of variable thickness and a labyrinthine Hartig net associated with root epidermal cells (**Massicotte et al. 1993**). However, arbutoid mycorrhizas differ from ectomycorrhizas mainly in the occurrence of hyphal penetration into the root epidermal cells.

Taxonomic diversity

The arbutoid mycorrhizas are mostly restricted to the host plant genera *Arbutus*, *Arctostaphylos*, and *Pyrola*, which are thought to be generally nonhost specific with respect to their mycorrhizal partners (Molina and Trappe 1982). The fungi involved in arbutoid mycorrhizas generally appear to overlap with those that form ectomycorrhizas in the same habitat (**Molina et al. 1992**).

Functional diversity

Because arbutoid mycorrhizas and ectomycorrhizas can share the same fungal species, arbutoid mycorrhizas may act as alternate hosts or "refuge plants" for ECM fungi. Therefore, after a disturbance such as fire or logging, the latter may survive in arbutoid symbioses prior to the reestablishment of ECM plants (Horton et al. 1999; **Hagerman et al. 2001**).

Ectendomycorrhizas

First categorized as "ectendotrophic mycorrhizas" (Melin 1923), the ectendomycorrhizas (or E-strain) are morphologically similar to ectomycorrhizas but are characterized by very thin to nonexistent fungal mantles. Anatomically, a Hartig net is formed as in ectomycorrhizas, but intracellular penetration of root epidermal and cortical cells also occurs to varying degrees (**Piché et al. 1986; Scales and Peterson 1991**).

Taxonomic diversity

The formation of ectendomycorrhizas appears to be restricted to particular combinations of ascomycetous fungi and species of *Pinus* and *Larix* (**Yu et al. 2001**). The most common fungi involved are species of *Wilcoxina* (*Wilcoxina mikolae* (Yang & Wilcox) Yang & Korf and *Wilcoxina rhemii* Yang & Korf) (**Egger and Fortin 1990; Egger et al. 1991**), although others, including *Sphaerosporella brunnea* (Alb. & Schwein.) Svrcek & Kubicka (**Danielson 1984**), *Phialophora finlandia* Wang & Wilcox and *Chloridium paucisporum* Wang & Wilcox (Wang and Wilcox 1985) have been identified. Symbioses between these fungi and tree species outside the genera *Pinus* and *Larix* (e.g., *Picea*, *Populus*, and *Betula*) result in the formation of ectomycorrhizas, rather than ectendomycorrhizas (**Yu et al. 2001**).

Functional diversity

Much of the early work on ectendomycorrhizas was conducted on forest nursery seedlings (Laiho 1965; Mikola 1965), where this type of symbiosis is relatively common. Indeed, ectendomycorrhizas tend to occur in disturbed (see Chapter 10), early successional habitats (Mikola 1988; **Danielson 1991**). The effect of ectendomycorrhizal colonization on seedling growth varies among fungal–host combinations. A positive response to an ectendomycorrhizal isolate was reported for jack pine (*Pinus banksiana* Lamb.) seedlings (**Danielson and Visser 1989**), whereas a negative response to inoculation with ectendomycorrhizal root pieces was reported in Scots pine (*Pinus sylvestris* L.; Levisohn 1954). Wilcox et al. (1983) found morphological variation among the root systems of red pine (*Pinus resinosa* Sol. ex Ait.) seedlings inoculated with a range of ectendomycorrhizal isolates.

Monotropoid mycorrhizas

In 1882, Franz Kamienski described the root–fungus association of *Monotropa hypopitys* L. as the most striking example of the mutualistic union of two vegetative organisms (Kamienski 1882; **Berch et al. 2005**). The monotropoid mycorrhiza consists of a multilayered fungal mantle, Hartig net between epidermal cells, and fungal pegs that penetrate the epidermal cells (Kamienski 1882; Duddridge and Read 1982; **Massicotte et al. 2005**).

Taxonomic diversity

Achlorophyllous plants in the clade Monotropoideae of the Ericaceae that form monotropoid mycorrhizas include *Allotropa*, *Cheilotheca*, *Hemitomes*, *Monotropa*, *Monotropantham*, *Monotropsis*, *Pityopus*, *Pleuricospora*, *Pterospora*, and *Sarcodes* (**Peterson et al. 2004**). For the relatively few Monotropoideae that have been studied, the associated fungi are highly specific. Using morphological and molecular characteristics, it has been determined that *M. hypopitys* forms mycorrhizas with fungi in the genus *Tricholoma* (Martin 1985; Bidartondo and Bruns 2001), *Monotropa uniflora* L. with fungi in the genus *Russula* and related Russulaceae (Martin 1986; Cullings et al. 1996; Bidartondo and Bruns

2001; **Young et al. 2002**), and *Pterospora andromedea* Nutt. and *Sarcodes sanguinea* Torrey with fungi in the genus *Rhizopogon* (Bruns and Read 2000).

Functional diversity

Leake (1994) described mycotrophic achlorophyllus vascular plants as "mycoheterotrophic," and **Brundrett (2002)** classified them as "exploitative" because the fungus seems to get little from the association. Björkman (1960) traced the movement of ^{14}C and ^{32}P from ECM host trees to neighbouring *M. hypopitys*; however, Duddridge (1980) did not find evidence of C transfer from *Salix* to this plant. Trudell et al. (2003) used the natural abundance of stable isotopes to determine that *M. hypopitys*, *M. unifora*, and *P. andromedea* receive both C and N from their associated fungi.

Final thoughts

The influence of Yolande Dalpé and J. André Fortin on mycorrhizal science and technology can be found in references used in this chapter and by tracing the scientific pedigree of scientists associated with many of these references, including the authors of this chapter. Their work and their enthusiasm have influenced many in the mycorrhiza world, and it is fair to say that their influence will continue into the next generations of mycorrhiza researchers. Theirs is a powerful legacy.

References

Abuzinadah, R.A., and Read, D.J. 1986. The role of proteins in the nitrogen nutrition of ectomycorrhizal plants. I. Utilization of peptides and proteins by ectomycorrhizal fungi. New Phytol. **103**: 481–493.

Addy, H.D., Miller, M.H., and Peterson, R.L. 1997. Infectivity of the propagules associated with extraradical mycelia of two AM fungi following winter freezing. New Phytol. **135**: 745–753.

Agerer, R. 1987–2006. Colour atlas of Ectomycorrhizae. Einhorn-Verlag, Schwabisch-Gmund, Germany.

Agerer, R. 1991. Characterization of ectomycorrhiza. *In* Methods in microbiology. Vol. 23. Techniques for the study of mycorrhiza. *Edited by* J.R. Norris, D.J. Read, and A.K. Varma. Academic Press, London. pp. 25–73.

Alexander, C., and Hadley, G. 1985. Carbon movement between host and mycorrhizal endophyte during the development of the orchid *Goodyera repens* Br. New Phytol. **101**: 657–665.

Alexander, C., Alexander, I.J., and Hadley, G. 1984. Phosphate uptake by *Goodyera repens* in relation to mycorrhizal infection. New Phytol. **97**: 401–411.

Arditti, J., Ernst, R., Yam, T.W., and Glabe, C. 1990. The contribution of orchid mycorrhizal fungi to seed germination: a speculative review. Lindleyana, **5**: 249–255.

Bago, B., Shachar-Hill, Y., and Pfeffer, P.E. 2000. Carbon metabolism and transport in arbuscular mycorrhizas. Plant Physiol. **124**: 949–957.

Bain, H.F. 1937. Production of synthetic mycorrhiza in the cultivated cranberry. J. Agric. Res. **55**: 811–835.

Bajwa, R., Abuarghub, S., and Read, D.J. 1985. The biology of mycorrhiza in the Ericaceae. X. The utilisation of proteins and the production of proteolytic enzymes by the mycorrhizal endophyte and by mycorrhizal plants. New Phytol. **101**: 469–486.

Bending, G.D., and Read, D.J. 1996. Nitrogen mobilization from protein–polyphenol complexes by ericoid and ectomycorrhizal fungi. Soil Biol. Biochem. **28**: 1603–1612.

Berch, S.M. 1977. Endomycorrhizae of southern Ontario ferns. M.Sc. thesis, University of Waterloo, Waterloo, Ont.

Berch, S.M., and Fortin, J.A. 1984. Some sporocarpic Endogonaceae from eastern Canada. Can. J. Bot. **62**: 170–180.

Berch, S.M., and Kendrick, B. 1982. Vesicular–arbuscular mycorrhizae of southern Ontario ferns and fern-allies. Mycologia, **74**: 769–776.

Berch, S.M., Deom, E., and Willingdon, T. 1991. Western red cedar growth and vesicular–arbuscular mycorrhizal colonization in fumigated and nonfumigated nursery beds. USDA For. Serv. Tree Plant. Notes, **42**: 14–16.

Berch, S.M., Allen, T.R., and Berbee, M.L. 2002. Molecular detection, community structure and phylogeny of ericoid mycorrhizal fungi. Plant Soil, **244**: 55–66.

Berch, S.M., Massicotte, H.B., and Tackaberry, L.E. 2005. Re-publication of a translation of 'The vegetative organs of *Monotropa hypopitys* L.' published by F. Kamienski in 1882, with an update on *Monotropa* mycorrhizas. Mycorrhiza, **15**: 323–332.

Bergero, R., Perotto, S., Girlanda, M., Vidano, G., and Luppi, A.M. 2000. Ericoid mycorrhizal fungi are common root associates of a Mediterranean ectomycorrhizal plant (*Quercus ilex*). Mol. Ecol. **9**: 1639–1649.

Beyrle, H.F., Smith, S.E., Peterson, R.L., and Franco, C.M.M. 1995. Colonization of *Orchis morio* protocorms by a mycorrhizal fungus: effects of nitrogen nutrition and glyphosate in modifying the responses. Can. J. Bot. **73**: 1128–1140.

Bidartondo, M.I., and Bruns, T.D. 2001. Extreme specificity in epiparasitic Monotropoideae (Ericaceae): widespread phylogenetic and geographical structure. Mol. Ecol. **10**: 2285–2295.

Bidartondo, M.I., Redecker, D., Hijri, I., Wiemken, A., Bruns, T.D., Dominguez, L., Sersic, A., Leake, J., and Read, D.J. 2002. Epiparasitic plants specialized on arbuscular mycorrhizal fungi. Nature, **419**: 389–392.

Björkman, E. 1960. *Monotropa hypopitys* L.—an epiparasite on tree roots. Physiol. Plant. **13**: 308–327.

Bonfante, P., and Perotto, S. 1995. Strategies of arbuscular mycorrhizal fungi when infecting host plants. New Phytol. **130**: 3–21.

Bradbury, S.M., Danielson, R.M., and Visser, S. 1998. Ectomycorrhizas of regenerating stands of lodgepole pine (*Pinus contorta*). Can. J. Bot. **76**: 218–227.

Brundrett, M.C. 2002. Co-evolution of roots and mycorrhizas of land plants. New Phytol. **154**: 275–304.

Brundrett, M.C., and Kendrick, B. 1988*a*. The mycorrhizal status, root anatomy and phenology of plants in a sugar maple forest. Can. J. Bot. **66**: 1153–1173.

Brundrett, M., and Kendrick, B. 1988*b*. Roots and mycorrhizae in a hardwood forest. *In* Canadian Workshop on Mycorrhizae in Forestry. *Edited by* M. Lalonde and Y. Piché. Centre de Recherche en Biologie Forestière, Faculté de Foresterie et de Géodésie, Université Laval, Québec, Que. pp. 89–95.

Brundrett, M.C., and Kendrick, B. 1990*a*. The roots and mycorrhizas of herbaceous woodland plants. I. Quantitative aspects of morphology. New Phytol. **114**: 457–468.

Brundrett, M.C., and Kendrick, B. 1990*b*. The roots and mycorrhizas of herbaceous woodland plants. II. Structural aspects of morphology. New Phytol. **114**: 469–479.

Bruns, T.D., and Read, D.J. 2000. *In vitro* germination of nonphotosynthetic, myco-heterotrophic plants stimulated by fungi isolated from the adult plants. New Phytol. **148**: 335–342.

Burgess, T.I., Malajczuk, N., and Grove, T.S. 1993. The ability of 16 ectomycorrhizal fungi to increase growth and phosphorus uptake of *Eucalyptus globulus* Labill and *E. diversicolor* F. Muell. Plant Soil, **153**: 155–164.

Cade-Menun, B.J., and Berch, S.M. 1997. Response of mycorrhizal western red cedar to organic P sources and benomyl. Can. J. Bot. **75**: 1226–1235.

Cairney, J.W.G., Jennings, D.H., and Agerer, R. 1991. The nomenclature of fungal multihyphal linear aggregates. Cryptogam. Bot. **2/3**: 246–251.

Cameron, D.D., Leake, J.R., and Read, D.J. 2006. Mutualistic mycorrhiza in orchids: evidence from plant–fungus carbon and nitrogen transfers in the green-leaved terrestrial orchid *Goodyera repens*. New Phytol. **171**: 405–416.

Cameron, D.D., Johnson, I., Leake, J.R., and Read, D.J. 2007. Mycorrhizal acquistion of inorganic phosphorus by the green-leaved terrestrial orchid *Goodyera repens*. Ann. Bot. (Lond.), **99**: 381–384.

Chalot, M., and Brun, A. 1998. Physiology of organic nitrogen acquisition by ectomycorrhizal fungi and ectomycorrhizas. FEMS Microbiol. Rev. **22**: 21–44.

Chambers, S.M., Liu, G., and Cairney, J.W.G. 2000. ITS rDNA sequence comparison of ericoid mycorrhizal endophytes from *Woollsia pungens*. Mycol. Res. **104**: 168–174.

Coughlan, A.P., Dalpé, Y., Lapointe, L., and Piché, Y. 2000. Soil pH modulates root colonization and reproduction of symbiotic arbuscular mycorrhizal fungi from healthy and declining maple forests. Can. J. For. Res. **30**: 1543–1554.

Courty, P.E., Pritsch, K., Schloter, M., Hartmann, A., and Garbaye, J. 2005. Activity profiling of ectomycorrhiza communities in two forest soils using multiple enzymatic tests. New Phytol. **167**: 309–319.

Couture, M., Fortin, J.A., and Dalpé, Y. 1983. *Oidiodendron griseum* (Robak): an endophyte of ericoid mycorrhiza in *Vaccinium* spp. New Phytol. **95**: 375–380.

Cullings, K.W., Szaro, T., and Bruns, T.D. 1996. Evolution of extreme specialization within a lineage of ectomycorrhizal epiparasites. Nature (London), **379**: 63–66.

Currah, R.S., and Sherburne, R. 1992. Septal ultrastructure of some fungal endophytes from boreal orchid mycorrhizas. Mycol. Res. **96**: 583–587.

Currah, R.S., and Van Dyk, M. 1986. A survey of some perennial vascular plant species native to Alberta for occurrence of mycorrhizal fungi. Can. Field-Nat. **100**: 330–342.

Currah, R.S., and Zelmer, C.D. 1992. A key and notes for genera of fungi mycorrhizal with orchids and a new species in the genus *Epulorhiza*. Rep. Tottori Mycol. Inst. **30**: 43–59.

Currah, R.S., Sigler, L., and Hambleton, S. 1987. New records and new taxa of fungi from the mycorrhizae of terrestrial orchids of Alberta. Can. J. Bot. **65**: 2473–2482.

Currah, R.S., Zettler, L.W., and McInnis, T.M. 1997. *Epulorhiza inquilina* sp. nov. from *Platanthera* (Orchidaceae) and a key to *Epulorhiza* species. Mycotaxon, **61**: 335–342.

Dalpé, Y. 1986. Axenic synthesis of ericoid mycorrhiza in *Vaccinium angustifolium* Ait. by *Oidiodendron* species. New Phytol. **103**: 391–396.

Dalpé, Y. 1989. Ericoid mycorrhizal fungi in the Myxotrichaceae and Gymnoascaceae. New Phytol. **113**: 523–527.

Dalpé, Y. 1991*a*. Inventaire de la flore endomycorhizienne des rivages et dunes maritimes du Québec, du Nouveau-Brunswick et de la Nouvelle-Écosse. Nat. Can. **116**: 219–236.

Dalpé, Y. 1991*b*. Statut mycorhizien du genre *Oidiodendron*. Can. J. Bot. **69**: 1712–1714.

Dalpé, Y., and Monreal, M. 2004. Arbuscular mycorrhizae inoculum to support sustainable cropping systems [online]. Crop Manage. doi:10.1094/CM-2004-0301-09-RV. http://www.plantmanagementnetwork.org/pub/cm/review/2004/amfungi/.

Dalpé, Y., Litten, W., and Sigler, L. 1989. *Scytalidium vaccinii* sp. nov. an ericoid endophyte of *Vaccinium angustifolium* roots. Mycotaxon, **35**: 371–377.

Dalpé, Y., Cranenbrouck, S., Séguin, S., and Declerck, S. 2005. Monoxenic culture of arbuscular mycorrhizal fungi as a tool for systematics and biodiversity. *In In vitro* culture of mycorrhizas. *Edited by* S. Declerck, D.G. Strullu, and A. Fortin. Springer-Verlag, Berlin. pp. 31–48.

Danielson, R.M. 1984. Ectomycorrhiza formation by the operculate discomycete *Sphaerosporella brunnea* (Pezizales). Mycologia, **76**: 454–461.

Danielson, R.M. 1991. Temporal changes and effects of amendments on the occurrence of sheathing (ecto-) mycorrhizas of conifers growing in oil sands tailings and coal spoil. Agric. Ecosyst. Environ. **35**: 261–281.

Danielson, R.M., and Visser, S. 1989. Host response to inoculation and behaviour of introduced and indigenous ectomycorrhizal fungi of jack pine grown on oil-sands tailings. Can. J. For. Res. **19**: 1412–1421.

Danielson, R.M., Zak, J.C., and Parkinson, D. 1984. Mycorrhizal inoculum in a peat deposit formed under a white spruce stand in Alberta. Can. J. Bot. **62**: 2557–2560.

DeBellis, T., Kernaghan, G., Bradley, R., and Widden, P. 2006. Relationships between stand composition and ectomycorrhizal community structure in boreal mixed-wood forests. Microb. Ecol. **52**: 114–126.

Declerck, S., and Dalpé, Y. 2001. GINCO. Mycorrhiza, **11**: 263.

Dighton, J., Mason, P.A., and Poskitt, J.M. 1990. Use of ^{32}P to measure phosphate uptake by birch mycorrhizas. New Phytol. **116**: 655–661.

Dixon, K.W., Kell, S.P., Barrett, R.L., and Cribb, P.J. 2003. Orchid conservation. Natural History Publications (Borneo), Kota Kinahalu, Malaysia.

Duclos, J.L., and Fortin, J.A. 1983. Effect of glucose and active charcoal on *in vitro* synthesis of ericoid mycorrhiza with *Vaccinium* spp. New Phytol. **94**: 95–102.

Duddridge, J.A. 1980. A comparative ultrastructural analysis of a range of mycorrhizal associations. Ph.D. thesis, University of Sheffield, Sheffield, England.

Duddridge, J.A., and Read, D.J. 1982. An ultrastructural analysis of the development of mycorrhizas in *Monotropa hypopithys* L. New Phytol. **92**: 203–214.

Durall, D.M., Jones, M.D., Wright, E.F., Kroeger, P., and Coates, K.D. 1999. Species richness of ectomycorrhizal fungi in cutblocks of different sizes in the Interior Cedar–Hemlock forests of northwestern British Columbia: sporocarps and ectomycorrhizas. Can. J. For. Res. **29**: 1322–1332.

Egger, K.N., and Fortin, J.A. 1990. Identification of taxa of E-strain mycorrhizal fungi by restriction fragment analysis. Can. J. Bot. **68**: 1482–1488.

Egger, K.N., and Sigler, L. 1993. Relatedness of the ericoid endophytes *Scytalidium vaccinii* and *Hymenoscyphus ericae* inferred from analysis of ribosomal DNA. Mycologia, **85**: 219–230.

Egger, K.N., Danielson, R.M., and Fortin, J.A. 1991. Taxonomy and population structure of E-strain mycorrhizal fungi inferred from ribosomal and mitochondrial DNA polymorphisms. Mycol. Res. **95**: 866–872.

Erland, S., and Taylor, A.F.S. 2002. Diversity of ecto-mycorrhizal communities in relation to the abiotic environment. *In* Mycorrhizal ecology. *Edited by* M.G.A. van der Heijden and I.R. Sanders. Springer, Berlin. Ecol. Stud. 157. pp. 163–200.

Finlay, R.D., Frostegård, Å., and Sonnerfeldt, A.-M. 1992. Utilization of organic and inorganic nitrogen sources by ectomycorrhizal fungi in pure culture and in symbiosis with *Pinus contorta* Dougl. ex Loud. New Phytol. **120**: 105–115.

Fortin, J.A., Bécard, G., Declerck, S., Dalpé, Y., St-Arnaud, M., Coughlan, A.P., and Piché, Y. 2002. Arbuscular mycorrhiza on root-organ cultures: a review. Can. J. Bot. **80**: 1–20.

Frank, A.B. 1885. Ueber die auf Wurzelsymbiose beruhende Ernährung gewisser Bäume durch unterirdische Pilze. Ber. Dtsch. Bot. Ges. **3**: 128–145. [English translation: 2005.] Mycorrhiza, **15**: 267–275.]

Gagnon, J., Langlois, C.G., and Fortin, J.A. 1987. Growth of containerized jack pine seedlings inoculated with different ectomycorrhizal fungi under a controlled fertilization schedule. Can. J. For. Res. **17**: 840–845.

Gavito, M.E., and Miller, M.H. 1998. Changes in mycorrhizal development in maize induced by crop management practices. Plant Soil, **198**: 185–192.

Girard, I. 1985. Écologie des mycorhizes: caractère mycorhizien des espèces végétales de forêts climatiques et réflexion sur la relation humus–mycorhize. M.Sc. thesis, Université Laval, Québec, Que.

Goodman, D.M., and Trofymow, J.A. 1998. Comparison of communities of ectomycorrhizal fungi in old-growth and mature stands of Douglas-fir at two sites on southern Vancouver Island. Can. J. For. Res. **28**: 574–581.

Goodman, D.M., Durall, D.M., Trofymow, J.A., and Berch, S.M. (*Editors*). 1996–1997. A manual of concise descriptions of North American ectomycorrhizae. Natural Resources Canada, Canadian Forest Service, British Columbia Ectomycorrhizal Research Network, Victoria, B.C.

Granger, R.L., Khanizadeh, S., Meheriuk, M., Bérard, L.S., and Dalpé, Y. 1995. Effects of simazine on the mycorrhizal population in soil beneath an apple canopy. Fruit Varieties J. **49**: 90–93.

Hagerman, S.M., Jones, M.D., Bradfield, G.E., Gillespie, M., and Durall, D.M. 1999. Effects of clear-cut logging on the diversity and persistence of ectomycorrhizae at a subalpine forest. Can. J. For. Res. **29**: 124–134.

Hagerman, S.M., Sakakibara, S.M., and Durall, D.M. 2001. The potential for woody understory plants to provide refuge for ectomycorrhizal inoculum at an interior Douglas-fir forest after clear-cut logging. Can. J. For. Res. **31**: 711–721.

Hambleton, S., and Currah, R.S. 1997. Fungal endophytes from the roots of alpine and boreal Ericaceae. Can. J. Bot. **75**: 1570–1581.

Hambleton, S., Egger, K.N., and Currah, R.S. 1998. The genus *Oidiodendron*: species delimitation and phylogenetic relationships based on nuclear ribosomal DNA analysis. Mycologia, **90**: 854–869.

Harley, J.L. 1989. The fourth benefactors' lecture: the significance of mycorrhiza. Mycol. Res. **92**: 129–139.

Harley, J.L., and Harley, E.L. 1987. A check-list of mycorrhiza in the British Flora. New Phytol. **105**(Suppl.): 1–102.

Horton, T.R., Bruns, T.D., and Parker, V.T. 1999. Ectomycorrhizal fungi associated with *Arctostaphylos* contribute to *Pseudotsuga menziesii* establishment. Can. J. Bot. **77**: 93–102.

Hung, L.L., and Trappe, J.M. 1983. Growth variation between and within species of ectomycorrhizal fungi in response to pH *in vitro*. Mycologia, **75**: 234–241.

Hutchison, L.J. 1990. Studies on the systematics of ectomycorrhizal fungi in axenic culture. V. Linear growth response to standard extreme temperatures used as a taxonomic character. Can. J. Bot. **68**: 2179–2184.

Hutchison, L.J. 1991. Description and identification of cultures of ectomycorrhizal fungi found in North America. Mycotaxon, **42**: 387–504.

Ingleby, K., Mason, P.A., Last, F.T., and Fleming, L.V. 1990. Identification of ectomycorrhizas. Her Majesty's Stationery Office, London. Inst. Terrestrial Ecol. Res. Publ. 5.

Jakobsen, I., and Rosendahl, L. 1990. Carbon flow into soil and external hyphae from roots of mycorrhizal cucumber plants. New Phytol. **115**: 77–83.

Kamienski, F.M. 1882. Les organes végétatifs du *Monotropa hypopithys* L. Mem. Soc. Natl. Sci. Nat. Math. Cherbourg, **24**: 5–40.

Kendrick, B. 1992. The fifth kingdom. 2nd ed. Focus, Newburyport/Mycologue Publications, Sidney.

Kerley, S.J., and Read, D.J. 1998. The biology of mycorrhiza in the Ericaceae. XX. Plant and mycorrhizal necromass as nitrogenous substrates for the ericoid mycorrhizal fungus

Hymenoscyphus ericae and its hosts. New Phytol. **139**: 353–360.

Kernaghan, G., and Currah, R.S. 1998. Ectomycorrhizal fungi at tree line in the Canadian Rockies. Mycotaxon, **69**: 39–79.

Kernaghan, G., and Harper, K.A. 2001. Community structure of ectomycorrhizal fungi across an alpine/subalpine ecotone. Ecography, **24**: 181–188.

Kernaghan, G., Hambling, B., Fung, M., and Khasa, D. 2002. *In vitro* selection of ectomycorrhizal fungi for reclamation of alkaline–saline habitats. Restor. Ecol. **10**: 43–51.

Kernaghan, G., Widden, P., Bergeron, Y., Légaré, S., and Paré, D. 2003. Biotic and abiotic factors affecting ectomycorrhizal diversity in boreal mixed-woods. Oikos, **102**: 497–505.

Kernan, M.J., and Finocchio, A.F. 1983. A new discomycete associated with roots of *Monotropa uniflora* (Ericaceae). Mycologia, **75**: 916–920.

Kittson, K., Bonti-Ankomah, S., Zafiriou, M., Gao, S., and Islam, N. 2007. An overview of the Canadian Agriculture and Agri-Food System. Agriculture and Agri-Food Canada, Strategic Policy Branch, Research and Analysis Directorate, Ottawa, Ont.

Kranabetter, J.M., and Friesen, J. 2002. Ectomycorrhizal community structure on western hemlock (*Tsuga heterophylla*) seedlings transplanted from forests into openings. Can. J. Bot. **80**: 861–868.

Kranabetter, J.M., and Wylie, T. 1998. Ectomycorrhizal community structure across forest openings on naturally regenerated western hemlock seedlings. Can. J. Bot. **76**: 189–196.

Kranabetter, J.M., Hayden, S., and Wright, E.F. 1999. A comparison of ectomycorrhizal communities from three conifer species planted on forest gap edges. Can. J. Bot. **77**: 1193–1198.

Laiho, O. 1965. Further studies on the ectendotrophic mycorrhiza. Acta For. Fenn. **79**: 1–35.

Last, F.T., Dighton, J., and Mason, P.A. 1987. Successions of sheathing mycorrhizal fungi. Trends Ecol. Evol. **2**: 157–161.

Leake, J.R. 1994. The biology of myco-heterotrophic ('saprophytic') plants. New Phytol. **127**: 171–216.

Leake, J.R., and Read, D.J. 1990. Chitin as a nitrogen source for mycorrhizal fungi. Mycol. Res. **94**: 993–1008.

Leake, J.R., Johnson, D., Donnelly, D., Muckle, G.E., Boddy, L., and Read, D.J. 2004. Networks of power and influence: the role of mycorrhizal mycelium in controlling plant communities and agro-ecosystem functioning. Can. J. Bot. **82**: 1016–1045.

Levisohn, I. 1954. Aberrant root infections of pine and spruce seedlings. New Phytol. **53**: 284–291.

Lilleskov, E.A., Fahey, T.J., Horton, T.R., and Lovett, G.M. 2002. Belowground ectomycorrhizal fungal community change over a nitrogen deposition gradient in Alaska. Ecology, **83**: 104–115.

Litten, W., Smagula, J.M., and Dalpé, Y. 1992. Growth of micropropagated lowbush blueberry with defined fungi in irradiated peat mix. Can. J. Bot. **70**: 2202–2206.

Lodge, D.J., and Wentworth, T.R. 1990. Negative associations among VA mycorrhizal fungi and some ectomycorrhizal fungi inhabiting the same root system. Oikos, **57**: 347–356.

Mah, K., Tackaberry, L.E., Egger, K.B., and Massicotte, H.B. 2001. The impacts of broadcast burning after clear-cutting on the diversity of ectomycorrhizal fungi associated with hybrid spruce seedlings in central British Columbia. Can. J. For. Res. **31**: 224–235.

Malloch, D., and Malloch, B. 1981. The mycorrhizal status of boreal plants: species from northeastern Ontario. Can. J. Bot. **59**: 2167–2172.

Martin, J.F. 1985. Sur la mycorhization de *Monotropa hypopithys* par quelques espèces du genre *Tricholoma*. Bull. Soc. Mycol. France, **101**: 249–256.

Martin, J.F. 1986. Mycorhization de *Monotropa uniflora* L. par des Russulaceae. Bull. Soc. Mycol. France, **102**: 155–159.

Massicotte, H.B., Melville, L.H., Molina, R., and Peterson, R.L. 1993. Structure and histochemistry of mycorrhizae synthesized between *Arbutus menziesii* (Ericaceae) and two basidiomycetes, *Pisolithus tinctorius* (Pisolithaceae) and *Piloderma bicolor* (Corticiaceae). Mycorrhiza, **3**: 1–11.

Massicotte, H.B., Melville, L.H., Peterson, R.L., and Luoma, D.L. 1998. Anatomical aspects of field ectomycorrhizas on *Polygonum viviparum* (Polygonaceae) and *Kobresia bellardii* (Cyperaceae). Mycorrhiza, **7**: 287–292.

Massicotte, H.B., Melville, L.H., and Peterson, R.L. 2005. Structural features of mycorrhizal associations in two members of the Monotropoideae, *Monotropa uniflora* and *Pterospora andromedea*. Mycorrhiza, **15**: 101–110.

Melin, E. 1923. Experimentelle Untersuchungen über die Konstitution und Ökologie der Mykorrhizen von *Pinus sylvestris* L. und *Picea abies* (L.) Karst. Mykol. Untersuchungen Ber. **2**: 73–331.

Meyer, F.H. 1973. Distribution of ectomycorrhizae in native and man-made forests. *In* Ectomycorrhizae. *Edited by* G.C. Marks and T.T. Kozlowski. Academic Press, New York. pp. 79–105.

Mikola, P. 1965. Studies on the ectendotrophic mycorrhiza of pine. Acta For. Fenn. **79**: 1–56.

Mikola, P. 1988. Ectendomycorrhiza of conifers. Silva Fenn. **22**: 19–27.

Miller, M.H., McGonigle, T.P., and Addy, H.D. 1995. Functional ecology of vesicular arbuscular mycorrhizas as influenced by phosphate fertilization and tillage in an agricultural ecosystem. Crit. Rev. Biotechnol. **15**: 241–255.

Molina, R., and Trappe, J.M. 1982. Lack of mycorrhizal specificity by the ericaceous hosts *Arbutus menziesii* and *Arctostaphylos uva-ursi*. New Phytol. **90**: 495–509.

Molina, R., Massicotte, H.B., and Trappe, J.M. 1992. Specificity phenomena in mycorrhizal symbiosis: community-ecological consequences and practical implications. *In* Mycorrhizal functioning—an integrated plant–fungus process. *Edited by* M.F. Allen. Chapman & Hall, New York. pp. 357–423.

Monreal, M., Berch, S.M., and Berbee, M. 1999. Molecular diversity of ericoid mycorrhizal fungi. Can. J. Bot. **77**: 1580–1594.

Moutoglis, P., and Widden, P. 1996. Vesicular–arbuscular mycorrhizal spore populations in sugar maple (*Acer saccharum* Marsh.) forests. Mycorrhiza, **6**: 91–97.

Pearson, V., and Read, D.J. 1973. The biology of mycorrhiza in the Ericaceae: I. The isolation of the endophyte and synthesis of mycorrhizas in aseptic cultures. New Phytol. **72**: 371–379.

Peterson, R.L., and Farquhar, M.L. 1994. Mycorrhizas—

integrated development between roots and fungi. Mycologia, **86**: 311–326.

Peterson, R.L., and Massicotte, H.B. 2004. Exploring structural definitions of mycorrhizas, with emphasis on nutrient-exchange interfaces. Can. J. Bot. **82**: 1074–1088.

Peterson, R.L., Uetake, Y., and Zelmer, C. 1998. Fungal symbiosis with orchid protocorms. Symbiosis, **25**: 29–55.

Peterson, R.L., Massicotte, H.B., and Melville, L.H. 2004. Mycorrhizas: anatomy and cell biology. NRC Research Press, Ottawa, Ont.

Piché, Y., Ackerley, C.A., and Peterson, R.L. 1986. Structural characteristics of ectendomycorrhizas synthesized between roots of *Pinus resinosa* and the E-strain fungus, *Wilcoxina mikolae* var. *mikolae*. New Phytol. **104**: 447–452.

Podeszfinski, C., Dalpé, Y., and Charest, C. 2002. *In situ* turfgrass establishment. I. Response to arbuscular mycorrhizae and fertilization. J. Sustain. Agric. **20**: 57–74.

Rasmussen, H.N. 1995. Terrestrial orchids: from seed to mycotrophic plant. Cambridge University Press, Cambridge, UK.

Read, D.J. 1974. *Pezizella ericae* sp. nov., the perfect state of a typical mycorrhizal endophyte of the Ericaceae. Trans. Br. Mycol. Soc. **63**: 381–383.

Read, D.J., Leake, J.R., and Perez-Moreno, J. 2004. Mycorrhizal fungi as drivers of ecosystem processes in heathland and boreal forest biomes. Can. J. Bot. **82**: 1243–1263.

Redhead, S.A. 1993. Macrofungi inventory requirements for British Columbia. Centre for Land and Biological Resources Research, Research Branch, Agriculture Canada, Ottawa, Ont.

Richardson, K.A., Peterson, R.L., and Currah, R.S. 1992. Seed reserves and early symbiotic protocorm development of *Platanthera hyperborea* (Orchidaceae). Can. J. Bot. **70**: 291–300.

Rufyikiri, G., Declerck, S., and Thiry, Y. 2004. Comparison of ^{233}U and ^{33}P uptake and translocation by the arbuscular mycorrhizal fungus *Glomus intraradices* in root organ culture conditions. Mycorrhiza, **14**: 203–207.

Sakakibara, S.M., Jones, M.D., Gillespie, M., Hagerman, S.M., Forrest, M.E., Simard, S.W., and Durall, D.M. 2002. A comparison of ectomycorrhiza identification based on morphotyping and PCR-RFLP analysis. Mycol. Res. **106**: 868–878.

Samson, J., and Fortin, J.A. 1986. Ectomycorrhizal fungi of *Larix laricina* and the interspecific and intraspecific variation in response to temperature. Can. J. Bot. **64**: 3020–3028.

Scagel, C.F. 2005. Inoculation with ericoid mycorrhizal fungi alters fertilizer use of highbush blueberry cultivars. HortScience, **40**: 786–794.

Scales, P.F., and Peterson, R.L. 1991. Structure and development of *Pinus banksiana* – *Wilcoxina* ectendomycorrhizae. Can. J. Bot. **69**: 2135–2148.

Schüßler, A., Schwarzott, D., and Walker, C. 2001. A new fungal phylum, the Glomeromycota: phylogeny and evolution. Mycol. Res. **105**: 1413–1421.

Sharples, J.M., Chambers, S.M., Meharg, A.A., and Cairney, J.W.G. 2000. Genetic diversity of root-associated fungal endophytes from *Calluna vulgaris* at contrasting field sites. New Phytol. **148**: 153–162.

Simon, L., Bousquet, J., Lévesque, R.C., and Lalonde, M. 1993. Origin and diversification of endomycorrhizal fungi and coincidence with vascular land plants. Nature (London), **363**: 67–69.

Smith, S.E., and Read, D.J. 1997. Mycorrhizal symbiosis. 2nd ed. Academic Press, London.

Stevens, C.M., Goulart, B.L., Dalpé, Y., Hancock, J.F., Demchak, K., and Yang, W.Q. 1996. Survey of ericoid mycorrhizae and isolation and characterization of fungal symbionts in native and commercial *Vaccinium* populations in central Pennsylvania. Acta Hortic. **446**: 411–420.

Taylor, D.L., and Bruns, T.D. 1997. Independent, specialized invasions of ectomycorrhizal mutualism by two non-photosynthetic orchids. Proc. Natl. Acad. Sci. U.S.A. **94**: 4510–4515.

Taylor, D.L., and Bruns, T.D. 1999. Population, habitat and genetic correlates of mycorrhizal specialization in the "cheating" orchids *Corallorhiza maculata* and *C. mertensiana*. Mol. Ecol. **8**: 1719–1732.

Thaxter, R. 1922. A revision of the Endogonaceae. Proc. Natl. Acad. Arts Sci. **57**: 289–350.

Trudell, S.A., Rygiewicz, P.T., and Edmonds, R.L. 2003. Nitrogen and carbon stable isotope abundances support the myco-heterotrophic nature and host-specificity of certain achlorophyllous plants. New Phytol. **160**: 391–401.

van der Heijdden, M.G.A., Klironomos, J.N., Ursic, M., Moutoglis, P., Streitwolf-Engel, R., Boller, T., Wiemken, A., and Sanders, I.R. 1998. Mycorrhizal fungal diversity determines plant biodiversity, ecosystem variability and productivity. Nature (London), **396**: 69–72.

Villeneuve, N., Grandtner, M.M., and Fortin, J.A. 1991. The coenological organization of ectomycorrhizal macrofungi in the Laurentide mountains of Quebec. Can. J. Bot. **69**: 2215–2224.

Visser, S. 1995. Ectomycorrhizal fungal succession in jack pine stands following wildfire. New Phytol. **129**: 389–401.

Wang, C.J.K., and Wilcox, H.E. 1985. New species of ectendomycorrhizal and pseudomycorrhizal fungi: *Phialophora finlandia*, *Chloridium paucisporum*, and *Phialocephala fortinii*. Mycologia, **77**: 951–958.

Weber, A., Karst, J., Gilbert, B., and Kimmins, J.P. 2005. *Thuja plicata* exclusion in ectomycorrhiza-dominated forests: testing the role of inoculum potential of arbuscular mycorrhizal fungi. Oecologia (Berl.), **143**: 148–156.

Wilcox, H.E., Yang, C.S., and LoBuglio, K.F. 1983. Responses of pine roots to E-strain ectendomycorrhizal fungi. Plant Soil, **71**: 293–297.

Xiao, G., and Berch, S.M. 1996. Diversity and abundance of ericoid mycorrhizal fungi of *Gaultheria shallon* on forest clearcuts. Can. J. Bot. **74**: 337–346.

Xiao, G., and Berch, S.M. 1999. Organic nitrogen use by salal ericoid mycorrhizal fungi from northern Vancouver Island and impacts on growth in vitro of *Gaultheria shallon*. Mycorrhiza, **9**: 145–149.

Young, B.W., Massicotte, H.B., Tackaberry, L.E., Baldwin, Q.F., and Egger, K.N. 2002. *Monotropa uniflora*: morphological and molecular assessment of mycorrhizae retrieved from sites in the Sub-Boreal Spruce biogeoclimatic zone in central British Columbia. Mycorrhiza, **12**: 75–82.

Yu, T.E.J.-C., Egger, K.N., and Peterson, R.L. 2001. Ectendomycorrhizal associations—characteristics and functions. Mycorrhiza, **11**: 167–177.

Zelmer, C.D., and Currah, R.S. 1995*a*. Evidence for a fungal liaison between *Corallorhiza trifida* (Orchidaceae) and *Pinus contorta* (Pinaceae). Can. J. Bot. **73**: 862–866.

Zelmer, C.D., and Currah, R.S. 1995*b*. *Ceratorhiza pernacatena* and *Epulorhiza calendulina* spp. nov.: mycorrhizal fungi of terrestrial orchids. Can. J. Bot. **73**: 1981–1985.

Zhang, Y.H., and Zhuang, W.Y. 2004. Phylogenetic relationships of some members in the genus *Hymenoscyphus* (Ascomycetes, Helotiales). Nova Hedwigia, **78**: 475–484.

Chapter 2
From a germinating spore to an established arbuscular mycorrhiza: signalling and regulation

José M. Garcia-Garrido and Horst Vierheilig

Introduction

The arbuscular mycorrhizal (AM) association is a symbiosis between root colonizing soilborne fungi and the majority of land plant species. Root colonization by AM fungi enhances plant growth. This is due to improved nutrient uptake (particularly of P) (Smith and Read 1997) and increased protection against soilborne pathogenic fungi (see review by St-Arnaud and Vujanovic (2007); see also Chapter 3). Briefly, the symbiosis can be divided into three phases (see also Chapter 3). The asymbiotic phase, when the AM fungal spore germinates and the hypha grows in absence of plant signals; the presymbiotic phase, during which AM fungal hyphal growth and differentiation occurs in the presence of signals exuded by plants; and the symbiotic phase, following colonization of the root, during which there is formation of intraradical structures (e.g., inter- and intra-radical hyphae, arbuscules and, in some species, vesicles) and an exchange of nutrients between the host and the symbiont.

The formation of the AM symbiosis is a complex developmental event requiring the coordination of the gene expression of both partners. Considerable progress has been made recently in the study of the pathway that results in AM formation, especially with regard to the host plant. This has been aided by the use of genetic tools (e.g., mutants affected in different stages of nodulation and AM formation) and, more recently, the use of DNA array techniques. The majority of these studies have been performed using leguminous plants, and the results obtained highlight the similarities and differences between the processes involved in nodulation and mycorrhization. Unfortunately, because of the difficulty of working with an obligate symbiont, information pertaining to AM fungal gene regulation and the production of fungal signals during AM formation is still scarce.

Signal exchanges have been reported during the presymbiotic phase, resulting in directional growth of existing AM fungal hyphae towards the host root and increased spore germination, hyphal growth, and hyphal branching. Together, these lead to the formation of appressoria, which are the AM fungal structures developed prior to root penetration (Fig. 2.1). Once the AM fungus has penetrated the root and formed intraradical structures, the plant defence system is triggered and a cascade of symbiotic events is initiated. Successful formation of the symbiosis seems to rely on a delicate balance of plant hormones and nutrients.

Recently, it has been suggested that, once the symbiotic phase is well established, the plant tries to limit further colonization by AM fungi. In analogy to the well-documented autoregulation in the *Rhizobium*–legume interaction, this regulatory process has been named mycorrhizal autoregulation. In the present work, we discuss the most recent data on signalling events and regulatory mechanisms during the presymbiotic (Fig. 2.1) and the symbiotic phases of the AM association.

J.M. Garcia-Garrido. Departamento de Microbiología de Suelos, Estación Experimental del Zaidín, CSIC, E-18008 Granada, Spain.

H. Vierheilig.[1,2] Institut für Pflanzenschutz (DAPP), Universität für Bodenkultur Wien, Peter Jordan-Strasse 82, A-1190 Wien, Austria.

[1]Corresponding author: (e-mail: horst.vierheilig@eez.csic.es).

[2]Present address: Estación Experimental del Zaidín, CSIC, Apdo. 419, E-18008 Granada, Spain.

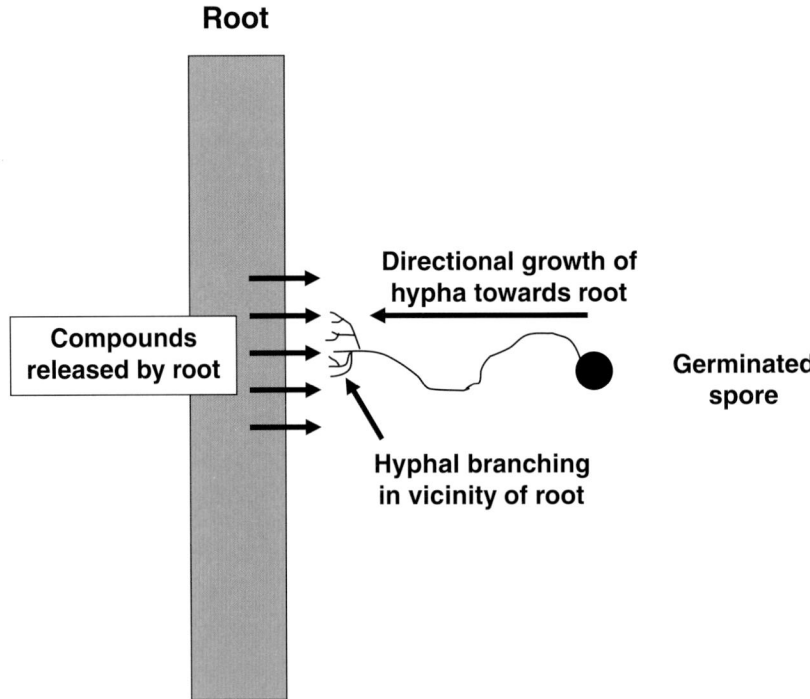

Fig. 2.1. Spore germination and hyphal growth is stimulated by root exudates. Hyphae show directional growth towards roots of host plants and a characteristic branching in the vicinity of roots. Hyphal branching is due to compounds in root exudates that may be a prerequisite for successful root colonization by arbuscular mycorrhizal fungi.

Presymbiotic phase

Root exudates, spore germination, and hyphal growth

Evidence from several plant–microbe interactions shows that root exudates act as primary plant signals (e.g., root exudates of legumes trigger a complex cascade of signals, from both the plant and bacteria of the genus *Rhizobium*) that finally result in nodule formation (Phillips and Tsai 1992). Root exudates also have an effect on AM fungi (see reviews by Morandi (1996) and Vierheilig et al. (1998*b*)). Root exudates of AM host plants stimulate spore germination (Gianinazzi-Pearson et al. 1989; Schreiner and Koide 1993*a*; Suriyapperuma and Koske 1995) and hyphal growth (e.g., Hepper and Mosse 1975; Graham 1982; Elias and Safir 1987; Bécard and Piché 1989; Gianinazzi-Pearson et al. 1989; Giovannetti et al. 1993*b*, 1994; Schreiner and Koide 1993*b*; Balaji et al. 1995; Pinior et al. 1999). Moreover, when root exudates are applied to plants inoculated with AM fungi (Tawaraya et al. 1998; Pinior et al. 1999), root colonization is clearly enhanced. Together, these data suggest that root exudates are the primary plant signals used during the establishment of the AM symbiosis.

A number of studies have shown that the concentration of P in plant tissues affects root colonization by AM fungi; however, the exact mechanisms involved are still unknown (Smith and Read 1997). Root exudates from plants grown in P-limited substrates stimulate hyphal growth (Elias and Safir 1987; Nagahashi et al. 1996; Tawaraya et al. 1996; Vierheilig et al. 1998*b*) and root colonization (Tawaraya et al. 1998; Pinior et al. 1999). By contrast, root exudates from plants with high P content stimulate hyphal growth to a lesser extent (Elias and Safir 1987; Tawaraya et al. 1996), and they do not enhance root colonization (Tawaraya et al. 1998). These results indicate that root exudates are involved, at least partially, in the regulation of root colonization by AM fungi.

The flavonoids in root exudates are known to play a key signalling role in a number of plant–microbe interactions. For example, in the *Rhizobium*–legume interaction, flavonoids act as chemoattractants for the bacteria and as specific inducers of their nodulation genes (*nod* genes), which are involved in the synthesis of lipochitooligosaccharide signals (Nod factors) (see review by Perret et al. (2000)). Nod factors induce the accumulation of flavonoids in roots of legumes, resulting in the secretion of even greater amounts of flavonoids by the root. In turn, this, further stimulates the production of Nod factors by the bacteria (Recourt et al. 1992; Dakora et al. 1993; Schmidt et al. 1994; Bolaños-Vasquéz and Werner 1997). A similar process seems to play an important role at the beginning of signal exchange between host plants and AM fungi.

Extensive data are available concerning the effect of flavonoids on hyphal growth and root colonization (see reviews by Morandi (1996) and Vierheilig et al. (1998b)). A number of flavonoids exhibit a clear stimulatory effect, and this effect seems to be dependent on the chemical structure of the molecule involved (Bécard et al. 1992; Chabot et al. 1992). The stimulatory effect of certain flavonoids is enhanced in the presence of CO_2 concentrations reflecting those found in the rhizosphere (Bécard et al. 1992; Chabot et al. 1992; Poulin et al. 1993). Recent studies by Scervino et al. (2005a, 2005b, 2005c) show that, during certain stages of the presymbiotic phase or when applied to plants inoculated with AM fungi, flavonoids exhibit an AM fungal genus- and even species-specific effect. However, although these data indicate that flavonoids are important signalling compounds for the establishment of the AM association, flavonoid patterns are relatively specific to each plant group. As a result, their role as general signalling compounds has been questioned (Bécard et al. 1995) and, only recently, has a general signalling compound from another chemical group present in root exudates been identified (see below).

Hyphal branching

As well as affecting hyphal growth, compounds released by host roots induce branching of AM fungal hyphae (Mosse and Hepper 1975; Powell 1976; Mosse 1988; Giovannetti et al. 1993a, 1994) (see also Chapter 3). Hyphal branching is considered to be an important event in host root recognition, and it precedes successful root colonization (Giovannetti and Sbrana 1998; Buée et al. 2000; Nagahashi and Douds 2000): the branching response increases "... the probability of encounter with a site on the root suitable for colonization ..." (Douds and Nagahashi 2000). Active branching signals have been reported from a wide range of host plants (Buée et al. 2000; Nagahashi and Douds 2000). Recently, Tamasloukht et al. (2003) determined the time course of a root-exuded branching factor. Application of the branching factor to a single hypha resulted in gene activation in *Gigaspora rosea* Nicol & Schenck after 0.5–1 h, alterations at the physiological level after 1.5–3 h, and hyphal branching after 5 h.

A hyphal branching effect has been reported for a number of flavonoids (Tsai and Phillips 1991; Phillips and Tsai 1992; Scervino et al. 2005a, 2005b, 2006); however, because the flavonoids that induce branching are only found in a limited number of plants, their role as general signalling compounds for hyphal branching can probably be discarded. A number of studies have attempted to identify the branching factor in root exudates of AM host plants (Nagahashi and Douds 1999, 2000, 2003; Buée et al. 2000); however, the molecules involved seem to occur at low concentrations and to be unstable. Nevertheless, a recent study has identified a group from the sesquiterpenes (strigolactones) in root exudates of *Lotus japonicus* (Regel) Larsen that has been clearly identified as the AM fungal hyphal branching factor (Akiyama et al. 2005). In addition, strigolactones also have a stimulatory effect on AM fungal spore germination (Besserer et al. 2006).

Although strigolactones are known to stimulate the germination of seeds of root parasites in the genera *Striga* and *Orobanche*, they have also been detected in the root exudates of plants that are not hosts to these parasites. Strigolactones have been reported from a wide range of plants (see review by Steinkellner et al. (2007)) including *Trifolium pratense* L. (Yokota et al. 1998), *Vigna unguiculata* Walp (Müller et al. 1992; Matusova et al. 2005), *Gossypium hirsutum* L. (Cook et al. 1972), *Sorghum bicolor* (L.) Moench (Hauck et al. 1992; Siame et al. 1993; Matusova et al. 2005), *Zea mays* L. and *Pennisetum glaucum* (L.) R. Br. (Siame et al. 1993), and *Solanum lycopersicum* L. (Yoneyama et al. 2004); thus, their role as a general signalling compound for the establishment of the AM symbiosis in all AM host plants has been suggested (Akiyama et al. 2005).

Recently, there have been more indications supporting the importance of strigolactones for the AM symbiosis. Because strigolactones are thought to be derived from the carotenoid pathway (Matusova et al. 2005), alterations of the carotenoid pathway could affect strigolactone production and, thus, alter root colonization by AM fungi. Gomez-Roldan et al. (2007), working with *Z. mays* mutants with a defect in the carotenoid pathway and *Z. mays* plants treated with an inhibitor of the carotenoid pathway, showed that root colonization by *G. rosea* was reduced in both treatments. Because the application of a strigolactone analogue to the inoculated mutant and inhibitor-treated plants restored AM root colonization to a similar levels to that of the control plants, it was concluded that the alterations of the carotenoid pathway resulted in reduced levels of strigolactones in the root exudates; in turn, this negatively affected root colonization by the AM fungus.

Infection of mycorrhizal *S. bicolor* and *Z. mays* by *Striga* spp. is lower than that of nonmycorrhizal plants (Lendzemo and Kuyper 2001; Gworgwor and Weber 2003; Lendzemo et al. 2005). Recent experiments (Fig. 2.2) have shown that root exudates of mycorrhizal *S. bicolor* and *Z. mays* reduce *Striga* seed germination when compared with exudates from nonmycorrhizal control plants (Lendzemo 2004; Matusova et al. 2005; Lendzemo et al. 2007). From these data, Matusova et al. (2005) hypothesized that AM formation might alter the level of strigolactones through a downregulation of the production of the mycorrhizal branching factor and, thus, affect the seed germination of parasitic plants. In vitro studies investigating the effect of root exudates from AM fungal colonized plants and noncolonized controls on the growth of AM fungi showed a reduction in branching in the presence of root exudates from mycorrhizal plants when compared with exudates from nonmycorrhizal controls (Pinior 1999; Pinior et al. 1999). Moreover, in several studies, it has been demonstrated that, once plants are mycorrhizal, further root colonization by AM fungi is slower than in nonmycorrhizal plants (Vierheilig et al. 2000b, 2000c; Catford et al. 2003; Vierheilig 2004b; Meixner et al. 2005, 2007). From these results, it is tempting to speculate that reduction in strigogalactone production reduces further root colonization by AM fungi (Fig. 2.3).

The hypothesis that branching factor(s) in root exudates might serve as a regulating mechanism for root colonization is further strengthened when the effect on hyphal branching of root exudates from plants with high tissue levels of P is considered. It has been reported that strigogalactone production is reduced in roots of plants grown in high P treatments (Yoneyama et al. 2001, 2007). Hyphal branching in the presence of root exudates from plants grown in low-P treatments has been reported in a number of studies (Nagahashi and Douds 1999, 2000, 2003; Buée et al. 2000). When the effect of root exudates collected from low- and high-P plants was compared, a clear stimulation of branching was observed with root exudates from low-P plants; however, root exudates collected from high-P plants exhibited no branching nor stimulatory activity (Nagahashi et al. 1996; Nagahashi and Douds 2000). To summarize, in general, a reduction in root colonization is observed in plants already colonized by AM fungi or in plants with high P levels. This is possibly linked with a reduction in the concentration of branching factor(s) in the exudates from the roots. We are convinced that, in the near future, further work in this field will allow many questions concerning signalling compounds for AM fungi in root exudates to be answered.

Recent work by Steinkellner et al. (2007) has shown that the role of strigolactones in the signalling between soilborne fungi and plants is possibly limited to AM fungi, because the application of the strigolactone analogue GR24, which has been shown to induce hyphal branching in the former group of fungi, did not induce branching in ectomycorrhizal fungi or soilborne pathogenic fungi and showed no effect on the germination of microconidia of *Fusarium oxysporum* Schltdl.

Chemotropism

Chemotropism is an important step for the establishment of soilborne pathogenic and symbiotic plant–fungus associations. Energy resources in fungal propagules are limited, and a directional growth response towards a potential host is likely important. A positive chemotropic response resulting in the formation of a symbiosis would allow the AM fungus to switch from presymbiotic growth, which depends on the energy resources of the spore, to symbiotic growth, which is driven by energy resources from the host.

Relatively few studies have treated chemotropism in AM fungi. However, in vitro studies have shown that, in the presence of volatiles from AM host plant roots, *Gigaspora gigantea* (Nicol. & Gerd.) Gerd. & Trappe germ tubes grow toward the root (Koske 1982; Gemma and Koske 1988; Suriyapperuma and Koske 1995). Working with a soil–sand substrate in a compartmental system, Vierheilig et al. (1998a) also showed directional hyphal growth of *Glomus mosseae* (Nicol. & Gerd.) Gerd. & Trappe toward a host plant root. This was recently confirmed in an elegant membrane-based experiment, where an attractional effect was observed at a distance of at least 910 µm from the root (Sbrana and Giovannetti 2005).

Very little is known concerning the nature of the host-derived signals involved in AM fungal chemotropism, and the data available are contradictory. In in vitro studies with *G. gigantea*, volatiles seemed to play a role in the observed chemotropism (Koske 1982; Gemma and Koske 1988; Koske and Gemma 1992; Suriyapperuma and Koske 1995). By contrast, in the study by Vierheilig et al. (1998a) using

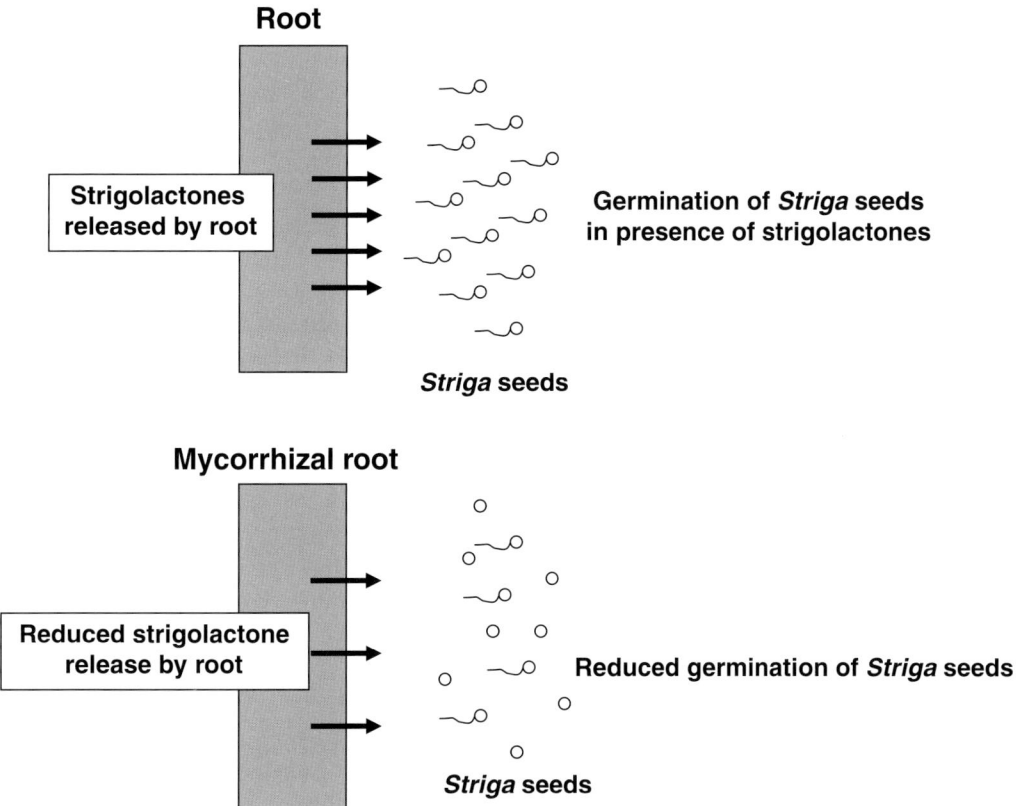

Fig. 2.2. *Striga* seed germination is induced in the presence of strigolactones that are released by roots. Root exudates of mycorrhizal plants induce a lower rate of germination than root exudates of nonmycorrhizal plants (Lendzemo 2004; Matusova et al. 2005; Lendzemo et al. 2007). This seems to be due to a reduction in the release of strigolactones by mycorrhizal roots.

Glomus mosseae (Nicol. & Gerd.) Gerd. & Trappe, the results suggested the involvement of water-soluble root exudates.

An experimental setup to test the effect of root exudates under in vitro conditions in the absence of a host root is a prerequisite before we can further characterize compound groups involved in AM fungal chemotropism. However, when root exudates from AM host plants were added to Petri dishes in the presence of germinated spores of *G. rosea* and *G. mosseae*, no attractional effect was observed (H. Vierheilig, unpublished results). Because of the difficulty of doing chemotropism studies with these obligate biotrophs, the exact mechanisms involved in the observed chemotropism are still unknown.

Arbuscular mycorrhizal fungal-to-plant signals

In analogy to the Nod factors in the *Rhizobium*–legume interaction, the existence of AM fungal signalling factors (possible Myc factors) that are perceived by the AM host plant prior to direct contact has been suggested (Albrecht et al. 1998; Blilou et al. 1999; Vierheilig and Piché 2002; Bécard et al. 2004; Vierheilig 2004a). The first indication of the existence of an AM fungal-derived signal comes from a study by Simoneau et al. (1994). The authors found that Ri T-DNA-transformed *S. lycopersicum* roots treated with a spore extract of *Glomus intraradices* Schenck & Smith produced a suite of new polypeptides. Further evidence for the presence of an AM fungal signalling factor was obtained from studies using nonhost plants within the family Brassicaceae. Briefly, alterations in the hydrolitic activity of β-1,3-glucanase and chitinase (Vierheilig et al. 1994) and the level of glucosinolates (Vierheilig et al. 2000a) were observed in the plant roots following inoculation with AM fungi. This hypothesis was further confirmed by Larose et al. (2002), who observed alterations in flavonoid levels in *Medicago sativa* L. roots treated with spores and hyphal fragments of *G. intraradices*. Recently, it was reported that a diffusible factor from different AM fungi growing in vitro not only induced symbiosis-spe-

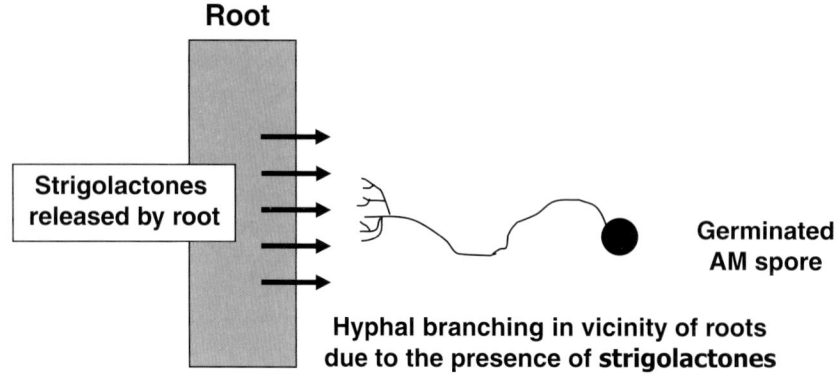
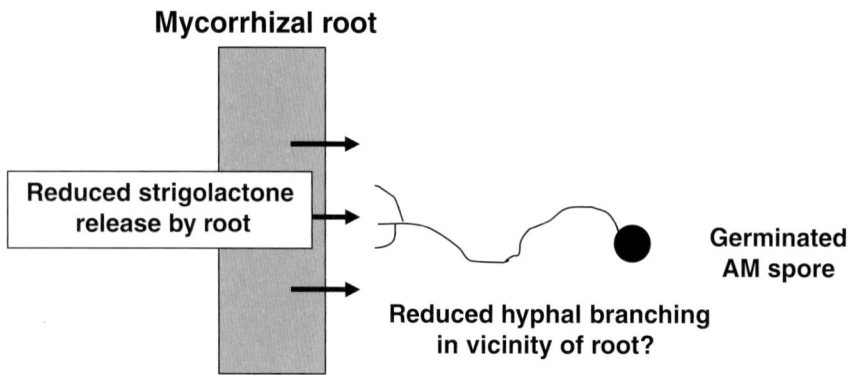

Fig. 2.3. Germination experiment with seeds of *Striga* (see Fig. 2) suggest a reduction in the release of strigolactones by mycorrhizal roots. These molecules have been shown to induce hyphal branching (Akiyama et al. 2005), which is a step thought to precede successful root colonization; thus, a reduction in strigolactone release by an already mycorrhizal plant would result in reduced hyphal branching in the vicinity of the root and could lead to a reduction in further colonization of the root by arbuscular mycorrhizal fungi (Meixner et al. 2005, 2007; Steinkellner et al. 2007).

cific MtENOD11 expression in roots (Kosuta et al. 2003) but also stimulated lateral root formation in *Medicago truncatula* Gaertn. (Olah et al. 2005).

Nothing is yet known about the chemical structure of the possible Myc factor. Furthermore, because some of the studies used hyphal fragments and spores (Simoneau et al. 1994; Larose et al. 2002) and others used diffusible molecules from AM fungi growing in vitro (Kosuta et al. 2003; Olah et al. 2005), it is not yet clear whether the effects observed with these different approaches are due to the same AM fungus-derived signal(s).

In the *Rhizobium*–legume symbiosis, roots secrete (iso)flavonoids, which induce bacterial genes involved in the synthesis of so-called Nod factors (lipochitooligosaccharides). These allow rhizobia to enter the root through infection threads (Perret et al. 2000). Recently, Catford et al. (2003) speculated that chitooligosaccharides released from cell walls of AM fungi might be the stimuli triggering regulatory events during mycorrhization. A certain functional similarity between the rhizobial and the mycorrhizal signal was reported by Vierheilig (2004a). The author treated *M. sativa* roots with AM fungal tissue or with Nod factor, and found similar changes in root flavonoid patterns in both cases.

Although the rhizobial and mycorrhizal signals might share certain similarities, they do not seem identical. In roots of *M. truncatula*, an AM fungus-derived signal induced the expression of the rhizobial symbiosis-specific gene MtENOD11. This was similar to the results obtained for roots treated with Nod factor; however, the localization of the gene was not identical. Working with *M. truncatula* mutants, Olah et al. (2005) also found certain functional differences. Briefly, not all the genes required for the stimulation of lateral root formation in the presence of Nod factor were required for stimulation of lateral root formation in the presence of AM fungal-derived signal(s). Moreover, by comparing the root formation in the presence of AM fungal-derived signal(s) with that observed at different auxin concentrations, the authors concluded that the observed fungal-induced branching effect was not due to auxins (Olah et al. 2005). Furthermore, the AM fungal factor seems specific to the Glomeromycota, because no MtENOD11 gene expression was observed in the presence of different fungal pathogens (Kosuta et al. 2003).

At the moment, we can only speculate as to the role of AM fungal-derived signal(s) during the establishment of the AM association. Larose et al. (2002) hypothesized that, in the presence of AM fungal signals, a more favourable environment for root penetration is created because of changes in the host's flavonoid pattern. Olah et al. (2005) proposed that the signal leads to the creation of "… new potential infection sites, e.g., by stimulating the formation of new (lateral) roots ..."; however, further studies are needed to fully understand the role of the early signalling events during the formation of the AM association.

Appressorium formation

The last step in the root – AM fungus interaction prior to colonization is the formation of an appressorium on the root surface; formation of this structure indicates that the fungus has recognized a potential host (Giovannetti et al. 1993b). The fact that appressoria may be formed on excised roots (Giovannetti et al. 1993b), on roots of host and nonhost plants grown in the dark (Vierheilig et al. 2002), and on fragments of *Daucus carota* L. root cell walls (Nagahashi and Douds 1997) led these authors to conclude that appressoria formation "... is a contact recognition process not requiring a chemical signal ... ." However, a specific signal seems necessary for further penetration (Douds and Nagahashi 2000).

Symbiotic stage

Regulation of mycorrhizal formation by induction or suppression of plant defence mechanisms

Certain events similar to those found in plant–pathogen interactions have been observed during plant–AM fungal interactions; these include signal perception, signal transduction, and defence gene activation. Two types of defence response can be distinguished during the development of AM fungi in roots: an initial response during the first stages of development with a slight and transitory character and a second defence response, which is stronger and more restricted to those cells that contain fungal arbuscules (see review by García-Garrido and Ocampo (2002)).

In a number of studies, it has been hypothesized that AM fungi possess surface molecules (e.g., chitin and glucans), which act as elicitors of the defence responses in plants (e.g., Lambais 2000; Salzer and Boller 2000). The concentration of these molecules could be particularly high at the arbuscular interface, where the fungal cell walls are thin and chitin is not highly polymerized (Bonfante-Fasolo et al. 1990; Balestrini and Bonfante 2005). Arbuscular mycorrhizal fungal elicitors seem to trigger plant defences at two scales: generally, at the initial stages of colonization and, locally, in cells containing arbuscules. Studies have revealed that three elements of the defence signal transduction pathway are activated in AM roots: (*i*) an oxidative burst (Salzer et al. 1999; Fester and Hause 2005), (*ii*) the activation of the metabolism of flavonoid compounds (Lambais 2000; Bonanomi et al. 2001*a*), and (*iii*) the accumulation of salicylic acid (SA) (Blilou et al. 2000*a*, 2000*b*). In mycorrhizal *Nicotiana tabacum* L. plants, transient increases of catalase and peroxidase activity (enzymes related to oxidative stress) coincided with appressoria formation and fungal penetration of the root (Blilou et al. 2000*a*). A similar pattern of peroxidase activity has been observed in *Allium porrum* L. and bean roots inoculated with AM fungi (Spanu and Bonfante-Fasolo 1988; Lambais 2000). The transient increase in catalase and peroxidase activity observed in mycorrhizal *N. tabacum* roots also coincided with the accumulation of SA (Blilou et al. 2000*a*). Members of different classes of plant defense genes, including genes encoding for the phenylpropanoid metabolism enzymes (Harrison and Dixon 1993, 1994; Volpin et al. 1994, 1995), for enzymes involved in the metabolism of reactive oxygen species (Blee and Anderson 2000), and for plant hydrolases (Lambais and Mehdy 1993, 1998; Blee and Anderson 1996; David et al. 1998; Salzer et al. 2000), have also been detected in plant cells containing arbuscules.

Recently, the use of transcriptomic profiling studies resulted in the discovery of a defence-related gene that was expressed in AM roots (Brechenmacher et al. 2004; Güimil et al. 2005; Hohnjec et al. 2005; Liu et al. 2007). In *Oryza sativa* L. roots, 45% of the mycorrhizal-induced *O. sativa* genes responded similarly to infection by fungal pathogens. This might be a molecular reflection of the high similarity between interactions relying on related infection strategies (Güimil et al. 2005).

The induction of defence gene expression during the formation and maintenance of the AM symbiosis could be considered to be a result of fungal elicitor recognition and the activation of signal transduction pathways. The relatively low level of defence gene activation, its transient nature, and its localization around arbuscules could be the result of a mechanism induced in the plant that suppresses or regulates the activated defence response, thus allowing fungal growth in roots. How AM fungi achieve the functional suppression of the plant's defence responses remains unclear; however, experimental evidence suggests that arbuscules may play a central role (Harrison 2005). This hypothesis is supported by two observations. Firstly, in associations with biotrophic fungal pathogens, defence responses are similarly regulated, and haustoria play a role in the suppression of the plant's defence responses. The arbuscules formed by AM fungi are structures analogous to the haustoria of pathogenic biotrophic fungi, and they could play a similar role in mediating defence suppression. Secondly, the spatial and temporal expression pattern of some defence genes, such as those encoding for basic endochitinase, β-1,3-endoglucanase, and genes whose products are involved in isoflavonoid phytoalexins synthesis, is consistent with a regulation mediated by the arbuscules. Briefly, during the initial stages of colonization, the expression of the above genes increases. This is followed by a downregulation that coincides with the onset of arbuscule formation. This negative effect on defence gene expression is probably maintained by the constant formation of new arbuscules by the AM fungus. At present, the mechanisms responsible for the suppression effect have not been determined.

Regardless of the origin of the suppressive effect, the defence response pattern activated in AM roots shows that a mechanism exists that regulates the suppression of the plant's defence system, thus allowing formation of the symbiosis. Two possible mechanisms that may attenuate the plant's defence response in the AM symbiosis must be considered: (*i*) the degradation of exogenous elicitor molecules produced by the AM fungus and (or) the prevention of endogenous elicitor release from the plant cell wall and (*ii*) the regulation mediated by alterations in the signal transduction pathway leading to the defence response.

Studies suggest that hydrolases may play a regulatory role in the AM symbiosis. Constitutive and mycorrhiza-specific chitinases, chitosanases, and β-1,3-glucanase isoforms (potential hydrolases of fungal elicitors) are differentially regulated during AM development (Dumas-Gaudot et al. 1992; Dassi et al. 1996; Pozo et al. 1998; Salzer et al. 2000). In situ analysis of a mycorrhiza-specific chitinase of *M. truncatula* (*Mtchit 3-3*) showed a specific pattern of transcript accumulation in cells containing arbuscules (Bonanomi et al. 2001*b*). Salzer and Boller (2000) proposed that, in the early stages of colonization, constitutively expressed plant chitinases are the hydrolases responsible for elicitor degradation (and, consequently, the attenuation of the plant's defence system), and that mycorrhiza-specific chitinase isoforms play a similar role in later developmental stages. However, Lambais (2000) suggested that the level of P probably regulated the factor responsible for the breakdown of the fungal elicitor molecules and that it is not related to chitinase activity in the root.

The intracellular root colonization by AM fungi creates a new interface compartment, which is composed of membranes from both partners separated by apoplastic material containing molecules common in the primary wall of plant cells: cellulose, pectin, xyloglucan, and hydroxyproline-rich glycoprotein (Bonfante 2001; Balestrini and Bonfante 2005). The lytic activity of plant and (or) fungal enzymes on these molecules could, in theory, generate oligomeric fragments that could act as elicitors for a localized defence response at the arbuscule level. However, the prevention of endogenous elicitor formation as a mechanism that mediates defence regulation should also be considered. The production of endogenous elicitors derived from the degradation of cell wall material does not seem to be an effective mechanism for the induction of the plant's defence response. Evidence for this hypothesis is provided by the fact that AM fungi produce little in the way of plant cell wall degrading enzymes (García-Romera et al. 1991; García-Garrido et al. 1992). Furthermore, host plants activate mycorrhiza-specific plant cell wall degrading enzyme genes (Maldonado-Mendoza et al. 2005) that alter the plant's cell wall structure without weakening it. This allows fungal growth but avoids lyses and the consequent production of endogenous elicitors.

A second mechanism that could attenuate the plant's defence response is the blockage of components of the signal transduction pathway that activate the response. Among these components, SA and reactive oxygen species (ROS) have been implicated as secondary messengers in AM associations. Direct and indirect evidence suggests that the levels of H_2O_2 in mycorrhizal associations are increased

(Salzer et al. 1999; Fester and Hause 2005). These include alterations in the production of antioxidative enzymes, such as catalase and peroxidase (Arines et al. 1994; Blilou et al. 2000*a*; Lambais 2000; Lambais et al. 2003), and increased amounts of jasmonates (Hause et al. 2002; Vierheilig and Piché 2002). In AM roots of *Z. mays*, the pattern of ROS accumulation in the cytoplasm of plant cells adjacent to fungal structures and on the surface of fungal hyphae suggests a relationship between ROS accumulation and arbuscule senescence (Fester and Hause 2005). Although the molecular significance of this localized cytoplasmic accumulation of ROS is not known, the oxidative compounds produced during the colonization process could activate the plant's defence response; a similar oxidative burst occurs in plant–pathogen interactions. The accumulation of antioxidative enzymes in cells containing arbuscules may be due to a localized regulation of the defence mechanism.

It is not known whether the regulation of SA levels in mycorrhizal plants is responsible for the suppression of plant defence; however, some experimental data support a potential role for SA in the regulation of root colonization. Blilou et al. (2000*a*, 2000*b*) observed transient increases of SA in AM *N. tabacum* and *O. sativa* roots. Furthermore, exogenous application of SA to *O. sativa* roots did not affect appressoria formation but delayed formation of the symbiosis (Blilou et al. 2000*b*). Both these results suggest that the regulation of the plant's defence response against AM fungi may be through the SA pathway. This is further supported by the fact that the SA content in the root affects colonization by AM fungi (Herrera-Medina et al. 2003). Experiments using transgenic *N. tabacum* plants showed that plants with reduced levels of SA (NahG) exhibited enhanced levels of root colonization, whereas plants with enhanced levels of SA (CSA) exhibited reduced levels of root colonization (Herrera-Medina et al. 2003). The hypothesis that the SA pathway participates in the regulation of mycorrhization is reinforced by the fact that enhanced SA levels were linked to the inability of Nod/Myc mutant *Pisum sativum* L. plants to form AM (Blilou et al. 1999).

Activation of a symbiotic genetic program

An important experimental approach has been the study of the interaction between the processes of AM development and nodule formation by bacteria in legume plants. Physiological, ecological, and biochemical data suggest a positive interaction between both processes in legume plant roots. Similarities at the molecular, cytological, and genetic level between both symbioses have been highlighted in a number of reviews (Gianinazzi-Pearson and Dénarié 1997; Hirsch and Kapulnik 1998; Parniske 2000; Guinel and Geil 2002; Kistner and Parniske 2002).

Molecular studies revealed that a number of genes are induced during both symbiotic interactions. Immunolocalization and hybridization showed the expression of early nodulin genes (van Rhijn et al. 1997), the leghaemoglobin gene *VFLb29* (Frühling et al. 1997), and the aquaporin related gene (Wyss et al. 1990) in both symbioses. Nevertheless, most scientific evidence for a conservative signal pathway in both symbioses has come from the characterization of SYM (symbiotic) mutant legume plants. The large number of these mutants allows the possibility of genetically identifying steps and genes that are essential for root symbioses with bacteria and fungi. The study of phenotypic alterations in the mycorrhizal colonization of these mutants, which are impaired in their ability to form nodules, has revealed the common *Sym* genes and the stage at which mycorrhizal colonization is blocked in each mutant (see Peterson and Guinel (2000) and Marsh and Schultze (2001)).

Two well-defined groups of phenotypic AM mutants have been identified. In the first and most common group, the process of fungal infection is blocked at the root epidermis. These mutants are unable to form infection threads in the interaction with *Rhizobium*. These mutants are well known at the cytological level, and some of the mutated genes have been cloned. The second group consists of mutants that are nodule defective (incapable of fixing nitrogen); these are also unable to form normal arbuscules in the symbiosis with AM fungi (Gianinazzi-Pearson et al. 1991; Resendes et al. 2001; Jacobi et al. 2003*a*, 2003*b*)

Symbiotic impaired mutants have been identified in *Melilotus alba* Medik., *Vicia faba* L., *Phaseolus vulgaris* L., *P. sativum*, *L. japonicus*, and *M. truncatula* (Marsh and Schultze 2001; Murray et al. 2006*a*, 2006*b*), but the clearest orthologous relationship between *Sym* genes has been established between *L. japonicus*, *P. sativum*, and *M. truncatula* (Parniske 2004). The group of extensively studied *Sym* genes consists of seven from *L. japonicus* (*LjSym2*, *LjSym3*, *LjSym4*, *LjSym6*, *LjSym15*, *LjSym23*, and *LjSym24*; Kistner et al. 2005), four from *P. sativum* (*PsSym8*, *PsSym19*, *PsSym30*, and *PsSym36*; Peterson and Guinel 2000), and three from *M. truncatula*

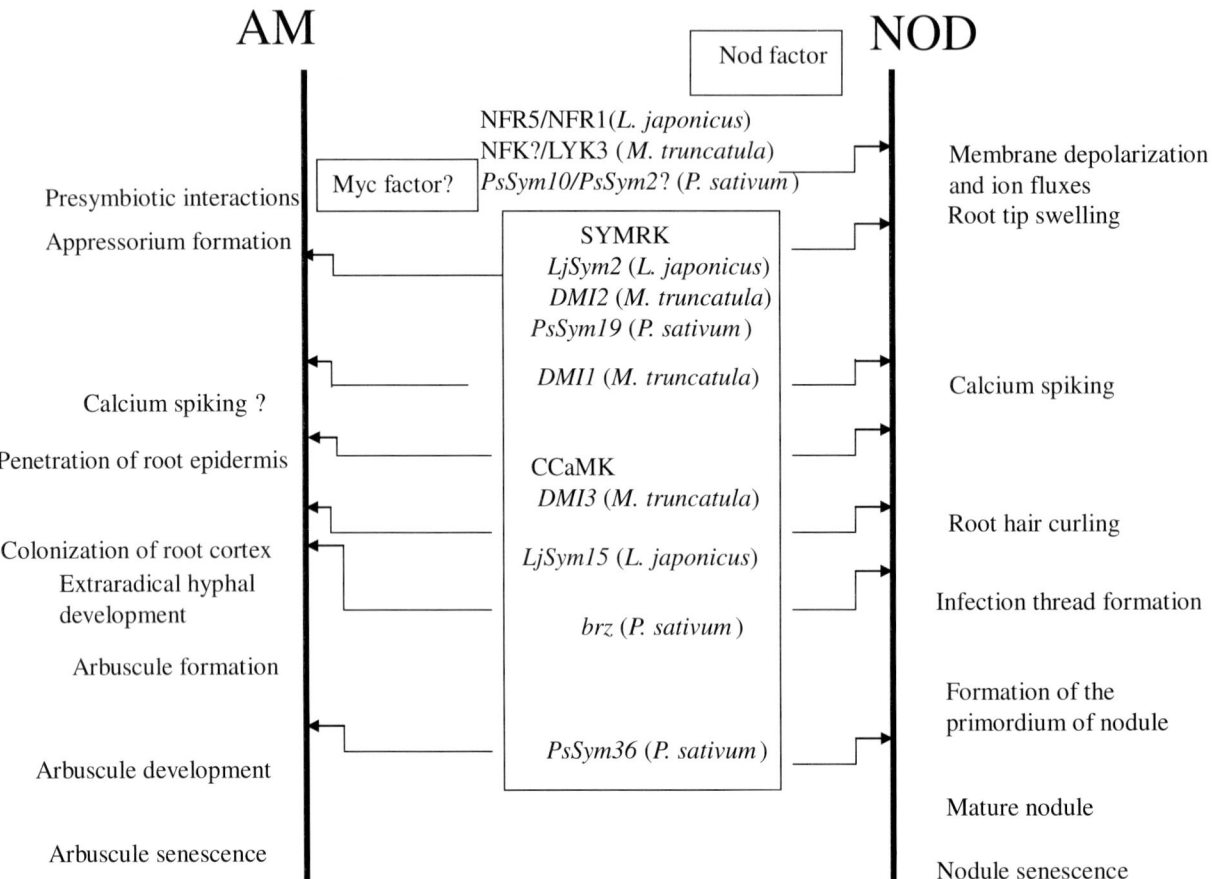

Fig. 2.4. Plant genes required for the formation of the arbuscular mycorrhizal (AM) symbiosis and the formation of nodules (NOD). Plant genes required for both symbioses are located in the central square. Genes exclusive to the rhizobial pathway are located above the central square. Arrows show the location of each gene in relation to well-defined events in the two symbioses.

(*DMI1*, *DMI2*, and *DMI3*; Catoira et al. 2000). A mycorrhizal mutant *S. lycopersicum* plant (nonleguminous) has also been isolated (Barker et al. 1998); however, at the moment, it is unknown if the mutated gene (reduced mycorrhizal colonization (*rmc*)) has an orthologue among the common *Sym* genes.

Results of physiological, biochemical, and genetic analyses of the above-mentioned SYM mutants show that the mutated genes are involved in the perception of microbial signals and (or) the transduction pathways initiated following perception. Figure 2.4 summarizes our current knowledge concerning the signals, structural and morphological cell adaptation events, and the common genetic loci that are involved in the processes of nodulation and AM fungal colonization. It has been proposed that several of the nodulation genes coding for plant kinases that have been cloned in *L. japonicus*, *M. truncatula*, and *P. sativum* (Limpens et al. 2003; Madsen et al. 2003; Radutoiu et al. 2003) may be bacterial Nod factor receptors (NFR) (see Parniske and Downie (2003)). These protein kinases contain LysM motifs in their extracellular domains, which are implicated in binding N-acetyl-glucosamine-containing molecules. These receptor kinases acts upstream of the common pathway. *Lotus* plant mutants in *NFR1* and *NFR5* lack all responses to Nod factor but are able to form normal AM (Radutoiu et al. 2003), which suggests that the hypothetical AM fungal signal(s) are quite different from the Nod factor.

The genes responsible for the first common step in the signal transduction cascade were identified in *L. japonicus*, *M. truncatula*, *M. sativa*, and *P. sativum* (Endre et al. 2002; Stracke et al. 2002). The coding proteins for these genes are kinases, known as symbiosis receptor-like kinase (SYMRK) or nodulation receptor kinase (NORK). They contain an extracellularly located leucine-rich repeat domain, which could be implicated in the perception of microbial signal molecules. In the case of the *Rhizobium* symbiosis, SYMRK plant mutants respond to early events provoked by the Nod factor, such as membrane depolarization and change in ion fluxes, but do not show calcium spiking (another initial event in colonized root cells). This evidence indicates a SYMRK/NORK placement downstream of the NFR, but how the Nod factor signal is transmitted after NFR recognition via SYMRK is not clear. It is also not clear how SYMRK/NORK can integrate fungal and bacterial signals.

As mentioned above, calcium spiking (rapid oscillations in cytoplasmic calcium levels) in root hair cells is a response observed in various legume species during the early stages of nodulation (Walker et al. 2000; Oldroyd and Downie 2004). This event has been used as a marker to differentiate between phenotypes of SYM mutants and to provide a framework for the location of genes in the signalling cascade (Oldroyd and Downie 2004). This is the case for *DMI* (does not make infections) mutants of *M. truncatula*. These mutants are unable to form rhizobial and AM associations and are affected in the early stages of interaction (Catoira et al. 2000). The *DMI1* and *DMI2* gene products act upstream of the calcium spiking, whereas *DMI3* acts downstream (Ané et al. 2002). The *DMI2* locus is the orthologue of *SYMRK/NORK* (Endre et al. 2002). The *DMI1* gene encodes for what is predicted to be membrane ion channel protein (Ané et al. 2004) and, theoretically, mediates one of the early ion fluxes that are well defined in the nodulation process but have not yet been experimentally verified in the AM symbiosis. The position of *DMI1* upstream of *DMI2* in the signalling chain is more theoretical than empirical (Parniske 2004). The signal may be transmitted via the intracellular kinase domain of *DMI2* to the *DMI1* ion channel.

The gene *DMI3* encodes a calcium- and calmodulin-dependent protein kinase (CCaMK) with domains implicated in calcium- and calmodulin-binding sites (Lévy et al. 2004; Mitra et al. 2004). It is reasonable to speculate that CCaMK perceives the calcium-spiking signal and transduces it as a phosphorylation event, leading to the activation of downstream responses. The product of *DMI3* is an essential component in the signalling cascade in both nodulation and in the formation of arbuscular mycorrhiza, and theoretically, this protein can perceive different calcium flux signals (Lévy et al. 2004; Mitra et al. 2004). This indicates that calcium spiking is also an essential component of the signalling cascade, leading to an effective AM symbiosis; however, experimental evidence for this does not exist. If rhizobia and AM fungi induce different calcium features, it is possible that *DMI3* could transduce these into different downstream signalling events (Harrison 2005).

The phenotypic description of a number of SYM mutants supports genetic evidence that the penetration of the root epidermis and colonization of the cortex are two key events in the development of both rhizobial and AM fungal colonization (Guinel and Geil 2002; Jacobi et al. 2003*a*, 2003*b*), revealing the importance of cell-type specificity in the development of the symbiotic program (Parniske 2004). In the case of the AM symbiosis, relationships between morphological cell-type events and genetic loci mutations have been established. These events have been summarized by Parniske (2004) and fall into three response classes that are mediated by distinct common SYM genes: (*i*) epidermal penetration; (*ii*) intracellular fungal accommodation in the outer cell layer of the cortex; and (*iii*) formation, development, and turnover of arbuscules. The penetration of the epidermis by fungal hyphae requires the formation of a cleft between the walls of two adjacent epidermal cells. The plant regulates this process, perhaps by mediation of hydrolytic enzymes; in the case of *L. japonicus*, the participation of *LjSym15* is essential (Demchenko et al. 2004). After the penetration of the epidermis, successful fungal establishment in the inner and outer layer of cortical cells is required. Cytological and genetic studies using *L. japonicus*, showed that the *SYMRK* and *LjSym4* genes are essential for this process (Bonfante et al. 2000; Novero et al. 2002; Demchenko et al. 2004). The LjSym4 phenotype is very similar to that of the *brz* (E107) gene mutant of *P. sativum*, and therefore, the *brz* gene probably has a similar function (Resendes et al. 2001). Moreover, host SYM genes have been identified that are required for arbuscule formation. The PsSym36 mutant forms stumpy branches instead of arbuscules (Gianinazzi-Pearson 1996; Guinel and Geil 2002) and some other *P. sativum* mutants (e.g., PsSym33 and PsSym40) show alterations in the control of arbuscule development and arbuscule turnover (Jacobi et al. 2003*a*, 2003*b*).

It is clear that the products of the SYM genes have a fundamental role in the establishment of the mycorrhizal and rhizobial symbioses and that a single mutation could result in incompatibility and resistance. Some of the identified genes that are important for both symbioses are involved in Nod factor perception, and putatively, the proteins that they code for contain domains capable of interacting with microbial signal molecules.

These findings reinforce the hypothesis that an AM fungus-derived factor activates the symbiotic programme in the root. Recent experiments have demonstrated an effective exchange of signals between AM fungi and roots before infection (see Hyphal branching above). The biochemical purification and characterization of these AM factors, along with the use of bioassays to investigate cross-talk and specificities in the Nod and Myc signalling pathways, together with the ability of genetic tools to identify new genes and reveal whole-transcriptome profiles, will provide exciting new data for the understanding of these processes in the near future.

Hormonal regulation of arbuscular mycorrhizal formation

Because of the fact that phytohormones coordinate processes of plant growth and morphogenesis, it is tempting to suggest that mycorrhizal interactions, in which processes of morphological and cytological adaptations are essential, could be hormonally regulated. A role for phytohormones in ectomycorrhizas and arbuscular mycorrhizas has been proposed in a number of reviews (e.g., Beyrle 1995; Ludwig-Müller 2000; Hause et al. 2007). In these and other publications, it was suggested that a fine balance between hormones and nutrient availability (including P, C, and N) is probably important for the regulation of mycorrhizal formation and functioning.

Despite increasing knowledge obtained about the genetic and molecular biology of AM, little is known about the biochemical and morphogenetical events mediated by phytohormones during AM formation. This is partially due to the intrinsic difficulty of working with this symbiosis coupled with the physiological and biochemical complexity of phytohormone action and the sophisticated methodology used for their identification and quantification. In the section below, the most relevant data about the involvement of plant hormones in AM formation are presented. Most of their putative functions are speculative and based on a few experiments involving hormone application or measurements of hormone levels in AM plants.

As in ectomycorrhizal (Barker and Tagu 2000) and *Rhizobium*–legume interactions (Hirsch et al. 1997), cytokinins and auxins have been suggested to play a general role in the enhancement of AM formation (and plant growth). Enhanced levels of cytokinins and gibberellin-like (GA-like) substances have been detected in AM plants (Allen et al. 1980, 1982; van Rhijn et al. 1997; Ginzberg et al. 1998). In addition, the production of cytokinins and GA-like substances by AM fungi has been shown (Barea and Azcón-Aguilar 1982). Furthermore, a complementary DNA of *G. intraradices* has been cloned that has sequence similarities with an *Arabidopsis* gene that is involved in the GA signal transduction pathway (Delp et al. 2000). Therefore, it is feasible that GA-like compounds are involved in the biology of AM fungi. The exogenous application of different doses of GA3 (gibberellic acid) affects arbuscule morphology and can inhibit colonization (El Ghachtouli et al. 1996). It is possible that GA3 triggers an interaction with the synthesis of polyamines, another group of plant-produced regulatory molecules that affect AM formation (El Ghachtouli et al. 1996).

Two important roles have been proposed for cytokinins in the AM symbiosis: (*i*) the suppression of plant defence responses and (*ii*) the regulation of mycorrhizal formation. Ginzberg et al. (1998) found a correlation between pathogenesis-related (PR) gene suppression and an increase in cytokinin in AM roots. Because of the reported role of cytokinin in PR inhibition (Shinshi et al. 1987), the authors speculated that cytokinins might suppress certain PR genes, which supports previous results concerning chitinase and β-1,3-endoglucanase regulation during AM formation (Spanu et al. 1989; Shaul et al. 2000). By contrast, Fang and Hirsch (1998), and Hirsch et al. (1997) proposed that the expression of the early nodulin genes *MsENOD2* and *MsENOD40* in *M. sativa* mycorrhizal roots be regulated by an increase of cytokinins in root tissues. These authors speculated that gene expression, in combination with enhanced hormone levels, determines the initiation of additional lateral root primordia in AM roots.

The level of free indol-3-acetic acid in roots of different plant species colonized by AM fungi does not differ significantly from that in roots of noncolonized plants (Ludwig-Müller 2000). However, the content of free and conjugated indol-3-butyric acid (IBA) seems to be regulated during AM colonization. An early increase in the free IBA content and IBA synthesis has been observed in AM *Z. mays* roots (Kaldorf and Ludwig-Müller 2000), and these events may facilitate AM fungal colonization by increasing lateral root formation (a mechanism similar to that observed for cytokinins). This is supported by the fact that exogenous applications of auxin have a positive effect on root colonization by AM fungi (see Beyrle (1995)). The formation of IBA and IBA conjugates was systemically induced in leaves of *Z. mays* plants during AM fungal colonization (Fitze et al. 2005), indicating a complex regulation of auxin metabolism in AM plants, including the regulation of synthesis and transport.

Studies on abscisic acid (ABA) implication in AM development have been conducted, mainly in experiments involving measurements of the hormone content in AM plants. Contradictory results have been obtained, and reports about changes in the endogenous level of ABA in mycorrhizal plants are controversial (Allen et al. 1982; Murakami-Mizukami et al. 1991; Danneberg et al. 1992; Meixner et al. 2005). Abscisic acid has also been detected in AM fungal hyphae at levels higher than in roots (Esch et al. 1994), and it was speculated that ABA in AM fungi could control the flux of water and mineral salts from the soil to the hyphae or from other fungal structures, such as arbuscules, to the root cells (Ruíz-Lozano 2003). Recently, by analyzing AM colonization in the *S. lycopersicum* mutant *sitiens*, in which ABA levels are reduced to 8% of those found in wild-type plants, Herrera-Medina et al. (2007) were able to show that ABA is necessary for successful root colonization and formation of arbuscules. In addition, their data suggest that the plant phytohormone ethylene plays an antagonistic role, regulating the action of ABA (Herrera-Medina et al. 2007). These experiments demonstrate that ABA plays an essential role in the establishment of a functional AM symbiosis.

Despite the key role of signalling compounds such as jasmonic acid (JA), SA (see above), and ethylene in plant–microbe interactions, their role in AM formation has received little attention (see reviews by Ludwig-Müller (2000) and Hause et al. (2007)). Two apparently contradictory results were obtained in experiments using exogenous foliar applications of JA to AM inoculated plants: Regvar et al. (1996) found increased colonization of *Allium sativum* L. plants, whereas Ludwig-Müller et al. (2002) observed a suppression of colonization in *Tropaeolum majus* L., *Carica papaya* L., and *Cucumis sativus* L. These differences are possibly due to slightly different methodologies, differences in the concentration of JA applied, and interspecific differences in JA sensitivity among plants. Recent experiments involving endogenous measures of JA in AM roots have shown positive correlations between jasmonates and AM fungal colonization in *Hordeum vulgare* L. (Hause et al. 2002), and in *C. sativus* (even in the nonmycorrhizal compartment of a split-root system of mycorrhizal cucumber) (Vierheilig and Piché 2002). Increase in jasmonate levels in AM *H. vulgare* roots was accompanied by the expression of a gene encoding for an enzyme involved in jasmonate synthesis and of a gene for a jasmonate-induced protein (Hause et al. 2002). In both cases, the gene expression was located in root cells containing arbuscules. The endogenous increase in jasmonate was found in advanced stages of mycorrhization, could be a consequence of fungal development, and may be linked to the stronger carbohydrate sink function of mycorrhizal roots compared to nonmycorrhizal ones. Genetic approaches, using *M. truncatula* hairy roots transformed with an antisense construction of allene oxide cyclase, an enzyme involved in JA biosynthesis, showed that the reduction in JA levels was accompanied by a delay in colonization (Isayenkov et al. 2005). The data obtained so far point to a specific function of jasmonates in the maintenance of the homeostasis between plants and AM fungi (Hause et al. 2007). The systemic increase of JA levels in mycorrhizal roots reported by Vierheilig and Piché (2002) could contribute to a systemic suppression of root colonization (autoregulation) and could also be implicated in the systemic enhancement of the defence status of mycorrhizal plants, because JA is known to play a role in defence gene induction (Kunkel and Brooks 2002).

In addition to its involvement in normal developmental processes such as fruit ripening and senescence, the gaseous phytohormone ethylene is involved in responses to multiple stresses, including mechanical damage and pathogen attack. Recently, significant advances have been made in the understanding of the physiology, biochemistry, and signalling pathway of this molecule (Alonso and Stepanova 2004; Guo and Ecker 2004). Nevertheless, few studies have investigated the involvement of ethylene in the AM symbiosis. Besmer and Koide (1999) and Cruz et al. (2000) showed that host ethylene production decreases following AM formation, presumably by 1-aminocyclopropane-1-carboxylate oxidase inhibition (McArthur and Knowles 1992). By contrast, other studies showed no effect (Vierheilig et al. 1994) or an enhancement of the ethylene level in mycorrhizal roots (Dugassa et al. 1996).

Alterations of AM development through exogenous ethylene application have been tested, and a general inhibitory effect has been demonstrated using ethrel (Azcón-Aguilar et al. 1981; Morandi 1989) or gaseous ethylene (Ishii et al. 1996; Geil et al. 2001; Geil and Guinel 2002). However, when the applied dose is low, ethylene may have a stimulatory effect on AM formation (Ishii et al. 1996). Taking into consideration the genetic and morphological similarity between the processes of nodulation and AM formation, and the central role of ethylene in the establishment of the *Rhizobium*–legume symbiosis, Guinel and Geil (2002) proposed a model for the development of rhizobial and mycorrhizal symbioses in legume plants. Basically, they suggest two potential regulatory roles for ethylene during AM formation that occur mainly at the epidermis–cortex interface: (*i*) the modulation of the signal transduction pathway at the calcium spiking level in plant cells and (*ii*) a putative implication in the plant's defence response. Nevertheless, direct evidence of a specific role of ethylene in the AM symbiosis does not exist, and further work is needed.

In conclusion, the role of plant hormones during AM formation is not clear. The results from experiments based on hormone levels in root tissues and the effect of exogenous application only allow speculation as to their involvement in AM regulation. Nevertheless, much evidence supports the hypothesis that these plant growth regulating substances play an important role in AM development. Future use of genetic tools, such as plant mutants defective in their capacity to produce or perceive hormonal signals, in combination with novel methods for simultaneous analysis of plant phytohormone content (Schmelz et al. 2003), should contribute to our understanding of the hormonal regulation of AM formation. This is supported by the fact that the combination of molecular and genetic approaches, principally through the use of mutant tomato plants altered in their hormonal synthesis or perception, has allowed us to demonstrate a negative and coordinated effect of SA and ethylene on AM formation (J.M. García-Garrido, unpublished results).

Mycorrhizal autoregulation

A nodulation regulation mechanism occurs in the rhizobial–legume interaction. Once a certain number of nodules has been formed, further nodulation is suppressed. In analogy to this autoregulation of nodulation (see Caetano-Anollés and Gresshoff (1991)), an autoregulation has been suggested for root colonization by AM fungi (Vierheilig and Piché 2002; Vierheilig 2004*a*). Root exudates of nonmycorrhizal *C. sativus* plants exhibit a stimulatory effect on AM fungal growth. When applied to roots of inoculated plants, they stimulate root colonization. By contrast, root exudates of mycorrhizal *C. sativus* plants have an inhibitory effect on root colonization by AM fungi (Pinior et al. 1999; Vierheilig et al. 2003). This indicates that, once plants are colonized, the root exudation pattern is altered, and this change could limit further root colonization by AM fungi.

Evidence for an autoregulation of AM fungal colonization has been provided by a number of studies working with split-root systems. Using such systems, precolonization of one compartment of the split-root system by AM fungi resulted in a clear reduction of root colonization in the other compartment (Vierheilig et al. 2000*b*). Furthermore, the greater the level of colonization in the first compartment, the greater was the level of suppression in the second (Vierheilig 2004*b*). However, adding P to one compartment of the split-root system did not result in a similar suppression, thus excluding P as the regulatory factor for the observed suppressional effect (Vierheilig et al. 2000*c*).

Carbon availability in mycorrhizal plants could also explain the reduction of further root colonization in already mycorrhizal plants. In theory, once a plant has been colonized by an AM fungus, carbon availability for another AM fungus is reduced. Different fungi have different C sink strengths in roots (Lerat et al. 2003*a*, 2003*b*) and, thus, should suppress further root colonization differently; however, in split-root systems, different species exhibit a similar suppressional effect on further root colonization. Therefore, the C sink strength of a given AM fungus can be discarded as the regulatory factor of the observed suppression on further root colonization in already mycorrhizal plants.

A number of signalling steps in the rhizobial and AM symbioses seem similar (Hirsch and Kapulnik 1998; Guinel and Geil 2002), and a certain similarity in the rhizobial and AM autoregulation has been suggested (Vierheilig and Piché 2002; Vierheilig 2004*a*). This similarity was confirmed by a recent study using *M. sativa* plants in which preinoculation of one side of a split-root system with an AM fungus resulted in suppression of nodulation on the other side, and the preinoculation of one side with *Rhizobium* resulted in suppression of mycorrhization on the other side (Catford et al. 2003). Moreover, recent experiments with supernodulating *Glycine max* (L.). Merr. mutants also pointed towards a similar rhizobial and mycorrhizal regulation. Supernodulating *G. max* are unable to autoregulate nodule formation and, thus, are characterized by the formation of a high number of nodules (Carrol et al. 1985). The supernodulating *G. max* mutant *nts1007* also lacked the autoregulatory mechanism for mycorrhization (Meixner et al. 2005, 2007). By using experimental systems to study the autoregulation of nodulation, we hope

Table 2.1. Clearly identified events of arbuscular mycorrhizal regulation, the molecular mechanisms associated with mycorrhizal regulation, and their site and (or) scale of impact.

Event	Molecular findings	Action
Blocked or altered mycorrhization	Mutation in symbiotic genetic loci	Generalized in roots
Alterations in root colonization	Hormonal and nutritional imbalances	Generalized in roots
Alterations in the degree of root colonization	Alterations of the salicylic acid content	Generalized in roots
Senescence of arbuscules	Accumulation of phenolics, jasmonic acid, phytoalexins, and reactive oxygen species	Localized in cells containing arbuscules
Autoregulation	Alterations of the salicylic acid and flavonoid content, and alterations in root exudates	Systemic suppression
Reduced stimulatory effect	Alterations in root exudates	Ecological

that more knowledge about AM autoregulation will become available.

Conclusion

Compared with our understanding of the signalling and regulation events in the *Rhizobium*–legume interaction, we know relative little about signalling and regulation in AM associations and only a few of the important events have been identified (Table 2.1). This is partially due to the obligately biotrophic nature of AM fungi. The recent identification of a branching factor, will allow detailed studies on the importance of this signalling molecule for AM root colonization. Increasing amounts of data are also available concerning a fungus-derived "Myc factor" and its activity on the plant host. We are certain that in the near future there will more exciting discoveries in this field.

Many questions are still unanswered regarding the symbiotic phase. The available data provide strong evidence that AM fungal penetration of the root and fungal growth in the root lead, in both symbionts, to the activation of "symbiotic programs" for a functional mycorrhiza. These result in important alterations in the morphology and physiology of the roots and the fungal hyphae. In addition, changes in the hormonal, nutritional, and defense pathways of the plant are necessary for the regulation and functioning of AM. Nevertheless, the importance of these changes for the AM formation and regulation are still unknown, and our knowledge on the molecular communication between both partners is still fragmentary and insufficient.

Hopefully, future research on the identification of new Myc mutants, together with the use of transcriptomic and metabolomic analysis, will provide answers to some of these questions and will help to elucidate which mechanisms and signalling molecules are decisive for the regulation and functioning of the AM symbiosis.

Acknowledgements

This work was partially funded by the "Acciones Integradas" program between Spain and Austria (HU2004-0019).

References

Akiyama, K., Matsuzaki, K., and Hayashi, H. 2005. Plant sesquiterpenes induce hyphal branching in arbuscular mycorrhizal fungi. Nature (Lond.), **435**: 824–827.

Albrecht, C., Geurts, R., Lapeyrie, F., and Bisseling, T. 1998. Endomycorrhizae and rhizobial nod factors both require SYM8 to induce the expression of the early nodulin genes *PsENOD5* and *PsENOD12A*. Plant J. **15**: 605–614.

Allen, M.F., Moore, T.S., Jr., and Christensen, M. 1980. Phytohormone changes in *Bouteloua gracilis* infected by vesicular–arbuscular mycorrhizae. II. Cytokinin increases in the host plant. Can. J. Bot. **58**: 371–374.

Allen, M.F., Moore, T.S., Jr., and Christensen, M. 1982. Phytohormone changes in *Bouteloua gracilis* infected by vesicular–arbuscular mycorrhizae. III. Altered levels of gibberellin-like substances and abscisic acid in the host plant. Can. J. Bot. **60**: 468–471.

Alonso, J.M., and Stepanova, A.N. 2004. The ethylene signaling pathway. Science (Washington, D.C.), **306**: 1513–1515.

Ané, J.-M., Lévy, J., Thoquet, P., Kulikova, O., de Billy, F.,

Penmetsa, V., Kim, D.J., Debellé, F., Rosenberg, C., Cook, D.R., Bisseling, T., Huguet, T., and Dénarié, J. 2002. Genetic and cytogenetic mapping of DMI1, DMI2, and DMI3 genes of *Medicago truncatula* involved in Nod factor transduction, nodulation, and mycorrhization. Mol. Plant Microbe Interact. **15**: 1108–1118.

Ané, J.-M., Kiss, G.B., Riely, B.K., Penmetsa, V., Oldroyd, G.E.D., Ayax, C., Lévy, J., Debellé, F., Baek, J.-M., Kalo, P., Rosendberg, C., Roe, B.A., Long, S.R., and Dénarié, J. 2004. *Medicago truncatula DMI1* required for bacterial and fungal symbioses in legumes. Science, **303**: 1364–1367.

Arines, J., Vilariño, A., and Palma, J. 1994. Involvement of the superoxide dismutase enzyme in the mycorrhization process. Agric. Food Sci. Finl. **3**: 303–306.

Azcón-Aguilar, C., Rodriguez-Navarro, D.N., and Barea, J.M. 1981. Effects of ethrel on the formation and responses to VA mycorrhiza in *Medicago* and *Triticum*. Plant Soil, **60**: 461–468.

Balaji, B., Poulin, M.J., Vierheilig, H., and Piché, Y. 1995. Responses of an arbuscular mycorrhizal fungus, *Gigaspora margarita*, to exudates and volatiles from the Ri T-DNA-transformed roots of nonmycorrhizal and mycorrhizal mutants of *Pisum sativum* L. Sparkle. Exp. Mycol. **19**: 275–283.

Balestrini, R., and Bonfante, P. 2005. The interface compartment in arbuscular mycorrhizae: a special type of plant cell wall? Plant Biosyst. **139**: 8–15.

Barea, J.M., and Azcón-Aguilar, C. 1982. Production of plant growth-regulating substances by the vesicular–arbuscular mycorrhizal fungus *Glomus mosseae*. Appl. Environ. Microbiol. **43**: 810–813.

Barker, S.J., and Tagu, D. 2000. The roles of auxins and cytokinins in mycorrhizal symbioses. J. Plant Growth Regul. **19**: 144–154.

Barker, S.J., Stummer, B., Gao, L., Dispain, I., O'Connor, P.J., and Smith, S.E. 1998. A mutant in *Lycopersicon esculentum* Mill with highly reduced VA mycorrhizal colonisation: isolation and preliminary characterization. Plant J. **15**: 791–797.

Bécard, G., and Piché, Y. 1989. Fungal growth stimulation by CO_2 and root exudates in the vesicular–arbuscular symbiosis. Appl. Environ. Microbiol. **55**: 2320–2325.

Bécard, G., Douds, D.D., and Pfeffer, P.E. 1992. Extensive *in vitro* hyphal growth of vesicular–arbuscular mycorrhizal fungi in the presence of CO_2 and flavonols. Appl. Environ. Microbiol. **68**: 1260–1264.

Bécard, G., Taylor, L.P., Douds, D.D., Jr., Pfeffer, P.E., and Doner, L.W. 1995. Flavonoids are not necessary plant signal compounds in arbuscular mycorrhizal symbiosis. Mol. Plant Microbe Interact. **8**: 252–258.

Bécard, G., Kosuta, S., Tamasloukht, M., Séjalon-Delmas, N., and Roux, C. 2004. Partner communication in the arbuscular mycorrhizal interaction. Can. J. Bot. **82**: 1186–1197.

Besmer, Y.L., and Koide, R.T. 1999. Effect of mycorrhizal colonization and phosphorus on ethylene production by snapdragon (*Antirrhinum majus* L.) flowers. Mycorrhiza, **9**: 161–166.

Besserer, A., Puech-Pàges, V., Kiefer, P., Gomez-Roldan, V., Jauneau, A., Roy, S., Portais, J.C., Roux, C., Bécard, G., and Séjalon-Delmas, N. 2006. Strigolactones stimulate arbuscular mycorrhizal fungi by activating mitochondria. PLoS Biol. **4**: 1239–1247.

Beyrle, H. 1995. The role of phytohormones in the function and biology of mycorrhizas. *In* Mycorrhiza: structure, function, molecular biology and biotechnology. Edited by A. Varma and B. Hock. Springer-Verlag, Berlin. pp. 365–390.

Blee, K.A., and Anderson, A.J. 1996. Defense-related transcript accumulation in *Phaseolus vulgaris* L. colonized by the arbuscular mycorrhizal fungus *Glomus intraradices* Schenk & Smith. Plant Physiol. **110**: 675–688.

Blee, K.A., and Anderson, A.J. 2000. Defense responses in plants to arbuscular mycorrhizal fungi. *In* Current advances in mycorrhizae research. Edited by G.K. Podila and D.D. Douds, Jr. American Phytopathological Society Press, St. Paul, Minn. pp. 27–44.

Blilou, I., Ocampo, J.A., and Garcia-Garrido, J.M. 1999. Resistance of pea roots to endomycorrhizal fungus or *Rhizobium* correlates with enhanced levels of endogenous salicylic acid. J. Exp. Bot. **50**: 1663–1668.

Blilou, I., Bueno, P., Ocampo, J.A., and García-Garrido, J.M. 2000a. Induction of catalase and ascorbate peroxidase activities in tobaccco roots inoculated with the arbuscular mycorrhizal fungus *Glomus mosseae*. Mycol. Res. **104**: 722–725.

Blilou, I., Ocampo, J.A., and García-Garrido, J.M. 2000b. Induction of *Ltp* (lipid transfer protein) and *Pal* (phenylalanine ammonia-lyase) gene expression in rice roots colonized by the arbuscular mycorrhizal fungus *Glomus mosseae*. J. Exp. Bot. **51**: 1969–1977.

Bolaños-Vasquéz, M.C., and Werner, D. 1997. Effect of *Rhizobium tropici*, *R.etli*, and *R. leguminosarum* bv. *phaseoli* on nod gene-inducing flavonoids in root exudates of *Phaseolus vulgaris*. Mol. Plant Microbe Interact. **10**: 339–346.

Bonanomi, A., Oetiker, J.H., Guggenheim, R., Boller, T., Wiemken, A., and Vögeli-Lange, R. 2001a. Arbuscular mycorrhizas in mini-mycorrhizotrons: first contact of *Medicago truncatula* roots with *Glomus intradices* induces chalcone synthase. New Phytol. **150**: 573–582.

Bonanomi, A., Wiemken, A., Boller, T., and Salzer, P. 2001b. Local induction of a mycorrhiza-specific class III chitinase gene in cortical root cells of *Medicago truncatula* containing developing or mature arbuscules. Plant Biol. **3**: 194–199.

Bonfante, P. 2001. At the interface between mycorrhizal fungi and plants: the structural organization of cell wall, plasma-membrane and cytoskeleton. *In* The Mycota IX. Edited by K. Esser and B. Hock. Springer-Verlag, Berlin, Heidelberg. pp. 45–61.

Bonfante, P., Genre, A., Faccio, A., Martini, I., Schauser, L., Stougaard, J., Webb, J., and Parniske, M. 2000. The *Lotus japonicus LjSym4* gene is required for the successful symbiotic infection of root epidermal cells. Mol. Plant Microbe Interact. **13**: 1109–1120.

Bonfante-Fasolo, P., Faccio, A., Perotto, S., and Schubert, A. 1990. Correlation between chitin distribution and cell wall morphology in the mycorrhizal fungus *Glomus versiforme*. Mycol. Res. **94**: 157–165.

Brechenmacher, L., Weidmann, S., van Tuinen, D., Chatagnier, O., Gianinazzi, S., Franken, P., and Gianinazzi-Pearson, V.

2004. Expression profiling of up-regulated plant and fungal genes in early and late stages of *Medicago truncatula – Glomus mosseae* interactions. Mycorrhiza, **14**: 253–262.

Buée, M., Rossignol, M., Jauneau, A., Ranjeva, R., and Bécard, G. 2000. The pre-symbiotic growth of arbuscular mycorrhizal fungi is induced by a branching factor partially purified from plant root exudates. Mol. Plant Microbe Interact. **13**: 693–698.

Caetano-Anollés, G., and Gresshoff, P.M. 1991. Plant genetic control of nodulation. Annu. Rev. Microbiol. **45**: 345–382.

Carrol, B.J., McNeil, D.L., and Gresshoff, P.M. 1985. A supernodulation and nitrate tolerant symbiotic (nts) soybean mutant. Plant Physiol. **78**: 34–40.

Catford, J.G., Staehelin, C., Lerat, S., Piché, Y., and Vierheilig, H. 2003. Suppression of arbuscular mycorrhizal colonization and nodulation in split-root systems of alfalfa after pre-inoculation and treatment with Nod factors. J. Exp. Bot. **54**: 1481–1487.

Catoira, R., Galera, C., de Billy, F., Varma Penmetsa, R., Journet, E.P., Maillet, F., Rosenberg, C., Cook, D., Gough, C., and Dénarié, J. 2000. Four genes of *Medicago truncatula* controlling components of a Nod factor transduction pathway. Plant Cell, **12**: 1647–1665.

Chabot, S., Bel-Rhlid, R., Chênevert, R., and Piché, Y. 1992. Hyphal growth promotion *in vitro* of the VA mycorrhizal fungus, *Gigaspora margarita* Becker & Hall, by the activity of structurally specific flavonoid compounds under CO_2-enriched conditions. New Phytol. **122**: 461–467.

Cook, C.E., Whichard, L.P., Wall, M.E., Egley, G.H., Coggon, P., Luhan, P.A., and McPhail, A.T. 1972. Germination stimulants. 2. The structure of strigol—a potent seed germination stimulant for witchweed (*Striga lutea* Lour.). J. Am. Chem. Soc. **94**: 6198–6199.

Cruz, A.F., Ishii, T., and Kadoya, K. 2000. Effects of arbuscular mycorrhizal fungi on tree growth, leaf water potential, and levels of 1-aminocyclopropane-1-carboxylic acid and ethylene in the roots of papaya under water-stress conditions. Mycorrhiza, **10**: 121–123.

Dakora, F.D., Joseph, C.M., and Phillips, D.A. 1993. Alfalfa (*Medicago sativa* L.) root exudates contain isoflavonoids in the presence of *Rhizobium meliloti*. Plant Physiol. **101**: 819–824.

Danneberg, G., Latus, C., Zimmer, W., Hundeshagen, B., Schneider-Poestsch, H.J., and Bothe, H. 1992. Influence of vesicular–arbuscular mycorrhiza on phytohormone balances in maize (*Zea mays* L.). J. Plant Physiol. **141**: 33–39.

Dassi, B., Dumas-Gaudot, E., Asselin, A., Richard, C., and Gianinazzi, S. 1996. Chitinase and β-1,3-glucanase isoforms expressed in pea roots inoculated with arbuscular mycorrhizal or pathogenic fungi. Eur. J. Plant Pathol. **102**: 105–108.

David, R., Itzhaki, H., Ginzberg, I., Gafni, Y., Galili, G., and Kapulnik, Y. 1998. Suppression of tobacco basic chitinase gene expression in response to colonization by the arbuscular mycorrhizal fungus *Glomus intraradices*. Mol. Plant Microbe Interact. **11**: 489–497.

Delp, G., Smith, S.E., and Barker, S.J. 2000. Isolation by differential display of three cDNAs coding for proteins from the VA mycorrhizal *G. intraradices*. Mycol. Res. **104**: 293–300.

Demchenko, K., Winzer, T., Stougaard, J., Parniske, A., and Pawlowski, K. 2004. Distinct roles of *Lotus japonicus SYMRK* and *SYM15* in root colonization and arbuscule formation. New Phytol. **163**: 381–392.

Douds, D.D., Jr., and Nagahashi, G. 2000. Signaling and recognition events prior to colonization of roots by arbuscular mycorrhiza fungi. *In* Current advances in mycorrhizae research. *Edited by* G.K. Podila and D.D. Douds, Jr. American Phytopatholgical Society Press, St. Paul, Minn. pp. 127–140.

Dugassa, G.D., von Alten, H., and Schönbeck, F. 1996. Effects of arbuscular mycorrhiza (AM) on health of *Linum usitatissimun* L. infected by fungal pathogens. Plant Soil, **185**: 173–182.

Dumas-Gaudot, E., Furlan, V., Grenier, J., and Asselin, A. 1992. New acidic chitinase isoform induced in tobacco roots by vesicular–arbuscular mycorrhizal fungi. Mycorrhiza, **1**: 133–136.

El Ghachtouli, N., Martin-Tangury, J., Paynot, M., and Gianinazzi, S. 1996. First report of the inhibition of arbuscular mycorrhizal infection of *Pisum sativum* by specific and irreversible inhibition of polyamine biosynthesis or by gibberellic acid treatment. Fed. Eur. Biochem. Soc., Lett. **385**: 189–192.

Elias, K.S., and Safir, G.R. 1987. Hyphal elongation of *Glomus fasciculatus* in response to root exudates. Appl. Environ. Microbiol. **53**: 1928–1933.

Endre, G., Kereszt, A., Kevei, Z., Mihacea, S., Kalo, P., and Kiss, G.B. 2002. A receptor kinase gene regulating symbiotic nodule development. Nature (Lond.), **417**: 962–966.

Esch, H., Hundeshagen, B., Schneider-Poetsch, H.J., and Bothe, H. 1994. Demonstration of abscisic acid in spores and hyphae of the arbuscular-mycorrhizal fungus *Glomus* and in the N_2-fixing cyanobacterium *Anabaena variabilis*. Plant Sci. **99**: 9–16.

Fang, Y., and Hirsch, A.M. 1998. Studying early nodulin gene *ENOD40* expression and induction by nodulation factor and cytokinin in transgenic alfalfa. Plant Physiol. **116**: 53–68.

Fester, F., and Hause, B. 2005. Accumulation of reactive oxygen species in arbuscular mycorrhizal roots. Mycorrhiza, **15**: 373–379.

Fitze, D., Wiepning, A., Kalforf, M., and Ludwig-Müller, J. 2005. Auxins in the development of an arbuscular mycorrhizal symbiosis in maize. J. Plant Physiol. **162**: 1210–1219.

Frühling, A., Roussel, H., Gianinazzi-Pearson, V., Phuler, A., and Perlick, A.M. 1997. The *Vicia faba* leghemoglobin gene *VfLb29* is induced in root nodules and in roots colonized by the arbuscular mycorrhizal fungus *Glomus fasciculatum*. Mol. Plant Microbe Interact. **10**: 124–131.

García-Garrido, J.M., and Ocampo, J.A. 2002. Regulation of the plant defence response in arbuscular mycorrhizal symbiosis. J. Exp. Bot. **53**: 1377–1386.

García-Garrido, J.M., García-Romera, I., and Ocampo, J.A. 1992. Cellulase production by the vesicular–arbuscular mycorrhizal fungus *Glomus mosseae* (Nicol. & Gerd.) Gerd. and Trappe. New Phytol. **121**: 221–226.

García-Romera, I., García-Garrido, J.M., and Ocampo, J.A. 1991. Pectolityc enzymes in the vesicular–arbuscular mycorrhizal fungus *Glomus mosseae*. Fed. Eur. Microb. Soc. Microbiol. Lett. **78**: 343–346.

Geil, R.D., and Guinel, F.C. 2002. Effects of elevated substrate-ethylene on colonization of *Allium porrum* (leek) roots by the vesicular–arbuscular mycorrhizal fungus, *Glomus versiforme*. Can. J. Bot. **80**: 114–119.

Geil, R.D., Peterson, R.L., and Guinel, F.C. 2001. Morphological alterations of pea (*Pisum sativum* cv. Sparkle) arbuscular mycorrhizas as a result of exogenous ethylene treatment. Mycorrhiza, **11**: 137–143.

Gemma, J.N., and Koske, R.E. 1988. Pre-infection interactions between roots and the mycorrhizal fungus *Gigaspora gigantea*: chemotropism of germ-tubes and root growth response. Trans. Br. Mycol. Soc. **91**: 123–132.

Gianinazzi-Pearson, V. 1996. Plant cell responses to arbuscular mycorrhizal fungi: getting to the roots of the symbiosis. Plant Cell, **8**: 1871–1883.

Gianinazzi-Pearson, V., and Dénarié, J. 1997. Red carpet genetic programmes for root endosymbioses. Trends Plant Sci. **2**: 371–372.

Gianinazzi-Pearson, V., Branzanti, B., and Gianinazzi, S. 1989. *In vitro* enhancement of spore germination and early hyphal growth of a vesicular–arbuscular mycorrhizal fungus by host root exudates and plant flavonoids. Symbiosis, **7**: 243–255.

Gianinazzi-Pearson, V., Gianinazzi, S., Guillemin, J.P., Trouvelot, A., and Duc, G. 1991. Genetic and cellular analysis of resistance to vesicular arbuscular (VA) mycorrhizal fungi in pea mutants. *In* Advances in molecular genetics of plant–microbe interactions. *Edited by* H. Hennecke and D.P.S. Verma. Kluwer Academic Publishers, Dordrecht, the Netherlands. pp. 336–342.

Ginzberg, I., David, R., Shaul, O., Elad, Y., Wininger, S., Ben-Dor, B., Badani, H., Fang, Y., van Rhijn, P., Li, Y., Hirsch, A.M., and Kapulnik, Y. 1998. *Glomus intraradices* colonization regulates gene expression in tobacco plants. Symbiosis, **24**: 145–157.

Giovannetti, M., and Sbrana, C. 1998. Meeting a non-host: the behaviour of AM fungi. Mycorrhiza, **8**: 123–130.

Giovannetti, M., Sbrana, C., Avio, L., Citernesi, A.S., and Logi, C. 1993*a*. Differential hyphal morphogenesis in arbuscular mycorrhizal fungi during pre-infection stages. New Phytol. **125**: 587–594.

Giovannetti, M., Avio, L., Sbrana, C., and Citernesi, A.S. 1993*b*. Factors affecting appressorium development in the vesicular–arbuscular mycorrhizal fungus *Glomus mosseae* (Nicol. & Gerd.) Gerd. & Trappe. New Phytol. **123**: 115–122.

Giovannetti, M., Sbrana, C., and Logi, C. 1994. Early processes involved in host recognition by arbuscular mycorrhizal fungi. New Phytol. **127**: 703–709.

Gomez-Roldan, V., Roux, C., Girard, D., Bécard, G., and Puech, V. 2007. Strigolactones: promising plant signals. Plant Signal. Behav. **2**: 163–164.

Graham, J.H. 1982. Effect of citrus root exudates on germination of chlamydospores of the vesicular–arbuscular mycorrhizal fungus *Glomus epigaeum*. Mycologia, **74**: 831–835.

Güimil, S., Chang, H.-S., Zhu, T., Sesma, A., Osbourn, A., Roux, C., Loannidis, V., Oakeley, E.J., Docquier, M., Descombes, P., Briggs, S.P., and Paszkowski, U. 2005. Comparative transcriptomics of rice reveals an ancient pattern of response to microbial colonization. Proc. Natl. Acad. Sci. U.S.A. **102**: 8066–8070.

Guinel, F.C., and Geil, R.D. 2002. A model for the development of the rhizobial and arbuscular mycorrhizal symbioses in legumes and its use to understand the roles of ethylene in the establishment of these two symbioses. Can. J. Bot. **80**: 695–720.

Guo, H., and Ecker, J.R. 2004. The ethylene signaling pathway: new insights. Curr. Opin. Plant Biol. **7**: 40–49.

Gworgwor, N.A., and Weber, H.C. 2003. Arbuscular mycorrhizal fungi–parasite–host interaction for the control of *Striga hermonthica* (Del.) Benth. in sorghum [*Sorghum bicolor* (L.) Moench]. Mycorrhiza, **13**: 277–281.

Harrison, M. 2005. Signaling in the arbuscular mycorrhizal symbiosis. Annu. Rev. Microbiol. **59**: 19–42.

Harrison, M., and Dixon, R. 1993. Isoflavonoid accumulation and expression of defense gene transcripts during the establishment of vesicular arbuscular mycorrhizal associations in roots of *Medicago truncatula*. Mol. Plant Microbe Interact. **6**: 643–659.

Harrison, M., and Dixon, R. 1994. Spatial patterns of expression of flavonoid/isoflavonoid pathway genes during interactions between roots of *Medicago truncatula* and the mycorrhizal fungus *Glomus versiforme*. Plant J. **6**: 9–20.

Hauck, C., Müller, S., and Schildknecht, H. 1992. A germination stimulant for parasitic flowering plants from *Sorghum bicolor*, a genuine host plant. J. Plant Physiol. **139**: 474–478.

Hause, B., Maier, W., Miersch, O., Kramell, R., and Strack, D. 2002. Induction of jasmonate biosynthesis in arbuscular mycorrhizal barley roots. Plant Physiol. **130**: 1213–1220.

Hause, B., Mrosk, C., Isayenkov, S., and Strack, D. 2007. Jasmonates in arbuscular mycorrhizal interactions. Phytochemistry, **68**: 101–110.

Hepper, C.M., and Mosse, B. 1975. Techniques used to study the interaction between *Endogone* and plant roots. *In* Endomycorrhizas. *Edited by* F.E. Sanders, B. Mosse, and B.P. Tinker. Academic Press, London. pp. 65–75.

Herrera-Medina, M.J., Gagnon, H., Piché, Y., Ocampo, J.A., García Garrido, J.M., and Vierheilig, H. 2003. Root colonization by arbuscular mycorrhizal fungi is affected by the salicylic acid content of the plant. Plant Sci. **164**: 993–998.

Herrera-Medina, M.J., Steinkellner, S., Vierheilig, H., Ocampo, J.A., and García-Garrido, J.M. 2007. Abscisic acid determines arbuscule development and functionality in the tomato arbuscular mycorrhiza. New Phytol. **175**: 554–564.

Hirsch, A.M., and Kapulnik, Y. 1998. Signal transduction pathways in mycorrhizal associations: comparisons with the *Rhizobium*–legume symbiosis. Fungal Genet. Biol. **23**: 205–212.

Hirsch, A.M., Fang, Y., Asad, S., and Kapulnik, Y. 1997. The role of phytohormones in plant–microbe symbioses. Plant Soil, **194**: 171–184.

Hohnjec, N., Vieweg, M.F., Pühler, A., Becker, A., and Küster, H. 2005. Overlaps in the transcriptional profiles of *Medicago truncatula* roots inoculated with two different *Glomus* fungi provide insights into the genetic program activated during arbuscular mycorrhiza. Plant Physiol. **137**: 1283–1301.

Isayenkov, S., Mrosk, C., Stenzel, I., Strack, D., and Hause, B. 2005. Suppression of allene oxide cyclase in hairy roots of *Medicago truncatula* reduces jasmonate levels and the degree of mycorhization with *Glomus intraradices*. Plant Physiol. **139**: 1401–1410.

Ishii, T., Shrestha, Y.H., Matsumoto, I., and Kadoya, K. 1996. Effect of ethylene on the growth of vesicular–arbuscular mycorrhizal fungi and on the mycorrhizal formation of trifoliate orange roots. J. Jpn. Soc. Hortic. Sci. **65**: 525–529.

Jacobi, L.M., Petreva, O.S., Tsyganov, V.E., Borisov, A.Y., and Tikhonovich, I.A. 2003*a*. Effect of mutations in the pea genes Sym33 and Sym40. I. Arbuscular mycorrhiza formation and function. Mycorrhiza, **13**: 3–7.

Jacobi, L.M., Zubkova, L.A., Barmicheva, E.M., Tsyganov, V.E., Borisov, A.Y., and Tikhonovich, I.A. 2003*b*. Effect of mutations in the pea genes *Sym33* and *Sym40*. II. Dynamics of arbuscule development and turnover. Mycorrhiza, **13**: 9–16.

Kaldorf, M., and Ludwig-Müller, J. 2000. AM fungi might affect the root morphology of maize by increasing indole-3-butyric acid biosynthesis. Physiol. Plant. **109**: 58–67.

Kistner, C., and Parniske, M. 2002. Evolution of signal transduction in intracellular symbiosis. Trends Plant Sci. **7**: 511–518.

Kistner, C., Winzer, T., Pitzschke, A., Mulder, L., Sato, S., Kaneko, T., Tabata, S., Sandal, N., Stougaard, J., Webb, K.J., Szczyglowski, K., and Parniske, M. 2005. Seven *Lotus japonicus* genes required for transcriptional reprogramming of the root during fungal and bacterial symbiosis. Plant Cell, **17**: 2217–2229.

Koske, R.E. 1982. Evidence for a volatile attractant from plant roots affecting germ tubes of a VA mycorrhizal fungus. Trans. Br. Mycol. Soc. **79**: 305–310.

Koske, R.E., and Gemma, J.N. 1992. Fungal reactions to plants prior to mycorrhizal formation. *In* Mycorrhizal functioning, an integrative plant–fungus process. *Edited by* M. Allen. Routledge, Chapman & Hall, New York. pp. 3–37.

Kosuta, S., Chabaud, M., Lougnon, G., Gough, C., Dénarié, J., Barker, D.G., and Bécard, G. 2003. A diffusible factor from arbuscular mycorrhizal fungi induces symbiosis-specific MtENOD11 expression in roots of *Medicago truncatula*. Plant Physiol. **131**: 952–962.

Kunkel, B.N., and Brooks, D.M. 2002. Cross talk between signaling pathways in pathogen defense. Curr. Opin. Plant Biol. **5**: 325–331.

Lambais, M.R. 2000. Regulation of plant defence-related genes in arbuscular mycorrhizae. *In* Current advances in mycorrhizae research. *Edited by* G.K. Podila and D.D. Douds, Jr. American Phytopathological Society Press, St. Paul, Minn. pp. 45–59.

Lambais, M.R., and Mehdy, M.C. 1993. Suppression of endochitinase, endoglucanase, and chalcone isomerase expression in bean vesicular–arbuscular mycorrhizal roots under different soil phosphate conditions. Mol. Plant Microbe Interact. **6**: 75–83.

Lambais, M.R., and Mehdy, M.C. 1998. Spatial distribution of chitinases and glucanase transcripts in bean arbuscular mycorrhizal roots under low and high soil phosphate conditions. New Phytol. **140**: 33–42.

Lambais, M.R., Ríos-Ruíz, W.F., and Andrade, M.R. 2003. Antioxidant responses in bean (*Phaseolus vulgaris*) roots colonized by arbuscular mycorrhizal fungi. New Phytol. **160**: 421–428.

Larose, G., Chênevert, R., Moutoglis, P., Gagné, S., Piché, Y., and Vierheilig, H. 2002. Flavonoid levels in roots of *Medicago sativa* are modulated by the developmental stage of the symbiosis and the root colonizing arbuscular mycorrhizal fungus. J. Plant Physiol. **159**: 1329–1339.

Lendzemo, V.W. 2004. The tripartite interaction between sorghum, *Striga hermonthica*, and arbuscular mycorrhizal fungi. Ph.D. thesis, Wageningen University, Wageningen, the Netherlands.

Lendzemo, V.W., and Kuyper, T.W. 2001. Effects of arbuscular mycorrhizal fungi on damage by *Striga hermonthica* on two contrasting cultivars of sorghum, *Sorghum bicolor*. Agric. Ecosyst. Environ. **87**: 29–35.

Lendzemo, V.W., Kuyper, T.W., Kropff, M.J., and van Ast, A. 2005. Field inoculation with arbuscular mycorrhizal fungi reduces *Striga hermonthica* performance on cereal crops and has the potential to contribute to integrated *Striga* management. Field Crops Res. **91**: 51–61.

Lendzemo, V., Kuyper, T.W., Matusova, R., Bouwmeester, H.J., and van Ast, A. 2007. Colonization by arbuscular mycorrhizal fungi of sorghum leads to reduced germination and subsequent attachment and emergence of *Striga hermonthica*. Plant Signal. Behav. **2**: 58–62.

Lerat, S., Lapointe, L., Gutjahr, S., Piché, Y., and Vierheilig, H. 2003*a*. Carbon partitioning in a split-root system of arbuscular mycorrhizal plants is fungal and plant species dependent. New Phytol. **157**: 589–595.

Lerat, S., Lapointe, L., Piché, Y., and Vierheilig, H. 2003*b*. Variable carbon sink strength of different *Glomus mosseae* strains colonizing barley roots. Can. J. Bot. **81**: 886–889.

Lévy, J., Bres, C., Geurts, R., Chalhoub, B., Kulikova, O., Duc, G., Journet, E.-P., Rosenberg, C., and Debellé, F. 2004. A putative Ca^{2+} and calmodulin-dependent protein kinase required for bacterial and fungal symbioses. Science, **303**: 1361–1364.

Limpens, E., Franken, C., Smit, P., Willemse, J., Bisseling, T., and Geurts, R. 2003. LysM domain receptor kinases regulating rhizobial Nod factor-induced infection. Science, **302**: 630–633.

Liu, J., Maldonado-Mendoza, I., Lopez-Meyer, M., Cheung, F., Town, C.D., and Harrison, M.J. 2007. Arbuscular mycorrhizal symbiosis is accompanied by local and systemic alterations in gene expression and an increase in disease resistance in the shoots. Plant J. **50**: 529–544.

Ludwig-Müller, J. 2000. Hormonal balance in plants during colonization by mycorrhizal fungi. *In* Arbuscular mycorrhizas: physiology and function. *Edited by* Y. Kapulnik and D.D. Douds, Jr. Kluwer Academic Publisher, Dordrecht, the Netherlands. pp. 263–286.

Ludwig-Müller, J., Bennett, R., García-Garrido, J.M., Piché, Y., and Vierheilig, H. 2002. Reduced arbuscular mycorrhizal root colonization in *Tropaeolum majus* and *Carica papaya* after jasmonic acid application cannot be attributed to increased glucosinolate levels. J. Plant Physiol. **159**: 517–523.

Madsen, E.B., Madsen, L.H., Radutoiu, S., Olbryt, M., Rakwalska, M., Szczyglowski, K., Sato, S., Kaneko, T., Tabata, S., Sandal, N., and Stougaard, J. 2003. A receptor kinase gene of the LysM type is involved in legume perception of rhizobial signals. Nature, **425**: 637–640.

Maldonado-Mendoza, I.E., Dewbre, G.R., Blaylock, L., and Harrison, M.J. 2005. Expression of a xyloglucan endotransglucoxylase/hydrolase gene, MtXTH1, from *Medicago truncatula* is induced systemically in mycorrhizal roots. Gene, **345**: 191–197.

Marsh, J.F., and Schultze, M. 2001. Analysis of arbuscular mycorrhizas using symbiosis-defective plant mutants. New Phytol. **150**: 525–532.

Matusova, R., Rani, K., Verstappen, F.W.A., Franssen, M.C.R., Beale, M.H., and Bouwmeester, H.J. 2005. The strigolactone germination stimulants of the plant-parasitic *Striga* and *Orobanche* spp. are derived from the carotenoid pathway. Plant Physiol. **139**: 920–934.

McArthur, D.A., and Knowles, N.R. 1992. Resistance response of potato to vesicular–arbuscular mycorrhizal fungi under varying abiotic phosphorus levels. Plant Physiol. **100**: 341–351.

Meixner, C., Ludwig-Müller, J., Miersch, O., Gresshoff, P., Staehelin, C., and Vierheilig, H. 2005. Lack of mycorrhizal autoregulation and phytohormonal changes in the supernodulating soybean mutant *nts1007*. Planta, **222**: 709–715.

Meixner, C., Vegvari, G., Ludwig-Müller, J., Gagnon, H., Steinkellner, S., Staehelin, C., Gresshoff, P., and Vierheilig, H. 2007. Two defined alleles of the LRR receptor kinase *GmNARK* in supernodulating soybean govern differing autoregulation of mycorrhization. Physiol. Plant. **130**: 261–270.

Mitra, R.M., Gleason, C.A., Edwards, A., Hadfield, J., Downie, J.A., Oldroyd, G.E., and Long, S.R. 2004. A Ca^{2+}/calmodulin-dependent protein kinase required for symbiotic nodule development: gene identification by transcript-based cloning. Proc. Natl. Acad. Sci. U.S.A. **101**: 4701–4705.

Morandi, D. 1989. Effect of xenobiotics on endomycorrhizal infection and isoflavonoid accumulation in soybean roots. Plant Physiol. Biochem. **27**: 697–701.

Morandi, D. 1996. Occurrence of phytoalexins and phenolic compounds on endomycorrhizal interactions, and their potential role in biological control. Plant Soil, **185**: 241–251.

Mosse, B. 1988. Some studies relating to "independent" growth of vesicular–arbuscular endophytes. Can. J. Bot. **66**: 2533–2540.

Mosse, B., and Hepper, C.M. 1975. Vesicular–arbuscular mycorrhizal infections in root organ cultures. Physiol. Plant Pathol. **5**: 215–223.

Müller, S., Hauck, C., and Schildknecht, H. 1992. Germination stimulants produced by *Vigna unguiculata* Walp cv. Saunders Upright. J. Plant Growth Regul. **11**: 77–84.

Murakami-Mizukami, Y., Yamamoto, Y., and Yamaki, S. 1991. Analyses of indole acetic acid and abscisic acid contents in nodules of soybean plants bearing VA mycorrhizas. Soil Sci. Plant Nutr. **37**: 291–298.

Murray, J., Geil, R., Wagg, C., Karas, B., Szczyglowski, K., and Peterson, R.L. 2006a. Genetic supressors of *Lotus japonicus har1-1* hypernodulation show altered interactions with *Glomus intraradices*. Funct. Plant Biol. **33**: 749–755.

Murray, J., Karas, B., Ross, L., Brachmann, A., Wagg, C., Geil, R., Perry, J., Nowakowski, K., MacGillivary, M., Held, M., Stougaard, J., Peterson, L., Parniske, M., and Szczyglowski, K. 2006b. Genetic suppressors of the *Lotus japonicus har1-1* hypernodulation phenotype. Mol. Plant Microbe Interact. **19**: 1082–1091.

Nagahashi, G., and Douds, D.D., Jr. 1997. Appressorium formation by AM fungi on isolated cell walls of carrot roots. New Phytol. **136**: 299–304.

Nagahashi, G., and Douds, D.D., Jr. 1999. Rapid and sensitive bioassay to study signals between root exudates and arbuscular mycorrhizal fungi. Biotechnol. Tech. **13**: 893–897.

Nagahashi, G., and Douds, D.D., Jr. 2000. Partial separation of root exudate components and their effects upon the growth of germinated spores of AM fungi. Mycol. Res. **104**: 1453–1464.

Nagahashi, G., and Douds, D.D., Jr. 2003. Action spectrum for the induction of hyphal branches of an arbuscular mycorrhizal fungus: exposure sites versus branching sites. Mycol. Res. **107**: 1075–1082.

Nagahashi, G., Douds, D.D., Jr., and Abney, G.D. 1996. Phosphorus amendment inhibits hyphal branching of the VAM fungus *Gigaspora margarita* directly and indirectly through its effect on root exudation. Mycorrhiza, **6**: 403–408.

Novero, M., Faccio, A., Genre, A., Stougaard, J., Webb, K.J., Mulder, L., Parniske, M., and Bonfante, P. 2002. Dual requirement of the *LjSym4* gene for mycorrhizal development in epidermal and cortical cells of *Lotus japonicus* roots. New Phytol. **154**: 741–749.

Olah, B., Brière, C., Bécard, G., Dénarié, J., and Gough, C. 2005. Nod factors and diffusible factors from arbuscular mycorrhizal fungi stimulate lateral root formation in *Medicago truncatula* via the DMI1/DMI2 signalling pathway. Plant J. **44**: 195–207.

Oldroyd, G.E.D., and Downie, J.A. 2004. Calcium, kinases and nodulation signaling in legumes. Nat. Rev. Mol. Cell Biol. **5**: 566–576.

Parniske, M. 2000. Intracellular accommodation of microbes by plants: a common developmental program for symbiosis and disease? Curr. Opin. Plant Biol. **3**: 320–328.

Parniske, M. 2004. Molecular genetics of the arbuscular mycorrhizal symbiosis. Curr. Opin. Plant Biol. **7**: 414–421.

Parniske, M., and Downie, J.A. 2003. Locks, keys and symbioses. Nature, **425**: 569–570.

Perret, X., Staehelin, C., and Broughton, W.J. 2000. Molecular basis of symbiotic promiscuity. Microbiol. Mol. Biol. Rev. **64**: 180–201.

Peterson, R.L., and Guinel, F.C. 2000. The use of plant mutants to study regulation of colonization by AM fungi. *In* Arbuscular mycorrhizas: physiology and function. *Edited by* Y. Kapulnik and D.D. Douds, Jr. Kluwer Academic Publishers, Dordrecht, the Netherlands. pp. 147–171.

Phillips, D.A., and Tsai, S.M. 1992. Flavonoids as plant signals to the rhizosphere microbes. Mycorrhiza, **1**: 55–58.

Pinior, A. 1999. Wurzelexsudate mykorrhizierter Pflanzen und deren regulierender Einfluss auf arbuskuläre Mykorrhizapilze. Masters thesis, Christian-Albrechts-Universität, Kiel, Germany.

Pinior, A., Wyss, U., Piché, Y., and Vierheilig, H. 1999. Plants colonized by AM fungi regulate further root colonization by AM fungi through altered root exudation. Can. J. Bot. **77**: 891–897.

Poulin, M.-J., Bel-Rhlid, R., Piché, Y., and Chênevert, R. 1993. Flavonoids released by carrot (*Daucus carota*) seedlings stimulate hyphal development of vesicular–arbuscular mycorrhizal fungi in the presence of optimal CO_2 enrichment. J. Chem. Ecol. **19**: 2317–2327.

Powell, C.L. 1976. Development of mycorrhizal infection from *Endogone* spores and infected root fragments. Trans. Br. Mycol. Soc. **66**: 439–445.

Pozo, M.J., Azcón-Aguilar, C., Dumas-Gaudot, E., and Barea, J.M. 1998. Chitosanase and chitinase activities in tomato roots during interactions with arbuscular mycorrhizal fungi or *Phytophthora parasitica*. J. Exp. Bot. **49**: 1729–1739.

Radutoiu, S., Madsen, L.H., Madsen, E.B., Felle, H.H., Umehara, Y., Gronlund, M., Sato, S., Nakamura, Y., and Sotougaard, J. 2003. Plant recognition of symbiotic bacteria requires two LysM receptor-like kinases. Nature, **425**: 585–592.

Recourt, K., van Tunen, A.J., Mur, L.A., van Brussel, A.A.N., Lugtenberg, B., and Kijne, J.W. 1992. Activation of flavonoid biosynthesis in roots of *Vicia sativa* subsp. *nigra* plants by inoculation with *Rhizobium leguminosarum* biovar *viciae*. Plant Mol. Biol. **19**: 411–420.

Regvar, M., Gogala, N., and Zalar, P. 1996. Effects of jasmonic acid on mycorrhizal *Allium sativum*. New Phytol. **134**: 703–707.

Resendes, C.M., Geil, R.D., and Guinel, F.C. 2001. Mycorrhizal development in a low nodulating pea mutant. New Phytol. **150**: 563–572.

Ruíz-Lozano, J.M. 2003. Arbuscular mycorrhizal symbiosis and alleviation of osmotic stress: new perspectives for molecular studies. Mycorrhiza, **13**: 309–317.

Salzer, P., and Boller, T. 2000. Elicitor-induced reactions in mycorrhizae and their suppression. *In* Current advances in mycorrhizae research. *Edited by* G.K. Podila and D.D. Douds, Jr. American Phytopathological Society Press, St. Paul, Minn. pp. 1–10.

Salzer, P., Corbière, H., and Boller, T. 1999. Hydrogen peroxide accumulation in *Medicago truncatula* roots colonized by the arbuscular mycorrhiza-forming fungus *Glomus mosseae*. Planta, **208**: 319–325.

Salzer, P., Bonanomi, A., Beyer, K., Vögeli-Lange, R., Aeschbacher, R.A., Lang, J., Wiemken, A., Kim, D., Cook, D.R., and Boller, T. 2000. Differential expression of eight chitinase genes in *Medicago truncatula* roots during mycorrhiza formation, nodulation, and pathogen infection. Mol. Plant Microbe Interact. **13**: 763–777.

Sbrana, C., and Giovannetti, M. 2005. Chemotropism in the arbuscular mycorrhizal fungus *Glomus mosseae*. Mycorrhiza, **15**: 539–545.

Scervino, J.M., Ponce, M.A., Erra-Bassels, R., Vierheilig, H., Ocampo, J.A., and Godeas, A. 2005*a*. Flavonoids exclusively present in mycorrhizal roots of white clover exhibit different effects on arbuscular mycorrhizal fungi than flavonoids exclusively present in non-mycorrhizal roots of white clover. J. Plant Interact. **15**: 22–30.

Scervino, J.M., Ponce, M.A., Erra-Bassels, R., Vierheilig, H., Ocampo, J.A., and Godeas, A. 2005*b*. Flavonoids exhibit fungal species and genus specific effects on the presymbiotic growth of *Gigaspora* and *Glomus*. Mycol. Res. **109**: 789–794.

Scervino, J.M., Ponce, M.A., Erra-Bassels, R., Vierheilig, H., Ocampo, J.A., and Godeas, A. 2005*c*. Arbuscular mycorrhizal colonization of tomato by *Gigaspora* and *Glomus* species in presence of roots flavonoids. J. Plant Physiol. **162**: 625–633.

Scervino, J.M., Ponce, M.A., Erra-Bassels, R., Vierheilig, H., Ocampo, J.A., and Godeas, A. 2006. Glycosidation of apigenin results in a loss of activity on different growth parameters of arbuscular mycorrhizal fungi from the genus *Glomus* and *Gigaspora*. Soil Biol. Biochem. **38**: 2919–2922.

Schmelz, E.A., Engelberth, J., Alborn, H.T., O'Donnell, P., Sammons, M., Toshima, H., and Tumlinson, J.H., III. 2003. Simultaneous analysis of phytohormones, phytotoxins, and volatile organic compounds in plants. Proc. Natl. Acad. Sci. U.S.A. **100**: 10552–10557.

Schmidt, P.E., Broughton, W.J., and Werner, D. 1994. Nod factors of B*radyrhizobium japonicum* and *Rhizobium* sp. NGR234 induce flavonoid accumulation in soybean root exudates. Mol. Plant Microbe Interact. **7**: 384–390.

Schreiner, R.P., and Koide, R.T. 1993*a*. Mustards, mustard oils and mycorrhizas. New Phytol. **123**: 107–113.

Schreiner, R.P., and Koide, R.T. 1993*b*. Stimulation of vesicular–arbuscular mycorrhizal fungi by mycotrophic and nonmycotrophic plant root systems. Appl. Environ. Microbiol. **59**: 2750–2752.

Shaul, O., David, R., Sinvani, G., Ginzberg, I., Ganon, D., Wininger, S., Ben-Dor, B., Badani, H., Ovdat, N., and Kapulnik, Y. 2000. Plant defense responses during arbuscular mycorrhiza symbiosis. *In* Current advances in mycorrhizae research. *Edited by* G.K. Podila and D.D. Douds. American Phytopathological Society, St. Paul, Minn. pp. 61–68.

Shinshi, H., Mohnen, D., and Meins, F., Jr. 1987. Regulation of a plant pathogenesis-related enzyme: inhibition of chitinase and chitinase mRNA accumulation in cultured tobacco tissues by auxin and cytokinin. Proc. Natl. Acad. Sci. U.S.A. **84**: 89–93.

Siame, B.A., Weerasuriya, Y., Wood, K., Ejeta, G., and Butler, L.G. 1993. Isolation of strigol, a germination stimulant for *Striga asiatica*, from host plants. J. Agric. Food Chem. **41**: 1486–1491.

Simoneau, P., Louisy-Louis, N., Plenchette, C., and Strullu, D.G. 1994. Accumulation of new polypeptides in Ri T-DNA-transformed roots of tomato (*Lycopersicon esculentum*) during the development of vesicular–arbuscular mycorrhizae. Appl. Environ. Microbiol. **60**: 1810–1813.

Smith, S., and Read, D. 1997. Mycorrhizal symbiosis. 2nd ed. Academic Press, London.

Spanu, P., and Bonfante-Fasolo, P. 1988. Cell wall-bound peroxidase activity in roots of mycorrhizal *Allium porrum*. New Phytol. **109**: 119–124.

Spanu, P., Boller, T., Ludwig, A., Wiemken, A., and Faccio, A. 1989. Chitinase in roots of mycorrhizal *Allium porrum*: regulation and localization. Planta, **177**: 447–455.

St-Arnaud, M., and Vujanovic, V. 2007. Effect of the arbuscular mycorrhizal symbiosis on plant diseases and pests. *In* Arbuscular mycorrhizae in crop production. *Edited by* C. Hamel and C. Plenchette. Haworth Press, Binghampton, N.Y. pp. 67–122.

Steinkellner, S., Lendzemo, V., Langer, I., Schweiger, P., Khaosaad, T., Toussaint, J.P., and Vierheilig, H. 2007. Flavonoids and strigolactones in root exudates as signals in symbiotic and pathogenic plant–fungus interactions. Molecules, **12**: 1290–1306.

Stracke, S., Kistner, C., Yoshida, S., Mulder, L., Sato, S., Kaneko, T., Tabata, S., Sandal, N., Stougaard, J., Szczyglowski, K., and Parniske, M. 2002. A plant receptor-like kinase required for both bacterial and fungal symbiosis. Nature (Lond.), **417**: 959–962.

Suriyapperuma, S.P., and Koske, R.E. 1995. Attraction of germ tubes and germination of spores of the arbuscular mycorrhizal fungus *Gigaspora gigantea* in the presence of roots of maize exposed to different concentrations of phosphorus. Mycologia, **87**: 772–778.

Tamasloukht, M.B., Séjalon-Delmas, N., Kluever, A., Jauneau, A., Roux, C., Bécard, G., and Franken, P. 2003. Root factors induce mitochondrial-related gene expression and fungal respiration during the developmental switch from asymbiosis to presymbiosis in the arbuscular mycorrhizal fungus *Gigaspora rosea*. Plant Physiol. **131**: 1468–1478.

Tawaraya, K., Watanabe, S., Yoshida, E., and Wagatsuma, T. 1995. Effect of onion (*Allium cepa*) root exudates on the hyphal growth of *Gigaspora margarita*. Mycorrhiza, **6**: 57–59.

Tawaraya, K., Hashimoto, K., and Wagatsuma, T. 1998. Effect of root exudate fractions from P-deficient and P-sufficient onion plants on root colonisation by the arbuscular mycorrhizal fungus *Gigaspora margarita*. Mycorrhiza, **8**: 67–70.

Tsai, S.M., and Phillips, D.A. 1991. Flavonoids released naturally from alfalfa promote development of symbiotic *Glomus* spores in vitro. Appl. Environ. Microbiol. **57**: 1485–1488.

van Rhijn, P., Fang, Y., Galili, S., Shaul, O., Atzmon, N., Wininger, S., Eshed, Y., Lum, M., Li, Y., To, V., Kapulnik, Y., and Hirch, A.M. 1997. Expression of early nodulin genes in alfalfa mycorrhizae indicates that signal transduction pathways used in forming arbuscular mycorrhizae and *Rhizobium*-induced nodules may be conserved. Proc. Natl. Acad. Sci. U.S.A. **94**: 5467–5472.

Vierheilig, H. 2004a. Regulatory mechanisms during the plant – arbuscular mycorrhizal fungus interaction. Can. J. Bot. **82**: 1166–1176.

Vierheilig, H. 2004b. Further root colonization by arbuscular mycorrhizal fungi in already mycorrhizal plants is suppressed after a critical level of root colonization. J. Plant Physiol. **161**: 339–341.

Vierheilig, H., and Piché, Y. 2002. Signalling in arbuscular mycorrhiza: facts and hypotheses. *In* Flavonoids in cell function. *Edited by* B. Buslig and J.A. Manthey. Kluwer Academic/Plenum Publishers, New York. pp. 23–39.

Vierheilig, H., Alt, M., Mohr, U., Boller, T., and Wiemken, A. 1994. Ethylene biosynthesis and activities of chitinase and β-1,3-glucanase in the roots of host and non-host plants of vesicular–arbuscular mycorrhizal fungi after inoculation with *Glomus mosseae*. J. Plant Physiol. **143**: 337–343.

Vierheilig, H., Alt-Hug, M., Engel-Streitwolf, R., Mäder, P., and Wiemken, A. 1998a. Studies on the attractional effect of root exudates on hyphal growth of an arbuscular mycorrhizal fungus in a soil compartment – membrane system. Plant Soil, **203**: 137–144.

Vierheilig, H., Bago, B., Albrecht, C., Poulin, M.-J., and Piché, Y. 1998b. Flavonoids and arbuscular-mycorrhizal fungi. *In* Flavonoids in the living system. *Edited by* J.A. Manthey and B.S. Buslig. Plenum Press, New York. pp. 9–33.

Vierheilig, H., Bennett, R., Kiddle, G., Kaldorf, M., and Ludwig-Müller, J. 2000a. Differences in glucosinolate patterns and arbuscular mycorrhizal status of glucosinolate-containing plant species. New Phytol. **146**: 343–352.

Vierheilig, H., Garcia-Garrido, J.M., Wyss, U., and Piché, Y. 2000b. Systemic suppression of mycorrhizal colonization of barley roots already colonized by AM fungi. Soil Biol. Biochem. **32**: 589–595.

Vierheilig, H., Maier, W., Wyss, U., Samson, J., Strack, D., and Piché, Y. 2000c. Cyclohexenone derivative- and phosphate-levels in split-root systems and their role in the systemic suppression of mycorrhization in precolonized barley plants. J. Plant Physiol. **157**: 593–599.

Vierheilig, H., Bago, B., Lerat, S., and Piché, Y. 2002. Shoot-produced, light-dependent factors are partially involved in the expression of the arbuscular mycorrhizal (AM) status of AM host and non-host plants. J. Plant Nutr. Soil Sci. **165**: 21–25.

Vierheilig, H., Lerat, S., and Piché, Y. 2003. Systemic inhibition of arbuscular mycorrhiza development by root exudates of cucumber plants colonized by *Glomus mosseae*. Mycorrhiza, **13**: 167–170.

Volpin, H., Elkind, Y., Okon, Y., and Kapulnik, Y. 1994. A vesicular arbuscular mycorrhizal fungus (*Glomus intraradix*) induces a defense response in alfalfa roots. Plant Physiol. **104**: 683–689.

Volpin, H., Phillips, D.A., Ocón, Y., and Kapulnik, Y. 1995. Suppression of an isoflavonoid phytoalexin defense response in mycorrhizal alfalfa roots. Plant Physiol. **108**: 1449–1454.

Walker, S.A., Viprey, V., and Downie, A. 2000. Dissection of nodulation signalling using pea mutants defective for calcium spiking induced by Nod factors and chitin oligomers. Proc. Natl. Acad. Sci. U.S.A. **97**: 13 413 – 13 418.

Wyss, P., Mellor, R.B., and Wiemken, J.A. 1990. Vesicular–arbuscular mycorrhizas of wild-type soybean and non-nodulating mutants with *Glomus mosseae* contain symbiosis-specific polypeptides (mycorrhizins), immunologically cross-reactive with nodulins. Planta, **182**: 22–26.

Yokota, T., Sakal, H., Okuno, K., Yoneyama, K., and Takeuchi, Y. 1998. Alectrol and orobanchol, germination stimulants for *Orobanche minor*, from its host red clover. Phytochemistry, **49**: 1967–1973.

Yoneyama, K., Takeuchi, Y., and Yokota, T. 2001. Production of clover broomrape seed germination stimulants by red clover roots requires nitrate but is inhibited by phosphate and ammonium. Physiol. Plant. **112**: 25–30.

Yoneyama, K., Takeuchi, Y., Sato, D., Sekimoto, H., and Yokota, T. 2004. Determination and quantification of strigolactones. *In* Proceedings of the 8th International Parasitic Weed Symposium, 24–25 June 2004, Durban, South Africa.

Edited by D.M. Joel. International Parasitic Plant Society, Amsterdam. p. 9.

Yoneyama, K., Xie, X.N., Kusumoto, D., Sekimoto, H., Sugimoto, Y., Takeuchi, Y., and Yoneyama, K. 2007. Nitrogen deficiency as well as phosphorus deficiency in sorghum promotes the production and exudation of 5-deoxystrigol, the host recognition signal for arbuscular mycorrhizal fungi and root parasites. Planta, **227**: 125–132.

Chapter 3
Growth and branching of asymbiotic, presymbiotic, and extraradical AM fungal hyphae: clarification of concepts and terminology

Christine Juge, Andrew P. Coughlan, J. André Fortin, and Yves Piché

Introduction

Arbuscular mycorrhizal (AM) fungi are obligate symbionts that have been coevolving with their host plants for the last 400 million years (Simon et al. 1993; Remy et al. 1994; Redecker et al. 2000). These coenocytic fungi play essential biological roles: they increase plant nutrient and water uptake (Cooper and Tinker 1978; Nelsen 1987; Marschner and Dell 1994), protect against pathogens (St-Arnaud et al. 1994) (see also Chapter 4), alter plant morphology (Plenchette et al. 1981), and dictate the outcome of competition between plants (Allen and Allen 1984). Despite the recognized ecological importance of the AM symbiosis and its omnipresence in terrestrial ecosystems (Brachmann and Parniske 2006), AM fungi are still somewhat of a biological enigma. For example, certain fundamental aspects of their genetics and physiology are still unknown. Advances in this field are complicated by the "primitive" nature of these fungi, which are apparently deprived of any classical sexual reproduction (Sanders 1999), and the obligate nature of the mycobionts. For many years, the early studies by Butler (1939), Godfrey (1957), and Mosse (1959a, 1959b) offered the only information on growth and branching of AM fungi outside the root. However, the development of root-organ cultures (Mosse and Hepper 1975; Mugnier and Mosse 1987; Bécard and Fortin 1988; Bécard and Piché 1992) now allows routine nondestructive observations of the growth of asymbiotic, presymbiotic, and extraradical hyphae (St-Arnaud et al. 1996; Fortin et al. 2002).

From the earliest studies on AM fungi (Butler 1939; Godfrey 1957; Mosse 1959a, 1959b) through to more recent works (Nagahashi and Douds 2004; Akiyama et al. 2005; Akiyama and Hayashi 2006; Besserer et al. 2006), the term branching has been extensively used in the description of the AM fungal mycelium. However, this term does not reflect the diversity of the branching phenomena observed throughout the life cycle of these fungi. According to Friese and Allen (1991), "The specialized hyphal architectures of these fungal endophytes seem to be linked to the unique function of each hyphal type." Moreover, the authors suggest that "this high degree of specialization may help explain why vesicular–arbuscular mycorrhizal fungi are so efficient in their beneficial role of nutrient and water transport to the host plant." Following several years of observation of AM fungi in vitro, we support the views of Friese and Allen (1991). Furthermore, we have observed that certain features of mycelial growth are common to all of the developmental phases that occur outside the root. After spore germination, there are three distinct growth phases: the short-lived asymbiotic phase, the establishment of the symbiosis (e.g., Holley and Peterson 1979; Alexander et al. 1989; Smith and Smith 1990; Bonfante and Perotto 1995), and the development of an extraradical mycelium. The latter is part of a persistent symbiotic phase, which influences the growth and development of the host plant. In this chapter, these phases are referred to as: (*i*) the germ tube, (*ii*) the presymbiotic hyphae (which corresponds to "M1," the spore-dependant stage described by Bécard and Piché (1989b)), and (*iii*) the extraradical mycelium. By presenting a comprehensive list of the different terms used to describe growth and branching of these fungi, we aim, for the first time, to clarify their nomenclature

C. Juge,[1] A.P. Coughlan, J.A. Fortin, and Y. Piché. Centre d'étude de la forêt, Pavillon C.-E.-Marchand, Université Laval, Québec, QC G1V 0A6, Canada.

[1]Corresponding author (e-mail: christine.juge@meidje.net).

and give a brief insight into their functional significance.

Growth and branching of the germ tube

In this section, the term germ tube refers to the initial hypha developing from a spore. Hyphal germination from intraradical propagules will be considered at the end of this review (see Sporulation from moribund root fragments below).

Prominent features of germ-tube growth

Germ-tube growth was recently reviewed by Dalpé et al. (2005). The authors reported that, independent of fungal species, the germ tube consists of a straight growing hypha. In this chapter, we refer to this growth pattern as "runner hypha(e)" (RH). At right angles from these aseptate straight hyphae, thinner (5–10 μm in diameter) hyphae develop, which are also aseptate and thick walled (Giovannetti 2000); these correspond to the germinative hyphae described by Godfrey (1957). However, the author occasionally observed a different growth pattern, where several branches of approximately the same size were produced shortly after spore germination. These latter hyphae were "fine and twisted"; in this chapter, we refer to this growth pattern as "fine branching" (FB).

Mosse (1959a), working with species of *Endogone* (former genus of AM fungi), also described hyphae resembling RH and FB hyphae. When hyphae developed under cellophane membranes, the submerged mycelium was characterized by coarse, straight hyphae with sparse, almost right-angled branches and strong apical dominance. These actively growing hyphae (3–4 mm / 24 h) were strongly translucent, showed rapid cytoplasmic streaming, and contained lipid droplets. By contrast, hyphae growing on the cellophane membrane (drier growing conditions) produced numerous short, thin-walled, often septate branches. These resemble the "rhizoid-like branches" described by Peyronel (1923), and the "fine, flattened, thread-like filaments" described by Butler (1939).

A recent study on the influence of cold stratification on the germination of spores of *Glomus intraradices* Schenck & Smith also revealed two types of hyphal growth (Juge et al. 2002). These showed a strong resemblance to those recorded by Mosse (1959a). The fine branching germ tube and associated coiled and anastomosed hyphae ("g-type pattern," Juge et al. 2002) seem to be produced when the fungus is stressed (Mosse 1959a) or when spore dormancy is incompletely broken (Juge et al. 2002). By contrast, the RH growth pattern is more frequently observed when the spore is under favorable conditions (Mosse 1959a) or when spore dormancy has been completely broken ("G-type pattern," Juge et al. 2002). The latter may potentially increase the chance of colonizing a new host root. Other environmental factors, such as pH, temperature, moisture, mineral or organic nutrient concentrations, host plants, and (or) soil microorganisms, have also been shown to influence spore germination (Giovannetti 2000) and may influence germination patterns too. Further studies may help elucidate the causes and mechanisms that determine the expression of different hyphal growth patterns.

The role of the germ tube and subsequent presymbiotic hyphae is to create new colonizations. However, in the absence of a suitable host root, germ-tube growth can only be maintained for two to four weeks. After this period, the germ tube undergoes a process of cytoplasm retraction and hyphal septation (Logi et al. 1998). This is considered to be a survival strategy for these obligate mycobionts. This mechanism appears to be a general feature in AM fungi and has been described by several other authors (Godfrey 1957; Mosse and Hepper 1975; Mosse 1988). However, in certain cases, this survival strategy does not function, and the spore becomes moribund (Mosse and Hepper 1975). The question of survival is possibly linked to the RH or FB germ-tube growth patterns. This idea is supported by Juge et al. (2002), who observed increased g-type growth and spore mortality when cold treatments where shorter than 14 days. Considering these results, further studies should investigate whether RH-type germinations can regerminate more frequently than fine branching ones and whether the abortion of the fine branching germination pattern is followed by spore death.

Morphologically distinct characteristics of germ tubes of the Gigasporacea

Gigaspora gigantea (Nicol. & Gerd.) Gerd. & Trappe and *Gigaspora rosea* Nicol. & Schenck (previously misidentified as *Gigaspora margarita* Becker & Hall) germ tubes show a strong negative geotropism (Watrud et al. 1978; Bécard and Fortin 1988; Mosse 1988), the physiology of which still remains unexplained. Mosse (1988), using inverted Petri dishes, observed that the main hyphae of *G. gigantea* spiraled towards the Petri dish base, where it continued its spiral growth, forming circular patterns against the base. By contrast, in noninverted Petri dishes, germ tubes grew out of the medium and into the air. Although this spiraling growth is probably an artifact of the in vitro growth conditions, the negative geotropism observed in species of *Gigaspora* is a natural phenomenon.

Species of this genus are frequent in sand dunes, and this feature could represent an environmental adaptation allowing the hyphae of germinating spores to growth up toward newly developing roots on partially buried sand dune plants. Furthermore, unlike other AM fungi, species of *Gigaspora* and *Scutellospora* form auxiliary cells on the extraradical mycelium (Trappe and Schenck 1982). These structures, possibly ancestral soilborne vesicles, first appear as rounded cells; as they mature, their surface becomes ornamented (Jabaji-Hare et al. 1986). The role of theses structures as infective propagules is controversial (Biermann and Linderman 1983; Pons and Gianinazzi-Pearson 1985; Morton and Benny 1990), and their biological significance is still unknown.

Production of secondary spores on germinative hyphae

Another morphological event observed during the asymbiotic phase is the production of new spores on germinative hyphae. For reasons that remain unknown, this phenomenon was more frequently recorded in earlier in vivo studies. Godfrey (1957), Mosse (1959a) and Powell (1976) observed small, solitary, and generally infrequent secondary spores (chlamydospores) on the germ tubes of certain *Endogone* species. In a later study, Chabot et al. (1992a), working with *G. intraradices* (previously also known as *Glomus intraradix*) in vitro, made a clear distinction between secondary spores (20–30 µm), which were observed as early as seven days after germination, and spores formed after root colonization (approximately 50 µm) based on their respective size.

In addition, secondary spores, also referred to as vesicle-like structures, have been observed on germinative hyphae of *Glomus caledonium* (Nicol. & Gerd.) Trappe & Gerd. (Karandashov et al. 2000), *Glomus clarum* Nicol. & Schenck (de Souza and Berbera 1999), and *Glomus mosseae* (Nicol. & Gerd.) Gerd. & Trappe (Mosse 1962) when grown with root organ cultures. Paula et al. (1990) and Hildebrandt et al. (2002) observed that the production of these spores was enhanced by the presence of soil bacteria. Secondary spores are always smaller than spores formed on the symbiotic extraradical mycelium. The generally smaller spores produced by *G. intraradices* after regrowth from fragments of moribund roots can colonize host roots (Diop et al. 1994; see also Sporulation from moribund root fragments); however, to the best of our knowledge, no studies have investigated the maturity or germination capacity of secondary spores produced on germinative hyphae. On the contrary, the observations of de Souza and Berbera (1999), suggest that they may not be viable.

To conclude this section, germ-tube growth can be divided into either the RH type, which is the most frequent and potentially the most efficient, or the FB type, probably expressed under conditions of stress. These two major types of germ-tube growth are illustrated in the Fig. 3.1. We propose to keep the terms "RH" and "FB" to refer to these two distinct germination patterns (Table 3.1).

Morphology of the presymbiotic hyphae

In the vicinity of a suitable host root, germinative hyphae undergo certain morphological changes. These are the first visible signs of a fungus–plant interaction. The induced branching pattern occurs both in the presence of a host root and in the presence of root exudates or certain molecules that they contain.

Morphological changes in the presence of stimulatory molecules

Several authors have successfully used root exudates and (or) volatiles to enhance AM fungal hyphal growth or branching (e.g., Graham 1982; Bécard and Piché 1989a, 1989b) (see also Chapter 2). Elias and Safir (1987), working with clover (*Trifolium* sp.), showed that only exudates from P-deficient seedlings stimulated hyphal elongation of *Glomus fasciculatus* (Thaxter) Gerd. & Trappe. This suggests that root exudates change according to the nutrient balance of the host plant. Moreover, Nagahashi and Douds (1999), working with minute quantities of root exudates, established a close correlation between exudate concentration and presymbiotic branching of *G. gigantea*. Following the above results, a number of studies have identified a range of molecules in root exudates that are respon-

Advances in Mycorrhizal Science and Technology

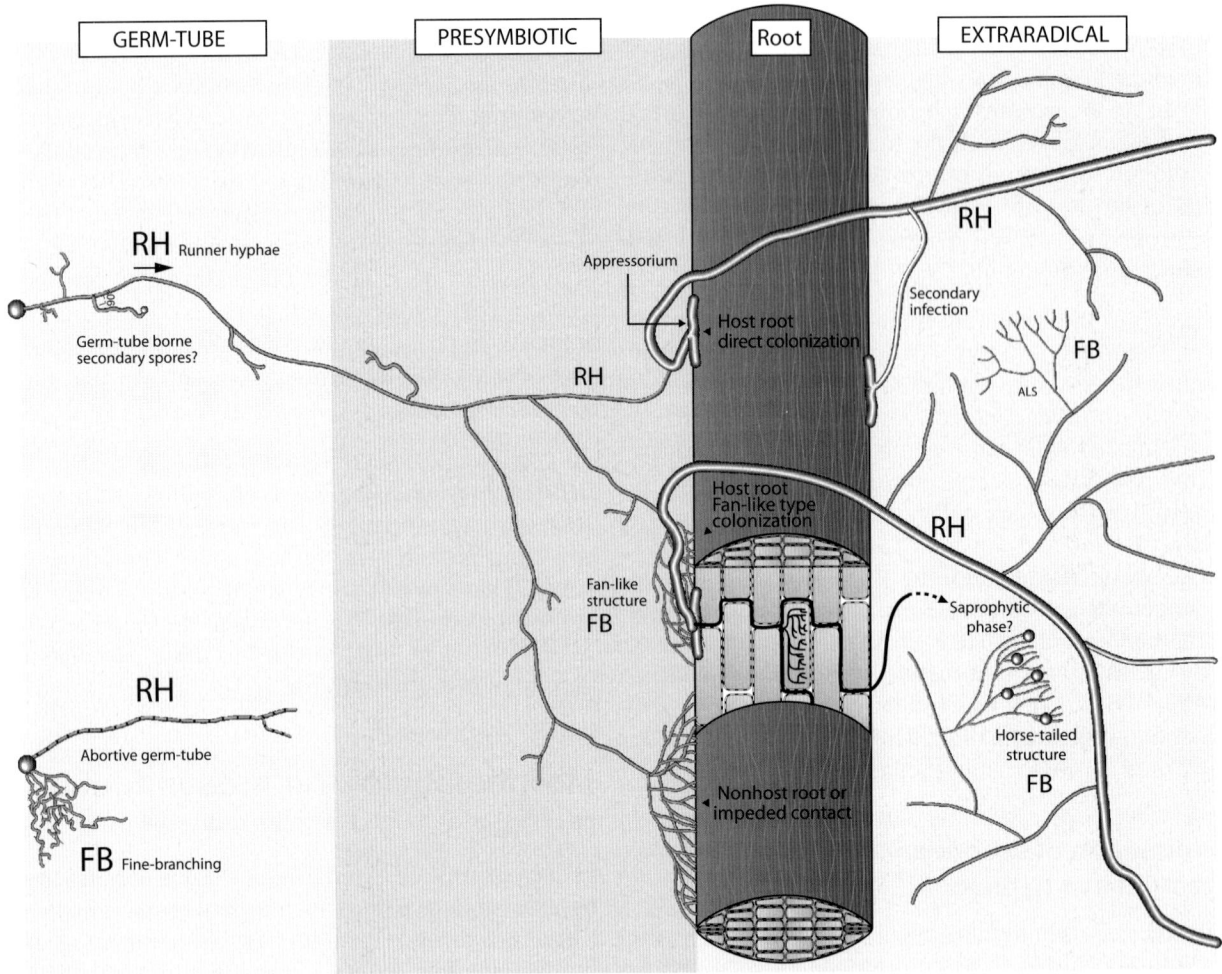

Fig. 3.1. Schematic representation of the morphological features of arbuscular mycorrhizal fungi through the three external phases of their life cycle, the asymbiotic germ-tube phase, the presymbiotic phase, and the extraradical symbiotic phase, illustrating the generalization of the runner hyphae (RH) – fine branching (FB) dimorphism.

sible for hyphal growth enhancement (Vierheilig and Piché 2002). However, these stimulatory molecules can have an inhibitory effect if present in too high a concentration.

Early studies showed that CO_2 from root respiration stimulated hyphal growth (Bécard and Piché 1989*a*). Subsequently, the optimal CO_2 concentration (2%) for hyphal elongation of *G. rosea* (previously misidentified as *G. margarita*) and *G. intraradices* was determined (Bécard et al. 1992; Chabot et al. 1992*b*). However, 5% CO_2 was found to detrimentally and irreversibly affect in vitro presymbiotic growth of *G. mosseae* (Le Tacon et al. 1983). These findings led Morandi (1996) to conclude that the lack of uniformity in CO_2 concentrations in different experiments could be one of the causes of divergent results observed by different research groups.

With regard to root exudates, one of the first groups of stimulatory molecules to be investigated were flavonoids (see reviews by Morandi (1996), Vierheilig and Piché (2002), Nagahashi and Douds (2005) and Vierheilig and Bago (2005)). However, a study by Bécard et al. (1995) demonstrated that flavonoids in root exudates were not necessarily the only molecules to stimulate AM fungal growth. The authors suggest that other root metabolites were also responsible for hyphal stimulation. Among the nonflavonoid compounds, Grandmaison et al. (1993) found that N-feruloyltyramine induced branching of hyphae of *G. intraradices* and *Glomus versiforme* (Karst.) Berch but reduced total hyphal length. More recently, Scervino et al. (2005), working with six flavonoid molecules and four species of AM fungi, found that flavonoids had different species- and genus-specific effects on hyphal growth.

Chapter 3. Growth and branching of asymbiotic, presymbiotic, and extraradical AM fungal hyphae

Table 3.1. Simplification under either runner hyphae (RH) or fine branching hyphae (FB) of the specific terms used to describe hyphal growth and branching of arbuscular mycorrhizal fungi during: the asymbiotic germ-tube phase, the presymbiotic phase, and the extraradical symbiotic phase.

Growth type	General features	Specific terms*	Species	Conditions of expression
Asymbiotic germ-tube phase				
RH	Straight, thick walled, 5–10 µm diameter, aseptate	RH (1,2,4)	*Endogone* spp. (1,2)	In vivo (1); in vitro, submerged hyphae (2)
		G-type (3)	*Glomus intraradices* (3)	In vitro, ≥14 days at 4 °C (3)
FB	Fine, twisted, 1–4 µm diameter, numerous branches	Rhizoid-like branches (5)	*Endogone* spp. (2,5,6)	In vivo (5,6)
		Thread-like filaments (6)		In vitro, on membranes (2)
		g-type (3)	*Glomus intraradices* (3)	In vitro, <14 days at 4 °C (3)
Presymbiotic phase				
RH	Direct colonization	RH (8)	*Endogone* spp. (7,8)	In vivo (7)
			Gigaspora rosea (8)	In vitro (8)
			Gigaspora gigantea (9)	Cell walls from carrot roots (9)
FB	Repeatedly branching, towards the root, cytoplasmic retraction, septation	Fan-like structures (7,10–12)	*Endogone* spp. (7)	In vivo, 1.6 mm from host root (7)
			Glomus mosseae (10,14), *Glomus intraradices* (11)	In vitro, root-organ cultures (10, 11)
			Gigaspora gigantea (12,13)	In vitro, three to five days after germination (12)
			Gigaspora rosea (12,13)	In vitro, near host root (13)
			Glomus intraradices (11)	In vivo, "sandwich" system (14)
Extraradical symbiotic phase				
RH	Straight, few or no branches, large (>20 µm diameter), thick-walled, angular projections (scars of extraradical FB)	Thick-walled hyphae (15)	*Endogone* spp. (15)	In vivo (15)
		RH (16)		In vivo, multiple secondary infections (hyphal bridges) and extraradical nets (16)
FB	Thin walled, fine (2–7 µm diameter), short lived, cytoplasmic retraction, septation	Thin-walled hyphae (15)	*Endogone* spp. (15,17)	In vivo (15)
		AHN† (16), ALS‡ (17), BAS§ (18)	*Glomus intraradices* (18)	In vitro (18)

*Numbers in parentheses refer to the following references: 1, Godfrey (1957); 2, Mosse (1959*a*); 3, Juge et al. (2002); 4, Dalpé et al. (2005); 5, Peyronel (1923); 6, Butler (1939); 7, Powell (1976); 8, Bécard and Fortin (1988); 9, Nagahashi and Douds (1997); 10, Mugnier and Mosse (1987); 11, Mosse (1988); 12, Bécard and Piché (1989*a*); 13, Bécard and Piché (1989*b*); 14, Giovannetti et al. (1993*b*); 15, Nicolson (1959); 16, Friese and Allen (1991); 17, Mosse (1959*b*); 18, Bago et al. (1998*a*, 1998*b*).
†Absorptive hyphal networks.
‡Arbuscule-like structures.

A recent scientific breakthrough on the early recognition between the two symbionts, showed that a nonflavonoid molecule, present in extremely low concentration in root exudates, was responsible for extensive hyphal branching (Buée et al. 2000). This molecule, referred to as "branching factor," was isolated from partially purified root exudates and induced a switch from asymbiotic to presymbiotic hyphal growth in *G. rosea*. A subsequent study by Tamasloukht et al. (2003), demonstrated that this branching factor was responsible for the elicitation of mitochondrial-related genes and fungal respiration. Recently, strigolactones, a group of plant sesquiterpenes, were shown to be the active molecules of the branching factor (Akiyama et al. 2005). Furthermore, Besserer et al. (2006) showed that strigolactones stimulated AM fungal mitochondria (see recent review by Akiyama and Hayashi (2006)).

Morphological changes in the close vicinity of the root

The fan-like structure

Presymbiotic hyphal structures in the vicinity of a root consist of a profusion of lateral and (or) apical branches proliferating from the germinative hypha. Powell (1976) was the first to use the term "fan-like structure" to describe this branching pattern. The author observed that all germinative hyphae of *G. mosseae*, and those of a second but unidentified AM fungus, first exhibited the RH-type growth pattern. These hyphae subsequently underwent a marked morphological change in close vicinity (1.6 mm) to onion (*Allium cepa* L.) roots. These hyphae branched dichotomously, forming fan-like structures of septate preinfection hyphae, which grew towards the root. As the hyphae approached the root surface, they became increasingly septate and of a more irregular form; however, the main subtending hypha remained aseptate. By contrast, the author observed that the germinative hyphae of *Acaulospora laevis* Gerd. & Trappe started to develop these structures at a greater distance (3.4 mm) from the root and the presymbiotic branching hyphae of this species were never septate.

Mugnier and Mosse (1987), Bécard and Fortin (1988), and Mosse (1988) observed similar branching when germ tubes of diverse *Glomus* and *Gigaspora* species came into contact with Ri T-DNA-transformed carrot (*Daucus carota* L.) roots. Working with *G. rosea* (previously misidentified as *G. margarita*), Bécard and Piché (1989*b*) proposed a spore-dependant mechanism, "M1," to explain this growth promotion in the absence of direct root contact. Briefly, the authors observed that when isolated carrot roots were added to Petri dishes three days after spore germination, there was a 20-fold increase in hyphal growth. Under these in vitro conditions, the hyphae exhibited rapid growth for the first three to six days. Following this, the growth rate decreased progressively and stopped after 15–28 days. As previously observed by Mosse (1988), this increased growth in the presence of a root ceased as soon as the root was removed.

In conclusion, germinative hyphae in the vicinity of a host root (presymbiotic hyphae) show a less pronounced apical dominance and developed a fine branching pattern, the fan-like structure.

Elicitation of fan-like structures in the absence of root contact

In an experiment using a system of membranes (0.45 µm), Giovannetti et al. (1993*b*) physically separated hyphae of *G. mosseae* and *G. intraradices* from host roots. However, this in vivo system allowed chemical exchanges between the two symbionts and the observation of hyphal morphogenesis elicited by the roots of several host species. Under these conditions, fan-like structures, similar to those described by Powell (1976), were formed in the absence of direct root contact. In the vicinity of the roots, the germinative hyphae branched profusely and produced irregular fine branches. These were heavily septate and exhibited a chaotic growth pattern. The modified hyphae grew rapidly (5 mm/day), giving rise to a thick hyphal network. The hyphal length in these networks was fourfold greater than the mycelium growing in the absence of roots, and branching was sevenfold greater. These observations were later validated in vitro (Giovannetti et al. 1996).

It should be noted that although fan-like branching is observed in the majority of cases, direct RH-type colonization, without the formation of fan-like structures, was observed by Powell (1976) and Bécard and Fortin (1988). This type of colonization is represented in Fig. 3.1 by a straight growing hypha and is referred to as RH (Table 3.1). In the case of direct colonization, absence of branching is not synonymous with an absence of root signals. Nevertheless, Nagahashi and Douds (1997) observed appressoria formation by *G. gigantea* and *G. rosea* (previously misidentified as *G. margarita*) on isolated cell walls obtained from carrot root organ cultures. The comparison of the molecular composition of root signals that induce direct colonization, with those that induce formation of fan-like structures, would probably give interesting new insights into presymbiotic interactions between the two symbiotic partners.

Appressorium formation: the last stage before root colonization

Once in contact with the root, the fungus forms a swollen appressorium from which develops the intraradical phase (Fig. 3.1). The appressorium, formed by a single hypha, develops two or three days after hyphal attachment to the root surface and often arises from a central hypha within the fan-like structure (Mugnier and Mosse 1987; Bécard and Fortin 1988; Giovannetti et al. 1993*a*). In the study by Giovannetti et al. (1993*a*), where hyphal–root contact was impeded by a series of membranes, "Elkhorn-like" structures were formed (not illustrated in Fig. 3.1). These structures were comprised of numerous enlarged hyphal tips separated by frequent septa. The authors considered these to be similar to appressoria and hypothesized that their formation was due to impeded contact with the root surface.

To conclude this section, several root-derived molecules have been shown to promote the formation of presymbiotic branching: these include CO_2, diverse phenolic compounds, and strigolactones. During the initial communication between both symbionts, these molecules trigger the morphological and physiological switch from a RH-type germinative hypha to a fine branching presymbiotic mycelium. In close proximity to the host root, the proliferation of the FB hyphae gives rise to a more specialized fan-like structure (Powell 1976). This structure, which is not dependent on direct root contact, probably improves exploration of the root surface to locate a suitable entry point. We recommend that the term "fan-like structure" be retained. The morphology of this presymbiotic structure is illustrated in Fig. 3.1, and the conditions of expression are summarized in Table 3.1.

Morphology of the extraradical symbiotic hyphae

Once the symbiosis is established, an extraradical mycelium develops. In general, AM fungal biomass production is greater during this phase than during the asymbiotic one (but see Godfrey (1957) and Powell (1976)). In natural ecosystems, mycorrhizal plants rely on the fungal networks produced for survival and reproductive fitness (Graham 1982). Moreover, the extraradical mycelium improves nutrient cycling and soil structure (see Chapter 5) and alters the microbial equilibrium in the mycorrhizosphere (see Chapter 4). Nevertheless, in spite of its recognized importance, the external phase of the AM fungus has been less well studied than the intraradical phase. This is mainly due to the difficulty of studying hyphal development within the soil matrix.

To the best of our knowledge, egress of hyphae from field-collected roots or colonized roots obtained from pot cultures has never been reported. This suggests that the extraradical mycelium typically develops from the hypha that colonized the root. This theory is supported by Brundrett and Juniper (1995), who used a nondestructive in vivo technique to observe hyphal growth of species of *Acaulospora*, *Glomus*, and *Scutellospora* and found "no evidence that any of the fungi produced several distinct systems of hyphae." However, recently Bago and Cano (2005) provided proof of egress of hyphae from colonized root-organ cultures. Our opinion is that the former strategy, based on branching of the colonizing hypha, is more probable during early establishment of the symbiosis, because it economizes both energy and time and means that the absorptive extraradical hyphae can begin to develop as soon as the symbiosis is formed (see Fig. 3.1). With regards to egress, our observations (unpublished) support those of Bago and Cano (2005); however, we only observed this phenomenon in older cultures. Egress of hyphae from old roots seems logical, because it would potentially allow the fungus to colonize younger roots.

Growth and branching of the extraradical hyphae

Runner hyphae (thick-walled hyphae)

Runner hyphae are large (20–27 µm diameter), thick-walled (1–3 µm) hyphae that resemble RH-type germinative hyphae (Nicolson 1959). These hyphae form the skeleton of the extraradical network (Friese and Allen 1991), and side branches develop at 30°–60° angles. The diameter of RH hyphae increases as the extraradical mycelium develops. These hyphae are able to form multiple secondary infections on a given root or to grow out several centimetres into the soil matrix to infect other roots. The result is the formation of hyphal bridges between adjacent roots of the same plant, two individual plants of the same species, or individual plants of two different species. Early studies mention the presence of "unilateral angular projections on these hyphae" (Peyronel 1923; McLuckie and Burges 1932; Butler 1939; Mosse 1959*b*; Nicolson 1959). These structures, which were considered to be a characteristic feature of the extraradical mycelium, have never been reported in in vitro studies (see following section for further details).

Fine branching (thin-walled hyphae)
Mosse (1959b) and Nicolson (1959) observed short, fine (2–7 µm diameter), thin-walled (<1 µm) hyphae that developed from runner hyphae. These typically became septate and according to the authors were more temporary structures. Mosse (1959b) underlined that, unlike the main hyphae that characteristically branched at 30°–60°, thin-walled secondary hyphae branched at right angles to the parent hypha. According to Nicolson (1959), these hyphae develop either from lateral branches of thick-walled hyphae or by the repeated branching of thick-walled hyphae. The septations observed in these structures resemble those occurring in aborted germinative hyphae and occlude the base of the thin-walled hyphae. Subsequent degeneration of the thin-walled hyphae in vivo leaves the above-mentioned conspicuous angular projections on the thick-walled hyphae, marking their former sites of attachment. Because the degeneration of thin-walled hyphae only occurs in the presence of soil microorganisms, these projections are not observed under in vitro conditions.

Functions of extraradical hyphal structures

Runner hyphae determine the architectural structure of the coenocytic network and serve as conduits, transporting soil nutrients to the plant hosts and carbohydrates to the developing extraradical hyphae. Hyphal architecture appears to be under the influence of both the AM fungal strain and the substrate. This is supported by a recent in vitro study by Bago et al. (2004), which showed that, for a giving individual AM fungus, the morphology of the extraradical mycelium was modified by the presence or absence of certain nutrients. Although it has not been demonstrated, an early study by Mosse (1959a) suggested that these "arbuscule-like structures" (ALS) have similar functions to intraradical arbuscules.

Following the original description of ALS by Mosse (1959a), several terms have been used to describe these extraradical fungal structures. For example, Friese and Allen (1991), working with several species of AM fungi and in soil, referred to these as "absorptive hyphal networks" (AHN). In a later series of detailed studies, Bago et al. (1998a, 1998b) referred to these structures as "branched absorbing structures" (BAS), because of the lack of a functional link between intraradical arbuscules and ALS.

In addition, BAS are probably important secretory sites; coupled with their rapid turnover, this probably influences soil microbial population dynamics (Linderman 1992).

Sporulation on the extraradical mycelium

Sporulation represents the completion of the fungal life cycle. Spore shape, size, ontogenesis, and final wall structure are highly species specific; until recently, these characteristics were the only basis for classification (Dalpé 1995). In this section, only the hyphal structures on which sporulation occurs are mentioned.

Mosse (1959b) documented the occurrence of spore formation on a repeatedly branching main hypha or on short lateral branches. These observations were supported by Bago et al. (1998c) and these structures were referred to as ALS spores. However, in a subsequent publication, the name was changed to BAS spores (Bago et al. 1998b), again for the reason outlined in the previous section. In addition, Bago et al. (1998c) occasionally observed hyphal structures, which they referred to as "horse-tailed," that appeared to be a preferential site for spore formation in *G. intraradices*. This structure may be similar to the repeatedly branching main hypha described by Mosse (1959b). The types of sporulation on FB extraradical structures vary from one species to the other but seem relatively uniform at the species level (Y. Dalpé, personal communication).

Sporulation from moribund root fragments

Strullu et al. (1997) observed hyphal growth from sections of AM roots obtained from pot cultures of *G. intraradices* and *G. versiforme*. Unlike germ tubes, this mycelium, which developed from intraradical hyphae and vesicles within the moribund root pieces, produced numerous viable spores (Diop et al. 1994). The existence of a saprophytic phase has not been proven; however, it is conceivable that the fungal mycelium issued from colonized root fragments could have acquired certain saprophytic properties from their passage within the host root. This might explain some of the observed differences between germinative hyphae and those growing from moribund root fragments. Although Strullu et al. (1997) recorded rapid outgrowth from colonized root fragments, no differences were observed in the colonization potential of root-segment inocula versus that of spores (Vimard et al. 1999) or between that of vesicles and spores (Nantais 1997). Nevertheless, in some cases, obtaining in vitro cultures is easier when done using a root segment than spores (Y. Dalpé, personal communication). However, it is not known whether this difference is due to differences in the physiological state of the fungus or a variation in the number of propagules present in the two types of inocula.

To conclude this section, the hyphae of the extraradical mycelium of AM fungi are, like the germ-tube and presymbiotic hyphae, dimorphic: either thick-walled RH, or thin-walled FB hyphae. We recommend that the original term "arbuscule-like structures" be retained for the distinctive FB hyphae of the extraradical mycelium. These structural features are illustrated in Fig. 3.1, and their general characteristics are summarized in Table 3.1. Thick-walled RH are responsible for the overall architecture of the external mycelium, and their large diameter permits rapid cytoplasmic streaming within the coenocytic hyphal network. The thin-walled ALS on the extraradical mycelium probably provide the main sites for bidirectional exchange between the mycorrhizal system and the soil matrix. However, the small size of these structures makes it extremely difficult to accurately determine their function.

General conclusion

To the best of our knowledge, this is the first comprehensive list of all the different terms previously used to describe the structural development of asymbiotic, presymbiotic, and extraradical AM fungal hyphae. A schematic representation of these three growth phases is shown in Fig. 3.1, and essential features and terms are given in Table 3.1. In each phase, the AM fungal hyphae are either elongating rapidly (RH) or extensively branching (FB). In the case of germinating spores, hyphae exhibit one or another of these growth patterns. However, in the two other growth phases, both patterns may cooccur (Fig. 3.1, Table 3.1).

In the absence of roots, straight, rapidly growing germinative hyphae should be referred to as "germinative RH," referring to the term "runner hyphae" first used by Nicolson (1959); the term "FB" should be used for the numerous smaller diameter germ-tubes produced under physical, chemical, and physiological stresses. In the spore-dependent presymbiotic phase, FB is stimulated by root exudates. Once in close proximity to a host root, FB hyphae mature to give fan-like structures, a term first used by Powell (1976). These structures probably enhance the initial establishment of the AM symbiosis. Nevertheless, germinative RH may occasionally colonize directly without the formation of these structures (Bécard and Piché 1989b).

In the last phase, the skeleton of the extraradical network is formed by the RH (Nicolson 1959; Friese and Allen 1991). From these RH, short-lived FB structures develop. We feel that these should be specifically referred to as ALS, rather than BAS or AHN, because the term refers to a morphological structure rather than to a precise function. The RH expand the mycelium into the soil matrix, and the fine branching ALS serve as probable bidirectionnal exchange sites. Together, these two hyphal growth patterns assure the efficient functioning of the established AM symbiosis.

Acknowledgments

The financial assistance provided by Natural Sciences and Engineering Research Council to the fourth author and scholarships from the Université Laval and the Canadian Oil Sands Network for Research and Development to the first author are gratefully acknowledged.

References

Akiyama, K., and Hayashi, H. 2006. Strigolactones: chemical signals for fungal symbionts and parasitic weeds in plant roots. Ann. Bot. (Lond.), **97**: 925–931.

Akiyama, K., Matsuzaki, K.I., and Hayashi, H. 2005. Plant sesquiterpenes induce hyphal branching in arbuscular mycorrhizal fungi. Nature (Lond.), **435**: 824–826.

Alexander, T., Toth, R., Meier, R., and Weber, H. 1989. Dynamics of arbuscule development and degeneration in onion, bean, and tomato with reference to vesicular–arbuscular mycorrhizae in grasses. Can. J. Bot. **67**: 2505–2513.

Allen, E., and Allen, M.F. 1984. Competition between plants of different successional stages: mycorrhizae as regulators. Can. J. Bot. **62**: 2625–2629.

Bago, B., and Cano, C. 2005. Breaking myths on arbuscular mycorrhizas *in vitro* biology. *In* In vitro culture of mycorrhizas. *Edited by* S. Declerck, D.G. Strullu, and J.A. Fortin. Springer-Verlag, Berlin. pp. 111–138.

Bago, B., Azcòn-Aguilar, C., Goulet, A., and Piché, Y. 1998a. Branched absorbing structures (BAS): a feature of the extraradical mycelium of symbiotic arbuscular mycorrhizal fungi. New Phytol. **139**: 375–388.

Bago, B., Azcòn-Aguilar, C., and Piché, Y. 1998b. Architecture and developmental dynamics of the external mycelium of the arbuscular mycorrhizal fungus *Glomus intraradices* grown under monoxenic conditions. Mycologia, **90**: 52–62.

Bago, B., Zipfel, W., Williams, R.M., Chamberland, H., Lafontaine, J.G., Webb, W.W., and Piché, Y. 1998c. *In vivo* studies on the nuclear behavior of the arbuscular mycorrhizal fungus *Gigaspora rosea* grown under axenic conditions. Protoplasma, **203**: 1–15.

Bago, B., Cano, C., Samson, J., Coughlan, A.P., and Piché, Y. 2004. Differential morphogenesis of the extraradical mycelium of an arbuscular mycorrhizal fungus grown monoxenically on spatially heterogeneous culture media. Mycologia, **96**: 452–462.

Bécard, G., and Fortin, J.A. 1988. Early events of vesicular–arbuscular mycorrhiza formation on Ri T-DNA transformed roots. New Phytol. **108**: 211–218.

Bécard, G., and Piché, Y. 1989*a*. Fungal growth stimulation by CO_2 and root exudates in vesicular–arbuscular mycorrhizal symbiosis. Appl. Environ. Microbiol. **55**: 2320–2325.

Bécard, G., and Piché, Y. 1989*b*. New aspects on the acquisition of biotrophic status by a vesicular–arbuscular fungus, *Gigaspora margarita*. New Phytol. **112**: 77–83.

Bécard, G., and Piché, Y. 1992. Establishment of vesicular–arbuscular mycorrhiza in root organ culture: review and proposed methodology. *In* Methods in microbiology. Vol. 24. Techniques for the study of mycorrhiza. *Edited by* J.R. Norris, D.J. Read, and A.K. Varma. Academic Press, London. pp. 89–108.

Bécard, G., Douds, D.D., Jr., and Pfeffer, P. 1992. Extensive *in vitro* hyphal growth of vesicular–arbuscular mycorrhizal fungi in the presence of CO_2 and flavonols. Appl. Environ. Microbiol. **58**: 821–825.

Bécard, G., Taylor, L.P., Douds, D.D., Jr., Pfeffer, P.E., and Doner, L.W. 1995. Flavonoids are not necessary plant signal compounds in arbuscular mycorrhizal symbioses. Mol. Plant Microbe Interact. **8**: 252–258.

Besserer, A., Puech-Pagès, V., Kiefer, P., Gomez-Roldan, V., Jauneau, A., Roy, S., Portais, J.C., Roux, C., Bécard, G., and Séjalon-Delmas, N. 2006. Strigolactones stimulate arbuscular mycorrhizal fungi by activating mitochondria. PLoS Biol. **4**: e226 doi: 10.1371/journal.pbio.0040226.

Biermann, B., and Linderman, R. 1983. Use of vesicular–arbuscular mycorrhizal roots, intraradical vesicles and extraradical vesicles as inoculum. New Phytol. **95**: 97–105.

Bonfante, P., and Perotto, S. 1995. Strategies of arbuscular mycorrhizal fungi when infecting host plants. New Phytol. **130**: 3–21.

Brachmann, A., and Parniske, M. 2006. The most widespread symbiosis on Earth. PLoS Biol. **4**: e239 doi: 10.1371/journal.pbio.0040239.

Brundrett, M., and Juniper, S. 1995. Non-destructive assessment of spore germination of VAM fungi and production of pot cultures from single spores. Soil Biol. Biochem. **27**: 85–91.

Buée, M., Rossignol, M., Jauneau, A., Ranjeva, R., and Bécard, G. 2000. The pre-symbiotic growth of arbuscular mycorrhizal fungi is induced by a branching factor partially purified from plant root exudates. Mol. Plant Microbe Interact. **13**: 693–698.

Butler, E. 1939. The occurrences and systematic position of the vesicular–arbuscular type of mycorrhizal fungi. Trans. Br. Mycol. Soc. **22**: 274–301.

Chabot, S., Bécard, G., and Piché, Y. 1992*a*. Life cycle of *Glomus intraradix* in root organ culture. Mycologia, **84**: 315–321.

Chabot, S., Bel-Rhlid, R., Chênevert, R., and Piché, Y. 1992*b*. Hyphal growth promotion *in vitro* of the VA mycorrhizal fungus, *Gigaspora margarita* Becker & Hall, by the activity of structurally specific flavonoid compounds under CO_2-enriched conditions. New Phytol. **122**: 461–467.

Cooper, K.M., and Tinker, P.B. 1978. Translocation and transfer of nutrients in vesicular–arbuscular mycorrhizas. II. Uptake and translocation of phosphorus, zinc, and sulphur. New Phytol. **81**: 43–52.

Dalpé, Y. 1995. Systématique des endomycorhizes à arbuscules: de la mycopaléontologie à la biochimie. *In* La symbiose mycorhizienne: état des connaissances. *Edited by* J.A. Fortin, C. Charest, and Y. Piché. Orbis, Québec, Que. pp. 1–20.

Dalpé, Y., Adriano de Souza, F., and Declerck, S. 2005. Life cycle of *Glomus* species in monoxenic culture. *In* In vitro culture of mycorrhizas. *Edited by* S. Declerck, D.G. Strullu, and J.A. Fortin. Springer-Verlag, Berlin. pp. 49–65.

de Souza, F.A., and Berbera, R.L.L. 1999. Ontogeny of *Glomus clarum* in Ri T-DNA transformed roots. Mycologia, **91**: 343–350.

Diop, T.A., Plenchette, C., and Strullu, D.G. 1994. *In vitro* culture of sheared mycorrhizal roots. Symbiosis, **17**: 217–227.

Elias, K., and Safir, G. 1987. Hyphal elongation of *Glomus fasciculatus* in response to root exudates. Appl. Environ. Microbiol. **53**: 1928–1933.

Fortin, J.A., Bécard, G., Declerck, S., Dalpé, Y., St-Arnaud, M., Coughlan, A.P., and Piché, Y. 2002. Arbuscular mycorrhiza on root-organ cultures. Can. J. Bot. **80**: 1–20.

Friese, C.F., and Allen, M.F. 1991. The spread of VA mycorrhizal fungal hyphae in the soil: inoculum types and external hyphal architecture. Mycologia, **83**: 409–418.

Giovannetti, M. 2000. Spore germination and pre-symbiotic mycelial growth. *In* Arbuscular mycorrhizas: physiology and function. *Edited by* Y. Kapulnick and D.D. Douds, Jr. Kluwer Academic Press, Dordrecht, the Netherlands. pp. 47–68.

Giovannetti, M., Avio, L., Sbrana, C., and Logi, C. 1993*a*. Factors affecting appressorium development in the vesicular–arbuscular mycorrhizal fungus *Glomus mosseae* (Nicol. & Gerd.) Gerd. & Trappe. New Phytol. **123**: 115–122.

Giovannetti, M., Sbrana, C., Avio, L., Citernesi, A., and Logi, C. 1993*b*. Differential hyphal morphogenesis in arbuscular mycorrhizal fungi during pre-infection stages. New Phytol. **125**: 587–593.

Giovannetti, M., Sbrana, C., Citernesi, A.S., and Avio, L. 1996. Analysis of factors involved in fungal recognition responses to host-derived signals by arbuscular-mycorrhizal fungi. New Phytol. **133**: 65–71.

Godfrey, R.M. 1957. Studies of British species of *Endogone*. III. Germination of spores. Trans. Br. Mycol. Soc. **40**: 203–210.

Graham, J.H. 1982. Effect of *Citrus* root exudates on germination of chlamydospores of the vesicular–arbuscular mycorrhizal fungus, *Glomus epigaeum*. Mycologia, **74**: 831–835.

Grandmaison, J., Olàh, G.M., Van Calsteren, M.R., and Furlan, V. 1993. Characterization and localization of plant phenolics likely involved in the pathogen resistance expressed by endomycorrhizal roots. Mycorrhiza, **3**: 155–164.

Hildebrandt, U., Janetta, K., and Bothe, H. 2002. Towards growth of arbuscular mycorrhizal fungi independant of a plant host. Appl. Environ. Microbiol. **68**: 1919–1924.

Holley, J.D., and Peterson, R.L. 1979. Development of a vesicular–arbuscular mycorrhiza in bean roots. Can. J. Bot. **57**: 1960–1978.

Jabaji-Hare, S.H., Piché, Y., and Fortin, J.A. 1986. Isolation and structural characterization of soil-borne auxiliary cells of *Gigaspora margarita* Becker & Hall, a vesicular–arbuscular mycorrhizal fungus. New Phytol. **103**: 777–784.

Juge, C., Samson, J., Bastien, C., Vierheilig, H., Coughlan, A.P., and Piché, Y. 2002. Breaking dormancy in spores of the arbuscular mycorrhizal fungus *Glomus intraradices*: a critical cold-storage period. Mycorrhiza, **12**: 37–42.

Karandashov, V., Kuzovkina, I., Hawkins, H.J., and George, E. 2000. Growth and sporulation of the arbuscular mycorrhizal fungus *Glomus caledonium* in dual culture with transformed carrot roots. Mycorrhiza, **10**: 23–28.

Le Tacon, F., Skinner, F.A., and Mosse, B. 1983. Spore germination and hyphal growth of a vesicular–arbuscular mycorrhizal fungus, *Glomus mosseae* (Gerdemann and Trappe), under decreased oxygen and carbon dioxide concentrations. Can. J. Microbiol. **29**: 1280–1285.

Linderman, R.G. 1992. Vesicular–arbuscular mycorrhizae and soil microbial interactions. *In* Mycorrhizae in sustainable agriculture. *Edited by* G.J. Bethlenfalvay and R.G. Linderman. American Society of Agronomy, Inc., Crop Science Society of America, Inc., and Soil Society of America, Inc., Madison, Wisconsin. ASA Spec. Publ. 54. pp. 45–70.

Logi, C., Sbrana, C., and Giovannetti, M. 1998. Cellular events involved in survival of individual arbuscular mycorrhizal symbionts growing in the absence of the host. Appl. Environ. Microbiol. **64**: 3473–3479.

Marschner, H., and Dell, B. 1994. Nutrient uptake in mycorrhizal symbiosis. Plant Soil, **159**: 89–102.

McLuckie, J., and Burges, A. 1932. Mycotropism in the Rutaceae. I. The mycorrhiza of *Eriostemon crowei*. Proc. Linn. Soc. N. S. W. **57**: 291–312.

Morandi, D. 1996. Occurrence of phytoalexins and phenolic compounds on endomycorrhizal interactions and their potential role in biological control. Plant Soil, **185**: 241–251.

Morton, J.B., and Benny, G.L. 1990. Revised classification of arbuscular mycorrhizal fungi (Zygomycetes): a new order, Glomales, two new suborders, Glomineae and Gigasporineae, and two new families, Acaulosporaceae and Gigasporaceae, with an emendation of Glomaceae. Mycotaxon, **37**: 471–491.

Mosse, B. 1959*a*. Observations on the extramatrical mycelium of a vesicular–arbuscular endophyte. Trans. Br. Mycol. Soc. **42**: 439–448.

Mosse, B. 1959*b*. The regular germination of resting spores and some observations on the growth requirements of an *Endogone* sp. causing vesicular–arbuscular mycorrhiza. Trans. Br. Mycol. Soc. **42**: 273–286.

Mosse, B. 1962. The establishment of vesicular–arbuscular mycorrhiza under aseptic conditions. J. Gen. Microbiol. **27**: 509–520.

Mosse, B. 1988. Some studies related to "independent" growth of vesicular–arbuscular endophytes. Can. J. Bot. **66**: 2533–2540.

Mosse, B., and Hepper, C. 1975. Vesicular–arbuscular mycorrhizal infections in root organ cultures. Physiol. Plant Pathol. **5**: 215–223.

Mugnier, J., and Mosse, B. 1987. Vesicular–arbuscular mycorrhizal infection in transformed root-inducing T-DNA roots grown axenically. Phytopathology, **77**: 1045–1050.

Nagahashi, G., and Douds, D.D., Jr. 1997. Appressorium formation by AM fungi on isolated cell walls of carrot roots. New Phytol. **136**: 299–304.

Nagahashi, G., and Douds, D.D., Jr. 1999. A rapid and sensitive bioassay to study signals between root exudates and arbuscular mycorrhizal fungi. Biotechnol. Tech. **13**: 893–897.

Nagahashi, G., and Douds, D.D., Jr. 2004. Isolated root caps, border cells, and mucilage from host roots stimulate hyphal branching of the arbuscular mycorrhizal fungus, *Gigaspora gigantea*. Mycol. Res. **108**: 1079–1088.

Nagahashi, G., and Douds, D.D., Jr. 2005. Environmental factors that affect presymbiotic hyphal growth and branching of arbuscular mycorrhizal fungi. *In In vitro* culture of mycorrhizas. *Edited by* S. Declerck, D.G. Strullu, and J.A. Fortin. Springer-Verlag, Berlin. pp. 95–110.

Nantais, L.M. 1997. Optimization of arbuscular mycorrhizal inoculum through selection and increased production of *Glomus intraradices* propagules. M.Sc. thesis, Département des sciences biologiques, Université de Montréal, Montréal, Que.

Nelsen, C. 1987. The water relations of vesicular–arbuscular mycorrhizal systems. *In* Ecophysiology of VA mycorrhizal plants. *Edited by* G.R. Safir. CRC Press, Boca Raton, Fla. pp. 71–91.

Nicolson, T. 1959. Mycorrhiza in the graminae I. Vesicular–arbuscular endophytes, with special reference to the external phase. Trans. Br. Mycol. Soc. **42**: 421–438.

Paula, M.A., dePinto, J.E.E., Siqueira, J.O., and Pasqual, M. 1990. Benefits of plant cell suspension to vesicular–arbuscular mycorrhizal fungi *in vitro*. 3. Effects of different culture medium. Pesq. Agropec. Bras. **25**: 1117–1124.

Peyronel, B. 1923. Prime ricerche sulle micorize endotrofiche e sulla microflora radicicola normale delle fanerogame. Riv. Biol. **5**: 463–485.

Plenchette, C., Furlan, V., and Fortin, J.A. 1981. Growth stimulation of apple trees in unsterilized soil under field conditions with VA mycorrhiza inoculation. Can. J. Bot. **59**: 2003–2008.

Pons, F., and Gianinazzi-Pearson, V. 1985. Observations on extramatrical vesicles of *Gigaspora margarita in vitro*. Trans. Br. Mycol. Soc. **84**: 168–170.

Powell, C. 1976. Development of mycorrhizal infections from *Endogone* spores and infected root segments. Trans. Br. Mycol. Soc. **66**: 439–445.

Redecker, D., Kodner, R., and Graham, L.E. 2000. Glomalean fungi from the Ordivician. Science (Washington, D.C.), **289**: 1920–1921.

Remy, W., Taylor, T.N., Hass, H., and Kerp, H. 1994. Four hundred-million-year-old vesicular arbuscular mycorrhizae. Proc. Natl. Acad. Sci. U.S.A. **91**: 11841–11843.

Sanders, I.R. 1999. Evolutionary genetics: no sex please, we are fungi. Nature (London), **399**: 737–739.

Scervino, J.M., Ponce, M.A., Erra-Bassels, R., Vierheilig, H., Ocampo, J.A., and Godeas, A. 2005. Flavonoids exhibit fungal species and genus specific effects on the presymbiotic growth of *Gigaspora* and *Glomus*. Mycol. Res. **109**: 789–794.

Simon, L., Bousquet, J., Lévesque, R., and Lalonde, M. 1993. Origin and diversification of endomycorrhizal fungi and coincidence with vascular land plants. Nature, **363**: 67–69.

Smith, S.E., and Smith, F.A. 1990. Structure and function of the interfaces in biotrophic symbioses as they relate to nutrient transport. New Phytol. **114**: 1–38.

St-Arnaud, M., Hamel, C., Caron, M., and Fortin, J.A. 1994. Inhibition of *Pythium ultimum* in roots and growth substrate of mycorrhizal *Tagetes patula* colonized with *Glomus intraradices*. Can. J. Plant Pathol. **16**: 187–194.

St-Arnaud, M., Hamel, C., Vimard, B., Caron, M., and Fortin, J.A. 1996. Enhanced hyphal growth and spore production of the arbuscular mycorrhizal fungus *Glomus intraradices* in an *in vitro* system in the absence of host roots. Mycol. Res. **100**: 328–332.

Strullu, D.G., Diop, T., and Plenchette, C. 1997. Réalisation de collections *in vitro* de *Glomus intraradices* (Schenck et Smith) et *Glomus versiforme* (Karsten et Berch) et proposition d'un cycle de développement. C. R. Acad. Sci. Paris Sci. Vie, **320**: 41–47.

Tamasloukht, M., Séjalon-Delmas, N., Kluever, A., Jauneau, A., Roux, C., Bécard, G., and Franken, P. 2003. Root factors induce mitochondrial-related gene expression and fungal respiration during the developmental switch from asymbiosis to presymbiosis in the arbuscular mycorrhizal fungus *Gigaspora rosea*. Plant Physiol. **131**: 1468–1478.

Trappe, J.M., and Schenck, N.C. 1982. Taxonomy of the fungi forming endomycorrhiza. A. Vesicular–arbuscular mycorrhizal fungi (Endogonales). *In* Methods and principles of mycorrhizal research. *Edited by* N.C. Schenck. American Phytopathological Society Press, St. Paul, Minn. pp. 1–9.

Vierheilig, H., and Bago, B. 2005. Host and non-host impact on the physiology of the AM symbiosis. *In In vitro* culture of mycorrhizas. *Edited by* S. Declerck, D.G. Strullu, and J.A. Fortin. Springer-Verlag, Berlin. pp. 139–158.

Vierheilig, H., and Piché, Y. 2002. Signalling in arbuscular mycorrhiza: facts and hypotheses. *In* Flavonoids in cell functions. *Edited by* B. Buslig and J. Manthey. American Chemical Society, Kluwer Academic/Plenum, New York. pp. 23–39.

Vimard, B., St-Arnaud, M., Furlan, V., and Fortin, J.A. 1999. Colonization potential of *in vitro*–produced arbuscular mycorrhizal fungus spores compared with a root-segment inoculum from open pot culture. Mycorrhiza, **8**: 335–338.

Watrud, L.S., Heithaus, J.J., and Jaworski, E.G. 1978. Geotropism in the endomycorrhizal fungus *Gigaspora margarita*. Mycologia, **70**: 449–452.

Chapter 4
Interactions between arbuscular mycorrhizal fungi and soil microorganisms

Laëtitia Lioussanne, Marie-Soleil Beauregard, Chantal Hamel, Mario Jolicoeur, and Marc St-Arnaud

Introduction

Bacteria, fungi, algae, nematodes, and protozoa are the principal components of the soil microflora and fauna. Most of these depend on organic matter as a source of C and generally proliferate in the top 10–15 cm of the soil profile, where plant roots and organic residues are most abundant. Soil microbial communities are essential for nutrient cycling, plant growth, and as a result, life on Earth. Thus, the extent to which these communities interact amongst themselves and with plant roots is of great importance. These interactions are complex and may fall anywhere along the continuum that ranges from wholly mutualistic to wholly pathogenic.

Soil microbial communities are species rich; however, they show an apparently high degree of functional redundancy, and the notion of functional groups is frequently employed when treating this subject. Arbuscular mycorrhizal (AM) fungi form one such group and have been coevolving symbiotically with plants for more than 400 million years. (Remy et al. 1994). These fungi act as root extensions that increase the volume of soil influenced and exploited by plants (Figs. 4.1 and 4.2). As a major interface between plant roots and the soil, they play a pivotal role in nutrient cycling (Dodd 2000; Hodge 2000) and ecosystem productivity (van der Heijden et al. 1998) and influence the outcome of interactions between plants (Klironomos 2002). Furthermore, they improve the physical quality of soil (Jastrow et al. 1998; Six et al. 2004), supply mineral nutrients and water to host plants (Subramanian and Charest 1998; Smith et al. 2001), and reduce the impact of root pathogens (St-Arnaud and Vujanovic 2007). Therefore, mycorrhizal fungi are of major ecological importance for plants and must be considered when developing management strategies for sustainable soil–plant systems and microbial-based biotechnologies for use in agriculture. Moreover, it is important to understand how soil microbial communities influence, and are influenced by, AM fungi. This chapter provides a summary of some of the recent advances in our understanding of microbial interactions in the mycorrhizosphere, and particular emphasis is placed on their role in nutrient cycling and plant health.

Arbuscular mycorrhizal fungi and general soil microbial diversity

The composition of the microbial community varies with the nature of the soil environment; however, its biomass is usually positively correlated with soil organic matter (SOM) content (Witter and Kanal 1998; Manjaiah et al. 2000; Bohme et al. 2005) or C availability (Campbell et al. 1999). Most of the 10^6–10^9 microbial units typically present in a single gram of soil are bacteria and fungi. The latter normally constitute the largest component of the soil microbial biomass (Domsch et al. 1980; Atlas and Bartha 1997; Brodie et al. 2003), and their hyphae

---

L. Lioussanne, M.-S. Beauregard, and M. St-Arnaud.[1] Institut de recherche en biologie végétale, Jardin botanique de Montréal, 4001, Sherbrooke Est, Montréal, QC H1X 2B2, Canada.

C. Hamel. Environmental Health / Water and Nutrients, Agriculture and Agri-Food Canada, 1 Airport Road, P.O. Box 1030, Swift Current, SK S9H 3X2, Canada.

M. Jolicoeur. Unité de recherche Bio-P2, Département de Génie Chimique, École Polytechnique de Montréal, Boîte postale 6079, succursale centre-ville, Montréal, QC H3C 3A7, Canada.

[1]Corresponding author (e-mail: marc.st-arnaud@umontreal.ca).

Fig. 4.1. *Glomus intraradices* on a carrot root-organ culture (scale bar: 1 mm). The arbuscular mycorrhizal fungal mycelium acts as an extension of the root increasing the volume of the substrate exploited by the plants, as well as enhancing the influence of plants on the soil environment.

may form extensive mycelial networks (Brodie et al. 2003; Leake et al. 2004) within the soil matrix. Of these fungi, those forming AM associations are typically highly abundant and may account for approximately 25% of the microbial biomass (Hamel et al. 1991; Olsson et al. 1999; Hamel 2007) and up to 80% of the fungal biomass (Kabir et al. 1997) in certain agricultural soils.

The distribution of the major mycorrhizal associations in ecosystems generally follows altitudinal and (or) latitudinal gradients (Read 1991; Smith and Read 1997), with arbuscular mycorrhiza being most abundant in temperate deciduous forests, grasslands, agricultural systems, and tropical forests, where P availability typically limits productivity. Therefore, arbuscular mycorrhizal fungi are an important component of Canada's agricultural and forest ecosystems (Dalpé 2003).

Biological interactions in soil

A soil microorganism may be affected by the presence of plant roots, other microorganisms, and diverse environmental conditions. Between 10% and 40% of the C fixed by plants through photosynthesis is released in the soil in forms readily available for soil microorganisms, thus stimulating microbial activity (Bowen and Rovira 1999; Uren 2000; Bertin et al. 2003). Consequently, rhizodeposition is an important factor coupling plant nutrition and microbial processes. Thus, much of the soil microbial activity is spatially organized around the roots, in the particularly active zone of soil called the rhizosphere. Active release of organic materials by roots occurs mostly at the root tips, where outer root cap cells secrete mucilage containing polysaccharides and proteins (Sievers and Braun 1996). The effect of plants on soil microorganisms depends on the plant species in question and its growth phase (Söderberg et al. 2002; Wamberg et al. 2003).

The potential of plants to influence soil microorganisms is highlighted by the occurrence of different microbial populations in the rhizosphere of different plant species in a given community (Ibekwe and Kennedy 1998; Marschner et al. 2001) or the presence of similar microbial communities in the rhizospere of a given plant species grown in different soils (Grayston et al. 1998; Miethling et al. 2000). Different AM fungal communities may occur in different plant communities (Zhang et al. 2004), on different plant species (Saito et al. 2004), and even on different parts of the same root (Scheublin et al. 2004). Although the reason for these differences is poorly understood, differences in the population of AM fungi with soil depth (Oehl et al. 2005) are likely a result of root distribution. However, assemblages of AM fungal taxa from the same soil sample were recently shown to differ markedly, depending on whether spores, intraradical mycelium, or extraradical mycelium were used to describe the community (Hempel et al. 2007).

Effect of arbuscular mycorrhizal fungi on rhizosphere microbiota

Mycorrhizal fungi effectively extend the influence of plants beyond the rhizosphere and interact directly and indirectly with other soil microorganisms. These interactions may lead to positive or negative changes in the populations of the different components of the soil microbial community and lead to the formation of the so-called mycorrhizosphere (Meyer and Linderman 1986a, 1986b; Paulitz and Linderman 1989; Calvet et al. 1992; St-Arnaud et al. 1995b; Rousseau et al. 1996; Filion et al. 1999; Vigo et al. 2000; Elsen et al. 2001; Talavera et al. 2001; Gryndler et al. 2002; St-Arnaud and Elsen 2005).

Microbial species differ in their ability to metabolize or access different nutrient sources (Baudoin et al. 2003). Formation of arbuscular mycorrhiza induces changes in root exudation patterns, modifying the plant's impact on the soil biota. Furthermore, the extraradical hyphae become an important source of exudates (Linderman 1992; Jakobsen et al. 2002), which consist of C of plant origin (Nakano et al.

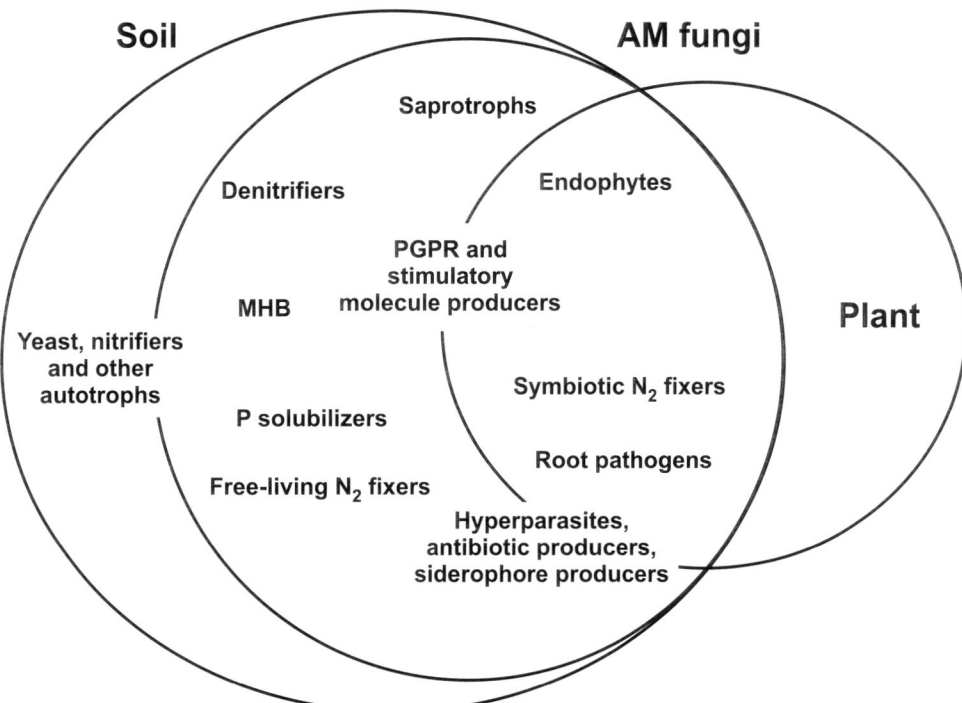

Fig. 4.2. Diagrammatic representation of the physicochemical environment in which interactions between arbuscular mycorrhizal (AM) fungi and soil microorganisms occur. The distance between a microbial group and the plant, AM fungi, or soil spheres indicates the presumed influence level of these spheres on the microbial group. Arbuscular mycorrhizal fungi enhance plant photosynthesis; because they are C sinks for plants, they export C belowground. Arbuscular mycorrhizal fungal hyphae growing thought the soil in the rooting zone modify nonhumic C distribution in soil. This C is an important microbial food source in soil, where microbial biomass is normally C limited. The production of inhibitory molecules by AM fungi has never been reported, but AM fungi may compete for soil resources with other soil organisms. Autotrophs are not stimulated by AM fungi as a C source but may be influenced by them indirectly through their impact on soil nutrient levels and interaction with other soil organisms. Interactions between AM fungi and members of the main functional groups of microorganisms illustrated here can be positive, negative, or neutral and likely result from the complex dynamic equilibrium between living organisms and environmental conditions that exists in soil. MHB, mycorrhiza helper bacteria; PGPRs, plant growth promoting rhizomicroorganisms.

1999; Johnson et al. 2002). This is supported by the changes both in the total microbial population and in the ratio between specific microbial taxa found in the hyphosphere of mycorrhizal root systems (the root-free zone of soil influenced by mycorrhizal structures) compared with the rhizosphere of non-AM plants (Posta et al. 1994; Andrade et al. 1998; Ravnskov et al. 1999; Marschner et al. 2001; Jeffries et al. 2003). Direct interactions between fungal hyphae and bacteria are diverse and involve mutualistic exudate-consuming bacteria that are associated with fungal surfaces as well as endosymbiotic and mycophagous species (Boer et al. 2005). The positive selection for mycorrhizal fungal-specific bacteria most likely occurs under a given plant (Linderman and Paulitz 1990; Olsson et al. 1996).

Several studies have shown qualitative, quantitative, and spatial shifts in bacterial communities due to mycorrhizal associations (Meyer and Linderman 1986b; Linderman and Paulitz 1990; Posta et al. 1994; Andrade et al. 1997). Furthermore, there are numerous reports of AM fungi influencing bacterial growth rate (Christensen and Jakobsen 1993; Marschner and Crowley 1996a, 1996b; Marschner et al. 1997). For example, growth of *Pseudomonas chlororaphis* (Guignard & Sauvageau) Bergey et al. was positively correlated with the concentration of unidentified components within the hyphosphere of an AM fungus. By contrast, growth of *Clavibacter michiganensis* corrig. (Smith) Davis et al. was not affected (Filion et al. 1999).

Arbuscular mycorrhizal fungi may also affect soil-borne fungi, but the outcome of interactions with saprophytic species is difficult to predict (Larsen et al. 1998; Olsson et al. 1998; Green et al. 1999). For example, negative, neutral, and positive effects of AM fungi on the population density of species of *Trichoderma* have been observed (Fracchia et al. 1998; Godeas et al. 1999; Green et al. 1999). Furthermore, recent results revealed that *Trichoderma pseudokoningii* Rifai influenced AM fungal development and function, thus potentially modifying the impact of AM fungi on soil and plants; however, the effect varied with the strain of saprophytic fungus used (Martinez et al. 2004). A number of studies also indicate that saprotrophic fungi influence presymbiotic AM fungi: positive and negative spore germination and germ-tube growth responses to volatiles and soluble exudates from saprotrophic fungi have been observed (McAllister et al. 1994, 1995; Fracchia et al. 1998). Furthermore, saprotrophic fungi may affect the functioning of AM fungi within host roots (McAllister et al. 1994, 1995; Fracchia et al. 1998; Garcia-Romera et al. 1998; Martinez et al. 2004). However, the nature of these interactions is still unclear.

Yeasts may also influence AM fungi. For example, *Rhodotorula mucilaginosa* (Jörgensen) Harrison, *Cryptococcus laurentii* (Kufferath) Skinner, and *Saccharomyces kunashirensis* James, Cai, Roberts & Collins, or their soluble and volatile exudates, were shown to stimulate spore germination, hyphal growth, and the development of the intraradical phase of *Glomus mosseae* (Nicol. & Gerd.) Gerd. & Trappe (Sampedro et al. 2004). Studies using *Saccharomyces cerevisiae* Meyen ex Hansen gave similar results (Larsen and Jakobsen 1996). Furthermore, when the yeast *Yarrowia lipolytica* (Wick. et al.) Van der Walt & Arx. was inoculated into a desertified soil, a 187% increase in AM fungal biodiversity was recorded (Medina et al. 2004a, 2004b).

Interspecific interactions may also occur between different AM fungi. In a microcosm experiment, plant productivity increased in the presence of several AM fungal species (Klironomos et al. 1998). However, inoculation of strawberry cultivars with a mixture of *Glomus intraradices* Schenck & Smith, *G. mosseae*, and *Glomus etunicatum* Becker & Gerd. did not increase yield and, sometimes, reduced mycorrhizal development when compared with inoculation with *G. intraradices* alone (Stewart et al. 2005).

Various microorganisms, including isolates of species of *Azotobacter*, *Azospirillum*, *Bacillus*, *Clostridium*, and *Pseudomonas*, improve plant growth by producing growth-enhancing substances and suppressing root pathogens (Glick 1995; Jeffries and Dodd 1996; Vazquez et al. 2000; Siddiqui and Shaukat 2002; Vessey 2003; Gamalero et al. 2004). Collectively, these organisms are referred to as plant growth promoting rhizomicroorganisms (PGPR). Certain PGPR also influence AM fungi. For example, it was shown that the inoculation of *Ficus benjamina* L. with two PGPR (*Bacillus coagulans* Hammer and *Trichoderma harzianum* Rifai) and an AM fungus (*G. mosseae*) not only positively affected plant biomass, but also enhanced root colonization by the AM fungus. This suggests a synergistic interaction between the three microorganisms (Srinath et al. 2003). By contrast, inoculation of strawberry (*Fragaria* sp.) plants with various combinations of *G. mosseae* and four PGPR, in an attempt to reduce the levels of infection caused by *Phytophthora* spp., gave varying results that ranged from inhibition of the pathogen to its stimulation (Vestberg et al. 2004). Interactions between soilborne plant pathogenic microorganisms and AM fungi are treated in greater depth below.

Little is currently known about the ecology of AM fungi under natural conditions (Hodge 2000). However, the above examples help to illustrate that the outcome of interactions between AM fungi and diverse soil microorganisms may be highly variable and difficult to predict. This suggests that the environmental conditions under which they occur may be important and that a better understanding of their ecology might be obtained by taking soil type into account. Different soil types offer different conditions for the soil biota. Rivera and Fernández (2003) found good relationships between AM fungal species and soil types. Crop yield increases were obtained mainly when AM fungal species were inoculated in their corresponding soil type. Because soils are more easily described than the soil microbial communities themselves, soil classification may be an important tool for the successful application of AM fungi to different agricultural systems.

Arbuscular mycorrhiza in nutrient mobilization

Arbuscular mycorrhizal fungi enhance plant uptake of poorly mobile nutrients (Barea et al. 2005) and those present in low concentrations. This is principally due to an increase in the volume of soil exploited and a more efficient mobilization of the nutrients than by roots alone. By removing nutrients from the soil solution, AM fungi disrupt the equilibrium that exists between the amount of a given nutrient present in solution and that in the solid phase, which triggers further release from the latter. Furthermore, AM fungi are able to influence the mobilization of soil nutrients via the transport of C into the soil matrix. This effect is indirect and occurs as a result of the stimulation of soil microorganisms involved in nutrient cycling. The subsequent biological activity causes the breakdown and mineralization of SOM and solubilization of mineral materials (Paul and Clark 1996).

Most soil microorganisms are involved in the mineralization of soil organic P and are considered to contribute significantly to the total soil phosphatase activity (Richardson 1994). To maintain their metabolism, soil microorganisms either use C released from plant roots and (or) mycorrhizal fungi (Sievers and Braun 1996; Bowen and Rovira 1999; Uren 2000; Bertin et al. 2003) or absorb dissolved organic molecules released during decomposition. In both cases, the molecules are oxidized and CO_2 released. Mineral nutrients resulting from the breakdown of SOM are released into the soil solution when their abundance exceeds the amount required for growth and maintenance of the microorganism involved. The CO_2 produced dissolves in soil water producing carbonic acid, which solubilizes soil mineral constituents together with the organic acids produced through microbial (and root) metabolism.

The amount of exudates produced by a root varies with its development, being more abundant in younger zones than in older ones. Bacteria behind the apex assimilate these substrates, creating successive turnover waves of diminishing amplitude along the root and an alternating immobilization and release of nutrients (Semenov et al. 1999; Zelenev et al. 2000). Considering the amount of nutrients released by plant root systems, microbial activity and succession in the root zone are certainly important steps in nutrient mobilization and cycling in soil (Azcón-Aguilar and Barea 1992; Toro et al. 1997; Stevenson and Cole 1999; Steinberg and Rillig 2003). There is growing interest in PGPR because of the role they play in plant health and soil fertility. These organisms are often divided into two groups based on their function: those involved in nutrient cycling, nutrient mobilization, and phytostimulation and those involved in the biocontrol of plant pathogens (see below). These two groups are not mutually exclusive.

Arbuscular mycorrhizal fungi and phosphorus availability

Phosphorus solubilization is a PGPR-mediated process (Richardson 2001). Phosphorus is very reactive in soils and as a result, the concentration of phosphate ions (predominantly HPO_4^{2-} and $H_2PO_4^-$) in the soil solution is very low, despite the relative abundance of total P in many soils. Because of the high reactivity of P, the diffusion of phosphate ions in soil is an extremely slow process, and P availability often limits plant growth in natural systems (Bieleski 1973; Richardson 2001). Different soil microorganisms have resolved the problem of P availability by solubilizing P sequestered in the soil mineral fraction, and various PGPR and other soil microorganisms exhibiting this capacity have been studied (Vassileva et al. 1998; Whitelaw 2000; Richardson 2001; Vivas et al. 2003; Gamalero et al. 2004; Barroso and Nahas 2005).

Mycorrhizal fungi also produce soil phosphatase enzymes (Koide and Kabir 2000) and, probably, excrete organic acids. However, the major contribution of AM fungi to plant P nutrition is probably related to the ability of their external mycelium to explore the soil intensively and to access microsites (Smith and Read 1997). The P taken up by the AM mycelium can be translocated relatively long distances within the hyphae to plant roots. In this manner, AM fungi give plants access to phosphate ions located beyond the P-depletion zone that normally develops around the roots.

Arbuscular mycorrhizal fungi and P-solubilizing microorganisms may have synergistic or complementary effects. For example, many PGPR are able to solubilize sparingly soluble P through the release of chelating organic acids, which may then be taken up by AM fungi and translocated to their host plant (Barea et al. 1983, 1997). Futhermore, inoculation with AM fungi can facilitate the establishment of both inoculated and indigenous P-solubilizing rhizobacteria (Barea et al. 2005). Moreover, AM fungi can enhance the P solubilization activity of PGPR. For example, the bacteria *Pseudomonas aeruginosa* (Schröter) Migula and *Pseudomonas putida* Trevisan solubilized more P when grown in the presence of *G. intraradices* than when grown alone (Villegas and Fortin 2001, 2002).

Arbuscular mycorrhizal fungi and nitrogen uptake

Nitrogen also occupies a key position among those elements essential for plant and microbial growth. Living organisms have a particularly large requirement for N, and it is often limiting in plant–soil systems. It appears that AM fungi have the potential to improve plant N uptake from soil. Uptake from organic sources may be partly due to the capacity of AM fungi to take up organic molecules (Hawkins et al. 2000); however, it is also likely related to the fact that AM fungi physically increase the absorbing capacity of plant root systems. Hyphae of an AM fungus proliferating in the vicinity of organic residues may increase its host plant's competitive ability to take up NH_4^+ released through mineralization (Hamel 2004). The ability of AM hyphae to take up and translocate NH_4^+ to plants has been clearly demonstrated (Subramanian and Charest 1999; Toussaint et al. 2004; Tanaka and Yano 2005). Recently, Govindarajulu et al. (2005) investigated N transfer in the AM symbiosis using mass spectrometry and quantitative reverse transcription polymerase chain reaction (PCR) approaches. The authors showed that inorganic N taken up by the AM fungus outside the roots is incorporated into amino acids, translocated from the extraradical to the intraradical mycelium as arginine, and probably, transferred to the plant as ammonium.

In natural ecosystems, N is mainly incorporated into the soil through chemical and biological N_2 fixation, the latter being performed by free-living bacteria and cyanobacteria or by plants in symbiosis with these organisms. In turn, organic forms of N are mineralized to ammonia by numerous bacteria and fungi and oxidized to nitrate by nitrifying bacteria (Stevenson and Cole 1999). Studies have suggested that the presence of AM fungi influences populations of N-transforming microorganisms and that these interactions may modify N availability in soils. For example, the occurrence of autotrophic nitrifying bacteria in the rhizosphere soil of maize (*Zea mays* L.) grown in pot cultures was significantly higher when the pots were inoculated with either *G. mosseae* or *Glomus fasciculatum* (Thaxter) Gerd. & Trappe, whereas ammonifying and denitrifying bacterial populations significantly decreased (Amora-Lazcano et al. 1998). By contrast, a recent study using soil under an organic farming regime revealed a diverse community of ammonia-oxidizing bacteria, and no significant differences were found between the rhizosphere soil of wild type tomato (*Solanum lycopersicum* L.) plants and that of a mycorrhiza-defective mutant (Cavagnaro et al. 2007).

Among those studies related to the influence of AM fungi on the organisms involved in N cycling, investigations concerning N_2 fixation are the most abundant. Phosphorus is involved in energy transport (ATP) and N_2 fixation is a very energy intensive process. Therefore, increased P uptake by a mycorrhizal plant would enhance N_2 fixation by diazotrophic bacteria. It has been shown using ^{15}N that N_2 fixation is greater in *Rhizobium*-inoculated mycorrhizal legumes than in nonmycorrhizal controls (Puppi et al. 1994; Jeffries et al. 2003). Several plant – *Rhizobium* – AM fungal combinations have been tested, and all have shown increased plant development, nutrient uptake, and N_2 fixation (Requena et al. 1997). Nitrogen fixation further promotes mycorrhizal development through improved plant nutrition and growth. Plants of tripartite symbioses generally contain higher N levels and, consequently, exhibit better root and mycorrhizal development (Puppi et al. 1994; Johansson et al. 2004).

An association among *G. mosseae*, *Rhizobium meliloti* Dangeard, a P-solubilizing rhizobacterium (*Enterobacter* sp.), and alfalfa plants (*Medicago sativa* L.) was investigated by Barea et al. (2002). In this experiment, the *Rhizobium* behaved as a mycorrhiza-helper bacterium (MHB), promoting establishment of the AM symbiosis. In turn, formation of arbuscular mycorrhiza increased N_2-fixation rates in *Rhizobium*-inoculated plants, compared with nonmycorrhizal controls. The dual inoculation of the AM fungus and the *Rhizobium* significantly increased microbial biomass and N and P accumulations in plant tissues. Furthermore, dual-inoculated plants also displayed lower specific $^{32}P/^{31}P$ activity in the presence of the *Enterobacter* sp. than their comparable controls. This suggests that the bacteria-inoculated mycorrhizal plants were using P sources made available through the activity of the P-solubilizing bacteria and, otherwise, unavailable to the plant. Therefore, it appears that these rhizosphere or mycorrhizosphere interactions contributed to the biogeochemical cycling of P and increased plant fitness.

Arbuscular mycorrhizal fungi and their role in controlling soilborne plant pathogens

Disease-suppressive soils were first described over 100 years ago. These soils show minimal disease development, even in the presence of virulent pathogens and susceptible plants. In these soils, propagules of the pathogens are subjected to fungistasis, and the germination of their propagules is reduced. The effects of such soils were first attributed to changes in pH and to SOM and clay contents. However, the fact that a soil's suppressive property can be eliminated by pasteurization or treatment with antibiotics and can be transmitted to a nonsuppressive soil by inoculation with a small amount of a suppressive soil, suggests that the responsible factor is microbial. Certain microorganisms seem to limit the development of certain pathogens, leading to the formation of specific suppressive soils (Cook and Baker 1983; Bruehl 1987; Mazzola 2002). The identification and study of these soil microorganisms led to the development of a new field of research termed biocontrol, bioprotection, or biological control. This was defined by Baker and Cook (1974), as "the reduction of inoculum density or disease-producing activities of a pathogen or parasite in its active or dormant state, by one or more organisms, accomplished naturally or through manipulation of the environment, host, or antagonist, or by mass introduction of one or more antagonists." Bioprotection is an environmentally friendly and economically viable means of controlling a number of plant diseases. However, to be used efficiently, the mechanisms by which biocontrol is achieved must be identified.

Much of the research investigating the interactions between AM fungi and other soilborne microorganisms has been related to plant pathogens and was performed to evaluate the potential use of AM fungi as biocontrol agents (St-Arnaud and Vujanovic 2007). Arbuscular mycorrhizal fungal-mediated bioprotection has been observed in the presence of parasitic stramenopiles, especially *Phytophthora nicotianae* Breda de Haan (Cordier et al. 1996; Trotta et al. 1996; Vigo et al. 2000; Pozo et al. 2002*a*), *Phytophthora fragariae* Hickman (Norman and Hooker 2000), *Pythium ultimum* Trow (St-Arnaud et al. 1994; Larsen et al. 2003), and *Aphanomyces euteiches* Drechsler (Slezack et al. 1999; Larsen and Bødker 2001; Thygesen et al. 2004), the pathogenic fungus *Fusarium oxysporum* Schltdl. (Caron et al. 1985, 1986*a*, 1986*b*, 1986*c*, 1986*d*), and nematodes (Borowicz 2001). The level of biocontrol provided by AM fungi largely depends on the species or isolate of the latter, on the plant species or cultivar, and on the variables considered (e.g., the biomass and the quantity and quality of the reproductive structures of both the plant and pathogen in question). However, the level of mycorrhizal colonization generally appears to be secondary. For the specific aspects of biocontrol induced by AM fungi, see reviews by Dehne (1982), Graham (1988), Perrin (1990), Hooker et al. (1994), Linderman (1994, 2000, 2001), St-Arnaud et al. (1995*a*), Azcón-Aguilar and Barea (1997), Borowicz (2001), Sharma and Johri (2002), St-Arnaud and Elsen (2005), Pozo and Azcón-Aguilar (2007), and St-Arnaud and Vujanovic (2007). Many different combinations of plants species and cultivars, pathogens, and AM fungi have been studied and in the large majority of cases, inoculation with AM fungi reduced disease development (St-Arnaud and Vujanovic 2007), confirming the importance of considering AM fungi in plant disease management.

Mechanisms of AM fungi-mediated biocontrol

The mechanism first advanced as being responsible for AM fungal-mediated bioprotection was the increased P assimilation of AM plants. According to this theory, mycorrhizal plants with greater nutrient uptake capacities and greater biomass would be better able to compensate for root damage caused by pathogens. However, as biocontrol was independent of soil P availability or plant P content (Caron et al. 1986*c*; St-Arnaud et al. 1994; Trotta et al. 1996), it became evident that biocontrol was also due to other phenomena (Fig. 4.3). Among these, the stimulation of host plant defence mechanisms appears to be an important factor. The synthesis of molecules involved in plant defence reactions (in particular chitinase, β-glucanase, peroxydase, and phytoalexin) is weakly elicited at the onset of mycorrhizal formation. Moreover, resistance reactions, including increased lignification and phenol metabolism stimulation, are enhanced by AM fungal colonization (Spanu and Bonfante-Fasolo 1988; Spanu et al. 1989; Grandmaison et al. 1993; Traquair 1995; Gianinazzi-Pearson et al. 1996; Garcia Garrido and Ocampo 2002). Arbuscular mycorrhizal fungal interactions with plants may activate or stimulate plant physiological pathways related to disease resistance and condition the roots to react faster when confronted with pathogens (Benhamou et al. 1994; Pozo et al. 1996, 1998, 2002*b*; Slezack et al. 2000; Zhu and Zao 2004). The observed biocontrol effect may either be localized or systemic (Cordier et al. 1998; Pozo et al. 2002*a*).

Colonization by AM fungi also changes the quantity and quality of root exudates (Graham et al. 1981; Bansal and Mukerji 1994; Azaizeh et al. 1995; Marschner et al. 1997; Gupta Sood 2003; Usha et al. 2004). Citric acid was detected in the soil solution collected from the hyphal compartment of plants

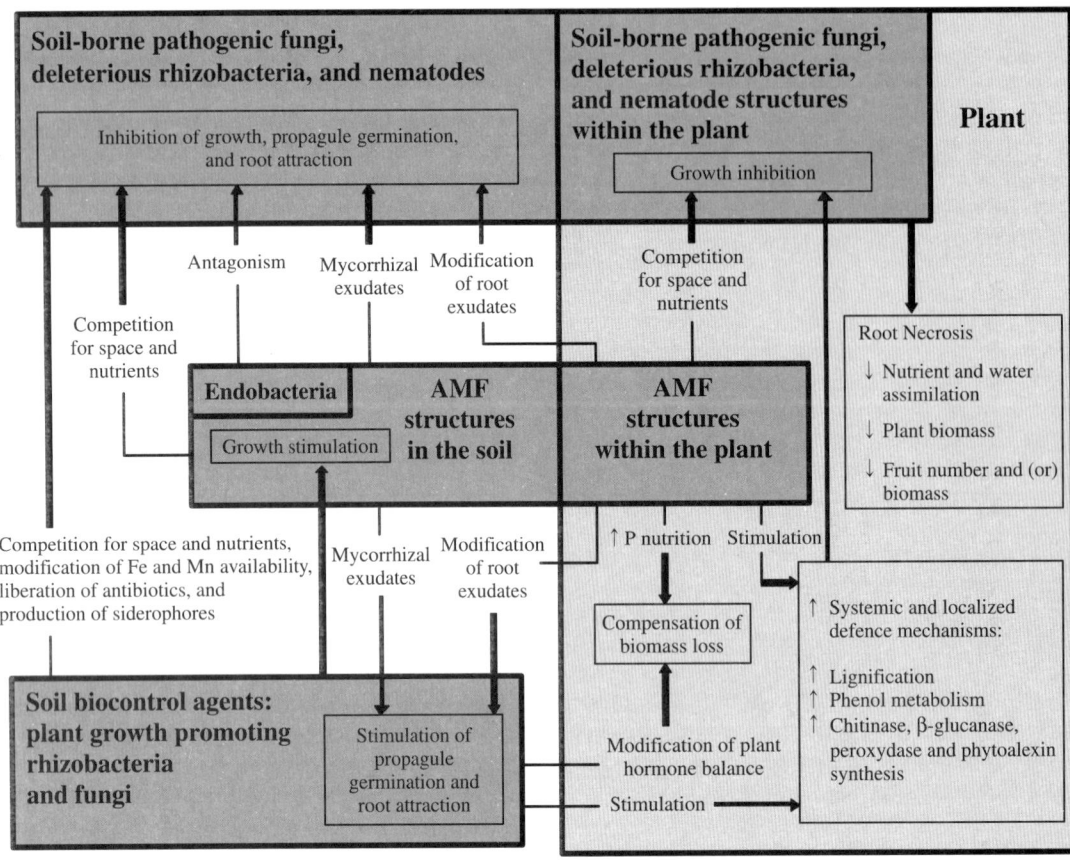

Fig. 4.3. Diagrammatic representation of the disease inhibiting interactions between arbuscular mycorrhizal (AM) fungi, soilborne pathogens, and other soil microorganisms. The plant, AM fungus, soilborne pathogens, and biocontrol-related microorganisms are represented by rectangles, and interactions are indicated with arrows. Arbuscular mycorrhizal fungi can provide bioprotection to plants directly by modifying the signalling- or defense-related biochemical pathways and through their influence on microbial populations in the mycorrhizosphere, which hampers the proliferation of pathogenic and deleterious fungi, bacteria, and nematodes in the vicinity of roots.

colonized with *Gigaspora margarita* Becker & Hall (Tawaraya et al. 2006). The potential role of such changes on plant bioprotection was investigated using a compartmented in vitro system (Lioussanne 2007, Lioussanne et al. 2008) (Fig. 4.4a). This approach permitted the collection of exudates from tomato root organs (colonized or not with *G. intraradices*) without having the confounding effects from other rhizospheric microorganisms. When *P. nicotianae* zoospores were exposed to exudates from mature mycorrhizal or nonmycorrhizal tomato roots (Fig. 4.4b), the former were less attractive to *P. nicotianae* zoospores than the latter. By contrast, exudates from mycorrhizal tomato roots grown in soil enhanced germination of *F. oxysporum* microconidia (Scheffknecht et al. 2006, 2007). Therefore, colonization by AM fungi modifies the composition of tomato exudates that, in turn, may lead to a change in root–microbial interaction in the soil. Filion et al. (1999) showed reduced *F. oxysporum* conidia germination in the presence of crude extracts of in vitro grown *G. intraradices* mycelium, whereas germination of conidia of the biocontrol agent *T. harzianum* was increased. Similar inhibitive effects on in vitro sporulation of the pathogen *P. fragariae* were observed in the presence of exudates from strawberry roots colonized by *G. etunicatum* and *Glomus monosporum* Gerd. & Trappe (Norman and Hooker 2000). These results suggest that changes in root exudate quality and quantity affect the growth of pathogens directly. Furthermore, exudates of tomato roots colonized by *G. fasciculatum* were also more attractive to *Pseudomonas fluorescens* (Flügge) Migula and *Azotobacter chroococcum* Beijerinck than exudates liberated by noncolonized roots (Gupta Sood 2003). Therefore, in vivo, changes may also occur in the proliferation of other microorganisms in the rhizosphere that are able to help reduce the proliferation of soilborne pathogens. The identification of the molecules responsible for the above effects

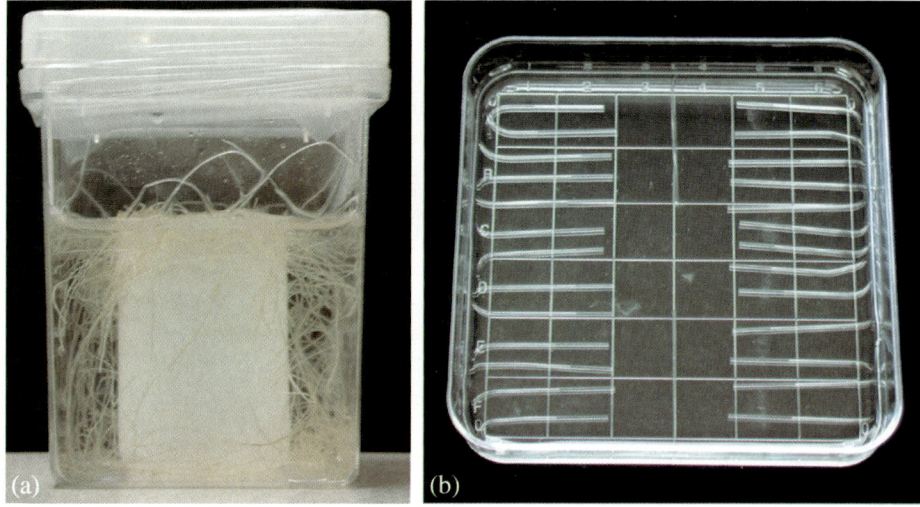

Fig. 4.4. In vitro system used to harvest root exudates and assess the effect of mycorrhizal colonization on zoospore chemotaxtic response. (a) Mycorrhizal and nonmycorrhizal transformed tomato roots were grown in two-compartment magenta box (Sigma, GA-7 Vessel: 7.7 cm wide) (Labour et al. 2003; Lioussanne et al. 2008) in a central compartment filled with solidified minimal medium. Roots and AM mycelia grew into the peripheral compartment containing a sugarless liquid minimal medium with three times the normal P concentration. Medium from the peripheral compartment containing the exudates was collected and used in biotests or analysed by high-performance liquid chromatography after filtration and freezing. (b) *Phytophthora nicotianae* zoospore suspensions were exposed to capillaries filled with either mycorrhizal tomato root exudates, nonmycorrhizal tomato root exudates, or sterilized pure water (control) in a square Petri dish (10 cm wide). The attraction or repulsion effect of exudates was estimated by counting the number of zoospores in each capillary.

would facilitate the development of effective AM-mediated bioprotection strategies. Although differences in the amino acid, organic acid, and sugar content of mycorrhizal and nonmycorrhizal root exudates have been described, the impact of these changes on other soil microorganisms has not been tested (Gupta Sood 2003). Modification in the production of signal molecules in mycorrhizal plants may also be involved in biocontrol. This is supported by studies showing that a mycorrhizal plant can inhibit further root colonization by the same or different AM fungi and that this is controlled by altered root exudation patterns (Pinior et al. 1999; Vierheilig and Piché 2002; Vierheilig et al. 2003; Vierheilig 2004) (see also Chapter 2) and systemic signalling (Ludwig-Muller 2000; Vierheilig et al. 2000; Vierheilig and Piché 2002; Herrera Medina et al. 2003; Vierheilig 2004). Blumenin (Fester et al. 1999, 2002), acacetin, and rhamnetin (Scervino et al. 2005b) are secondary metabolic compounds found only in colonized roots. These molecules may be involved in the regulation of the symbiosis and in the functioning of arbuscules. Moreover, AM fungal genus-, species-, and development-specific effects of the flavonoids chrysin, isorhamnetin, kaempferol, luteolin, morm, and rutin were recently shown (Scervino et al. 2005a). However, this has prevented the establishment of any generalization concerning the effect of root flavonoids on AM fungi.

Similarities in the systemic regulatory mechanism of the AM and *Rhizobium* symbioses were demonstrated using a split-root system (Catford et al. 2003) in which *G. mosseae* colonization suppressed *Sinorhizobium meliloti* (Dangeard) De Lajudie et al. nodulation and vice versa. Because root diseases can also be reduced systemically in mycorrhizal root systems (Pozo et al. 2002a), pathogen development may also be affected by the same regulatory mechanisms. Secondary metabolites induced by AM fungal colonization may play a significant role in the observed AM-mediated biocontrol. This hypothesis is strongly supported by the accumulation of the flavonoids rishitin and solavetivone in potato (*Solanum tuberosum* L.) roots colonized by *G. etunicatum*. These flavonoids, which inhibit *Rhizoctonia solani* Kühn growth, only accumulated in potato roots when they were challenged by the pathogen (Yao et al. 2003).

Arbuscular mycorrhizal fungal mediated biocontrol may also be partly due to a competition for space and nutrients between AM fungi and root pathogens. Cordier et al. (1996) showed that *P. nicotianae* and *G. mosseae* never simultaneously occupied the same tomato root tissues. Furthermore, fatty acid signatures have been used to show the reduction in biomass and energy reserves of *G. mosseae* and *A. euteiches* cooccupying pea (*Pisum sativum* L.) roots (Larsen and Bødker 2001). In most studies to date, plants have been precolonized with AM fungi prior to being inoculated with the pathogen; however, this has been shown to be unnecessary for obtaining the biocontrol effect (Caron et al. 1986*b*; St-Arnaud et al. 1994, 1997). By contrast, different plant pathogens may reduce the extent of mycorrhizal colonization (Davis and Menge 1980; Bååth and Hayman 1983; Krishna and Bagyaraj 1983). This again supports the occurrence of competitive interactions. Inhibition of the proliferation of deleterious rhizobacteria by mycorrhizal colonization has also reported (Garcia-Garrido and Ocampo 1988, 1989).

Synergistic interactions between AM fungi and PGPR or other soil microorganisms may enhance bioprotection. For example, Siddiqui and Mahmood (1998) demonstrated that the combined inoculation of *G. mosseae* and *P. fluorescens* caused a greater reduction in galling and nematode reproduction than when they were used alone. Diedhiou et al. (2003) also showed a similar interaction between *Glomus coronatum* Giovannetti and the nonpathogenic *F. oxysporum* strain Fo162 in the control of *Meloidogyne incognita* (Kofoid & White) Chitwood on tomato.

The mechanisms used by PGPR to protect plants against pathogens are well known: competition for space and nutrients, modification of Fe and Mn availability, liberation of antibiotics and HCN, plant growth promotion by modification of plant hormone balance, and stimulation of systemic and localized plant defence mechanisms (Azcón-Aguilar and Barea 1996; Nehl et al. 1997; M'piga et al. 1997; Bowen and Rovira 1999; Larkin and Fravel 1999). Plant growth promoting rhizomicroorganisms may act in concert or be stimulated by AM fungal colonization. *Paenibacillus* sp. strain B2 isolated from the mycorrhizosphere of *G. mosseae* colonized *Sorghum bicolor* (L.) Moench had an antagonistic effect on *P. nicotianae* in vitro and in vivo (Budi et al. 1999, 2000). This suggests that the AM-mediated biocontrol of this pathogen (Cordier et al. 1996; Trotta et al. 1996; Vigo et al. 2000; Pozo et al. 2002*a*) is due to the AM fungus and its associated bacteria. A species of *Paenibacillus* was also frequently found in the hyphosphere of cucumber (*Cucumis sativus* L.) plants colonized by *G. intraradices* (Mansfeld-Giese et al. 2002), indicating that bacteria of this genus may live in close association with the AM fungal mycelium. The increase in AM fungal biomass caused by the presence of MHB would intensify the biocontrol effect. Nonetheless, mycoparasitism by the biocontrol fungus *Trichoderma* (a PGPR) has been observed (Brimner and Boland 2003), especially on *G. intraradices* and *G. mosseae*. Moreover, Ravnskov et al. (2002) and Larsen et al. (2003) reported a reduction of biomass of the biocontrol agent *Burkholderia cepacia* (Palleroni & Holmes) Yabuuchi et al. in the presence of *G. intraradices*. This highlights the need for controlled experiments prior the use of biocontrol agents in the field.

Nevertheless, management of microbial resources in the field (in particular AM fungi) is likely to be a promising avenue towards the control of plant diseases. Recently, in a large scale study on *Fusarium* crown and root rot disease in asparagus (*Asparagus officinalis* L.), a reduction in AM fungal biomass was shown to be one of the most significant factors associated with disease outbreaks (Hamel et al. 2005*a*, 2005*b*). In this study, the *Fusarium* community structure was shown to include at least 16 species that varied in relation to climatic geographical regions, soil types, cultivars, and plant tissues but not with plant health (Vujanovic et al. 2006). Sampling site and plant age significantly influenced the AM fungal community structure, whereas only sampling site consistently influenced the *Fusarium* community. Diseased and healthy plants hosted similar *Fusarium* and AM fungal communities (Yergeau et al. 2006), but the distributions of some *Fusarium* and AM fungal strains were largely mutually exclusive. This observation may reflect antagonism between some AM fungal and *Fusarium* isolates, and this issue is worth further investigation. Our understanding of the ecology of microorganisms within the rhizosphere has been hampered by the lack of reliable, fast, and low-cost high-throughput approaches required to process the number of samples in large-scale ecological studies. The increasing availability of culture-independent PCR-based technologies has greatly improved the direct detection, identification, and characterization of microorganisms within soil. Various methods such as denaturating gradient gel electrophoresis (Duineveld et al. 2001; Marschner and Baumann 2003; de Souza et al. 2004; Yergeau et al. 2005), single strand conformation polymorphism (Lee et al. 1996; Brandao et al. 2002), terminal restriction fragment length polymorphism (Tiedje et al. 1999), and real-time PCR (Filion et al. 2003*a*, 2003*b*; Harms et al. 2003) are now widely available, making it possible to investigate the highly complex multitrophic interactions of AM fungi with other soil microorganisms. These techniques are likely to refine our classical view of the roles and functions of soil microbes, and allow the optimization of the use and management of soil-inhabiting microorganisms in plant production.

Conclusion

The effects of AM fungi on soil microorganisms influence soil fertility and plant health. This influence is largely due to modifications in the quantity, quality, and distribution of plant-derived C in soil. This varies with the host plant, the AM fungus or fungi involved, the other microorganisms present, and the prevailing environmental conditions. Arbuscular mycorrhizal fungi can influence those soil microorganisms mineralizing SOM, solubilizing minerals, chelating metals, and fixing N. They also provide bioprotection to plants directly by modifying the signalling or defense-related biochemical pathways and through their influence on microbial populations in the mycorrhizosphere, which hampers the proliferation of pathogenic and deleterious fungi, bacteria, and nematodes in the vicinity of roots. The latter probably constitutes a large part of the bioprotective capability of the AM symbiosis. The successful management of AM fungi in agriculture could reduce the need for agrochemicals and improve the sustainability of these production systems.

Commercial AM fungal inocula exist in several countries, including the United States and Canada (see Chapter 8). They are primarily marketed as biofertilizers for home gardening, landscaping, and production of certain ornamental plants. However, because of regulation specificities, they cannot currently be registered as biopesticides (Whipps 2004). Although AM fungi are rarely managed in crop production systems, they almost certainly contribute to crop yield (see Chapter 5). Public demand for sustainable systems increases, and biotechnology evolves. Consideration of AM fungal resources in crop management can be expected to increase with the development of easy to use, effective, and inexpensive inocula. Not only will the bioproducts that are currently being tested become valuable tools for the management of soil microbial communities in plant production systems, they will allow a more environmentally friendly approach to agriculture.

References

Amora-Lazcano, E., Vazquez, M.M., and Azcón, R. 1998. Response of nitrogen-transforming microorganisms to arbuscular mycorrhizal fungi. Biol. Fertil. Soils, **27**: 65–70.

Andrade, G., Mihara, K.L., Linderman, R.G., and Bethlenfalvay, G.J. 1997. Bacteria from rhizosphere and hyphosphere soils of different arbuscular-mycorrhizal fungi. Plant Soil, **192**: 71–79.

Andrade, G., Mihara, K.L., Linderman, R.G., and Bethlenfalvay, G.J. 1998. Soil aggregation status and rhizobacteria in the mycorrhizosphere. Plant Soil, **202**: 89–96.

Atlas, R.M., and Bartha, R. 1997. Microbial ecology: fundamentals and applications. Benjamin Cummings Science Publishing, Menlo Park, Calif.

Azaizeh, H.A., Marschner, H., Romheld, V., and Wittenmayer, L. 1995. Effects of a vesicular–arbuscular mycorrhizal fungus and other soil microorganisms on growth, mineral nutrient acquisition and root exudation of soil-grown maize plants. Mycorrhiza, **5**: 321–327.

Azcón-Aguilar, C., and Barea, J.M. 1992. Interactions between mycorrhizal fungi and other rhizosphere microorganisms. *In* Mycorrhizal functioning. *Edited by* M.F. Allen. Chapman & Hall, London. pp. 163–198.

Azcón-Aguilar, C., and Barea, J.M. 1997. Arbuscular mycorrhizas and biological control of soil-borne plant pathogens—an overview of the mechanisms involved. Mycorrhiza, **6**: 457–464.

Bååth, E., and Hayman, D.S. 1983. Plant growth responses to vesicular–arbuscular mycorrhizae. XIV. Interactions with Verticillium wilt on tomato plants. New Phytol. **95**: 419–426.

Baker, K.F., and Cook, R.J. 1974. Biological control of plant pathogens. W.H. Freeman, San Francisco, Calif.

Bansal, M., and Mukerji, K.G. 1994. Positive correlation between VAM-induced changes in root exudation and mycorrhizosphere mycoflora. Mycorrhiza, **5**: 39–44.

Barea, J.M., Azcón, R., and Azcón-Aguilar, C. 1983. Interactions between phosphate solubilizing bacteria and VA mycorrhiza to improve plant utilization of rock phosphate in non-acidic soils. *In* Proceedings, 3rd International Congress on Phosphorus Compounds, 4–6 Oct. 1983 Brussels, Belgium. pp. 127–144.

Barea, J.M., Azcón-Aguilar, C., and Azcón, R. 1997. Interactions between mycorrhizal fungi and rhizosphere microorganisms within the context of sustainable soil–plant systems. *In* Multitrophic interactions in terrestrial systems. *Edited by* A.C. Gange and V.K. Brown. Blackwell Science Publishers, Osney, UK. pp. 65–77.

Barea, J.M., Toro, M., Orozco, M.O., Campos, E., and Azcón, R. 2002. The application of isotopic (^{32}P and ^{15}N) dilution techniques to evaluate the interactive effect of phosphate-solubilizing rhizobacteria, mycorrhizal fungi and *Rhizobium* to improve the agronomic efficiency of rock phosphate for legume crops. Nutr. Cycl. Agroecosyst. **63**: 35–42.

Barea, J.M., Pozo, M.J., Azcón, R., and Azcón-Aguilar, C. 2005. Microbial co-operation in the rhizosphere. J. Exp. Bot. **56**: 1761–1778.

Barroso, C.B., and Nahas, E. 2005. The status of soil phosphate fractions and the ability of fungi to dissolve hardly soluble phosphates. Appl. Soil Ecol. **29**: 73–83.

Baudoin, E., Benizri, E., and Guckert, A. 2003. Impact of artificial root exudates on the bacterial community structure in bulk soil and maize rhizosphere. Soil Biol. Biochem. **35**: 1183–1192.

Benhamou, N., Fortin, J.A., Hamel, C., St-Arnaud, M., and Shatilla, A. 1994. Resistance responses of mycorrhizal Ri T-DNA-transformed carrot roots to infection by *Fusarium oxysporum* f. sp. *chrysanthemi*. Phytopathology, **84**: 958–968.

Bertin, C., Yang, X., and Weston, L.A. 2003. The role of root exudates and allelochemicals in the rhizosphere. Plant Soil, **256**: 67–83.

Bieleski, R.L. 1973. Phosphate pools, phosphate transport and phosphate availability. Annu. Rev. Plant Physiol. **24**: 225–252.

Boer, W.D., Folman, L.B., Summerbell, R.C., and Boddy, L. 2005. Living in a fungal world: impact of fungi on soil bacterial niche development. FEMS Microbiol. Rev. **29**: 795–811.

Bohme, L., Langer, U., and Bohme, F. 2005. Microbial biomass, enzyme activities and microbial community structure in two European long-term field experiments. Agric. Ecosyst. Environ. **109**: 141–152.

Borowicz, V.A. 2001. Do arbuscular mycorrhizal fungi alter plant–pathogen relations? Ecology, **82**: 3057–3068.

Bowen, G.D., and Rovira, A.D. 1999. The rhizosphere and its management to improve plant growth. Adv. Agron. **66**: 1–102.

Brandao, P.F.B., Torimura, M., Kurane, R., and Bull, A.T. 2002. Dereplication for biotechnology screening: PyMS analysis and PCR–RFLP–SSCP (PRS) profiling of 16S rRNA genes of marine and terrestrial actinomycetes. Appl. Microbiol. Biotechnol. **58**: 77–83.

Brimner, T.A., and Boland, G.J. 2003. A review of the non-target effects of fungi used to biologically control plant diseases. Agric. Ecosyst. Environ. **100**: 3–16.

Brodie, E., Edwards, S., and Clipson, N. 2003. Soil fungal community structure in a temperate upland grassland soil. FEMS Microbiol. Ecol. **45**: 105–114.

Bruehl, G.W. 1987. Soilborne plant pathogens. Macmillan Publishing Co., New York.

Budi, S.W., van Tuinen, D., Martinotti, G., and Gianinazzi, S. 1999. Isolation from the *Sorghum bicolor* mycorrhizosphere of a bacterium compatible with arbuscular mycorrhiza development and antagonistic towards soilborne fungal pathogens. Appl. Environ. Microbiol. **65**: 5148–5150.

Budi, S.W., van Tuinen, D., Arnould, C., Dumas-Gaudot, E., Gianinazzi-Pearson, V., and Gianinazzi, S. 2000. Hydrolytic enzyme activity of *Paenibacillus* sp. strain B2 and effects of the antagonistic bacterium on cell integrity of two soil-borne pathogenic fungi. Appl. Soil Ecol. **15**: 191–199.

Calvet, C., Barea, J.M., and Pera, J. 1992. *In vitro* interactions between the vesicular–arbuscular mycorrhizal fungus *Glomus mosseae* and some saprophytic fungi isolated from organic substrates. Soil Biol. Biochem. **24**: 775–780.

Campbell, C.A., Biederbeck, V.O., Wen, G., Zentner, R.P., Schoenau, J., and Hahn, D. 1999. Seasonal trends in soil biochemical attributes: effects of crop rotation in the semiarid prairie. Can. J. Soil Sci. **79**: 73–84.

Caron, M., Fortin, J.A., and Richard, C. 1985. Influence of substrate on the interaction of *Glomus intraradices* and *Fusarium oxysporum* f. sp. *radicis-lycopersici* on tomatoes. Plant Soil, **87**: 233–239.

Caron, M., Fortin, J.A., and Richard, C. 1986*a*. Effect of *Glomus intraradices* on infection by *Fusarium oxysporum* f.sp. *radicis-lycopersici* in tomatoes over a 12-week period. Can. J. Bot. **64**: 552–556.

Caron, M., Fortin, J.A., and Richard, C. 1986*b*. Effect of inoculation sequence on the interaction between *Glomus intraradices* and *Fusarium oxysporum* f. sp. *radicis-lycopersici* in tomatoes. Can. J. Plant Pathol. **8**: 12–16.

Caron, M., Fortin, J.A., and Richard, C. 1986*c*. Effect of phosphorus concentration and *Glomus intraradices* on *Fusarium* root rot of tomatoes. Phytopathology, **76**: 942–946.

Caron, M., Richard, C., and Fortin, J.A. 1986*d*. Effect of preinfestation of the soil by a vesicular–arbuscular mycorrhizal fungus, *Glomus intraradices*, on *Fusarium* crown and root rot of tomatoes. Phytoprotection, **67**: 15–19.

Catford, J.G., Staehelin, C., Lerat, S., Piché, Y., and Vierheilig, H. 2003. Suppression of arbuscular mycorrhizal colonization and nodulation in split-root systems of alfalfa after pre-inoculation and treatment with Nod factors. J. Exp. Bot. **54**: 1481–1487.

Cavagnaro, T.R., Jackson, L.E., Scow, K.M., and Hristova, K.R. 2007. Effects of arbuscular mycorrhizas on ammonia oxidizing bacteria in an organic farm soil. Microb. Ecol. **54**: 618–626.

Christensen, H., and Jakobsen, I. 1993. Reduction of bacterial growth by a vesicular–arbuscular mycorrhizal fungus in the rhizosphere of cucumber (*Cucumis sativus* L). Biol. Fertil. Soils, **15**: 253–258.

Cook, R., and Baker, K. 1983. The nature and practice of biological control of plant pathogens. American Phytopatholgical Society Press, St-Paul, Minn.

Cordier, C., Gianinazzi, S., and Gianinazzi-Pearson, V. 1996. Colonisation patterns of root tissues by *Phytophthora nicotianae* var. *parasitica* related to reduced disease in mycorrhizal tomato. Plant Soil, **185**: 223–232.

Cordier, C., Pozo, M.J., Barea, J.M., Gianinazzi, S., and Gianinazzi-Pearson, V. 1998. Cell defense responses associated with localized and systemic resistance to *Phytophthora parasitica* induced in tomato by an arbuscular mycorrhizal fungus. Mol. Plant Microbe Interact. **11**: 1017–1028.

Dalpé, Y. 2003. Mycorrhizal fungi biodiversity in Canadian soils. Can. J. Soil Sci. **83**: 321–330.

Davis, R.M., and Menge, J.A. 1980. Influence of *Glomus fasciculatus* and soil phosphorus on *Phytophthora* root rot of *Citrus*. Phytopathology, **70**: 447–452.

Dehne, H.-W. 1982. Interaction between vesicular–arbuscular mycorrhizal fungi and plant pathogens. Phytopathology, **72**: 1115–1119.

de Souza, F.A., Kowalchuk, G.A., Leeflang, P., van Veen, J.A., and Smit, E. 2004. PCR-denaturing gradient gel electrophoresis profiling of inter- and intraspecies 18S rRNA gene sequence heterogeneity is an accurate and sensitive method to assess species diversity of arbuscular mycorrhizal fungi of the genus *Gigaspora*. Appl. Environ. Microbiol. **70**: 1413–1424.

Diedhiou, P.M., Hallmann, J., Oerke, E.C., and Dehne, H.W. 2003. Effects of arbuscular mycorrhizal fungi and a non-pathogenic *Fusarium oxysporum* on *Meloidogyne incognita* infestation of tomato. Mycorrhiza, **13**: 199–204.

Dodd, J.C. 2000. The role of arbuscular mycorrhizal fungi in agro- and natural ecosystems. Outlook Agric. **29**: 55–62.

Domsch, K.H., Gams, W., and Anderson, T.H. 1980. Compendium of soil fungi. Academic Press, London.

Duineveld, B.M., Kowalchuk, G.A., Keijzer, A., van Elsas, J.D., and van Veen, J.A. 2001. Analysis of bacterial communities in the rhizosphere of *Chrysanthemum* via denaturing gradient gel electrophoresis of PCR-amplified 16S rRNA as well as DNA fragments coding for 16S rRNA. Appl. Environ. Microbiol. **67**: 172–178.

Elsen, A., Declerck, S., and De Waele, D. 2001. Effects of *Glomus intraradices* on the reproduction of the burrowing nematode (*Radopholus similis*) in dixenic culture. Mycorrhiza, **11**: 49–51.

Fester, T., Maier, W., and Strack, D. 1999. Accumulation of secondary compounds in barley and wheat roots in response to inoculation with an arbuscular mycorrhizal fungus and co-inoculation with rhizosphere bacteria. Mycorrhiza, **8**: 241–246.

Fester, T., Hause, B., Schmidt, D., Halfmann, K., Schmidt, J., Wray, V., Hanse, G., and Strack, D. 2002. Occurrence and localization of apocarotenoids in arbuscular mycorrhizal plant roots. Plant Cell Physiol. **43**: 256–265.

Filion, M., St-Arnaud, M., and Fortin, J.A. 1999. Direct interaction between the arbuscular mycorrhizal fungus *Glomus intraradices* and different rhizosphere microorganisms. New Phytol. **141**: 525–533.

Filion, M., St-Arnaud, M., and Jabaji-Hare, S.H. 2003*a*. Direct quantification of fungal DNA from soil substrate using real-time PCR. J. Microbiol. Methods, **53**: 67–76.

Filion, M., St-Arnaud, M., and Jabaji-Hare, S.H. 2003*b*. Quantification of *Fusarium solani* f. sp. *phaseoli* in mycorrhizal bean plants and surrounding mycorrhizosphere soil using real-time polymerase chain reaction and direct isolations on selective media. Phytopathology, **93**: 229–235.

Fracchia, S., Mujica, M.T., Garcia-Romera, I., Garcia-Garrido, J.M., Martin, J., Ocampo, J.A., and Godeas, A. 1998. Interactions between *Glomus mosseae* and arbuscular mycorrhizal sporocarp-associated saprophytic fungi. Plant Soil, **200**: 131–137.

Gamalero, E., Trotta, A., Massa, N., Copetta, A., Martinotti, M.G., and Berta, G. 2004. Impact of two fluorescent pseudomonads and an arbuscular mycorrhizal fungus on tomato plant growth, root architecture and P acquisition. Mycorrhiza, **14**: 185–192.

Garcia-Garrido, J.M., and Ocampo, J.A. 1988. Interaction between *Glomus mosseae* and *Erwinia carotovora* and its effects on the growth of tomato plants. New Phytol. **110**: 551–555.

Garcia-Garrido, J.M., and Ocampo, J.A. 1989. Effect of VA mycorrhizal infection of tomato on damage caused by *Pseudomonas syringae*. Soil Biol. Biochem. **21**: 165–167.

Garcia-Garrido, J.M., and Ocampo, J.A. 2002. Regulation of the plant defence response in arbuscular mycorrhizal symbiosis. J. Exp. Bot. **53**: 1377–1386.

Garcia-Romera, I., Garcia-Garrido, J.M., Martin, J., Fracchia, S., Mujica, M.T., Godeas, A., and Ocampo, J.A. 1998. Interactions between saprotrophic *Fusarium* strains and arbuscular mycorrhizas of soybean plants. Symbiosis, **24**: 235–245.

Gianinazzi-Pearson, V., Dumas-Gaudot, E., Gollotte, A., Tahiri Alaoui, A., and Gianinazzi, S. 1996. Cellular and molecular defence-related root responses to invasion by arbuscular mycorrhizal fungi. New Phytol. **133**: 45–57.

Glick, B.R. 1995. The enhancement of plant growth promotion by free living bacteria. Can. J. Microbiol. **41**: 109–117.

Godeas, A., Fracchia, S., Mujica, M.T., and Ocampo, J.A. 1999. Influence of soil impoverishment on the interaction between *Glomus mosseae* and saprobe fungi. Mycorrhiza, **9**: 185–189.

Govindarajulu, M., Pfeffer, P.E., Jin, H., Abubaker, J., Douds, D.D., Allen, J.W., Bücking, H., Lammers, P.J., and Shachar-Hill, Y. 2005. Nitrogen transfer in the arbuscular mycorrhizal symbiosis. Nature (Lond.), **435**: 819–823.

Graham, J.H. 1988. Interaction of mycorrhizal fungi with soilborne plant pathogens and other organisms: an introduction. Phytopathology, **78**: 365–366.

Graham, J.H., Leonard, R.T., and Menge, J.A. 1981. Membrane-mediated decrease in root exudation responsible for phosphorus inhibition of vesicular–arbuscular mycorrhiza formation. Plant Physiol. **68**: 548–552.

Grandmaison, J., Olah, G.M., Van Calsteren, M.-R., and Furlan, V. 1993. Characterization and localization of plant phenolics likely involved in the pathogen resistance expressed by endomycorrhizal roots. Mycorrhiza, **3**: 155–164.

Grayston, S.J., Wang, S., Campbell, C.D., and Edwards, A.C. 1998. Selective influence of plant species on microbial diversity in the rhizosphere. Soil Biol. Biochem. **30**: 369–378.

Green, H., Larsen, J., Olsson, P.A., Jensen, D.F., and Jakobsen, I. 1999. Suppression of the biocontrol agent *Trichoderma harzianum* by mycelium of the arbuscular mycorrhizal fungus *Glomus intraradices* in root-free soil. Appl. Environ. Microbiol. **65**: 1428–1434.

Gryndler, M., Vosatka, M., Hrselova, H., Catska, V., Chvatalova, I., and Jansa, J. 2002. Effect of dual inoculation with arbuscular mycorrhizal fungi and bacteria on growth and mineral nutrition of strawberry. J. Plant Nutr. **25**: 1341–1358.

Gupta Sood, S. 2003. Chemotactic response of plant-growth-promoting bacteria towards roots of vesicular–arbuscular mycorrhizal tomato plants. FEMS Microbiol. Ecol. **45**: 219–227.

Hamel, C. 2004. Impact of arbuscular mycorrhizal fungi on N and P cycling in the root zone. Can. J. Soil Sci. **84**: 383–395.

Hamel, C. 2007. Extraradical arbuscular mycorrhizal mycelia: shadowy figures in the soil. *In* Mycorrhizae in crop production. *Edited by* C. Hamel and C. Plenchette. Haworth Press, Binghampton, N.Y. pp. 1–36.

Hamel, C., Neeser, C., Barrantes-Cartín, U., and Smith, D.L. 1991. Endomycorrhizal fungal species mediate ^{15}N-transfer from soybean to corn in non-fumigated soil. Plant Soil, **138**: 41–47.

Hamel, C., Vujanovic, V., Jeannotte, R., Nakano-Hylander, A., and St-Arnaud, M. 2005a. Negative feedback on a perennial crop: Fusarium crown and root rot of asparagus is related to changes in soil microbial community structure. Plant Soil, **268**: 75–87.

Hamel, C., Vujanovic, V., Nakano-Hylander, A., Jeannotte, R., and St-Arnaud, M. 2005b. Factors associated with Fusarium crown and root rot of asparagus outbreaks in Quebec. Phytopathology, **95**: 867–873.

Harms, G., Layton, A.C., Dionisi, H.M., Gregory, I.R., Garrett, V.M., Hawkins, S.A., Robinson, K.G., and Sayler, G.S. 2003. Real-time PCR quantification of nitrifying bacteria in a municipal wastewater treatment plant. Environ. Sci. Technol. **37**: 343–351.

Hawkins, H.J., Johansen, A., and George, E. 2000. Uptake and transport of organic and inorganic nitrogen by arbuscular mycorrhizal fungi. Plant Soil, **226**: 275–285.

Hempel, S., Renker, C., and Buscot, F. 2007. Differences in the species composition of arbuscular mycorrhizal fungi in spore, root and soil communities in a grassland ecosystem. Environ. Microbiol. **9**: 1930–1938.

Herrera Medina, M.J., Gagnon, H., Piché, Y., Ocampo, J.A., Garcia-Garrido, J.M., and Vierheilig, H. 2003. Root colonization by arbuscular mycorrhizal fungi is affected by the salicylic acid content of the plant. Plant Sci. **164**: 993–998.

Hodge, A. 2000. Microbial ecology of the arbuscular mycorrhiza. FEMS Microbiol. Ecol. **32**: 91–96.

Hooker, J.E., Jaizme-Vega, M., and Atkinson, D. 1994. Biocontrol of plant pathogens using arbuscular mycorrhizal fungi. *In* Impact of arbuscular mycorrhizas on sustainable agriculture and natural ecosystems. *Edited by* S. Gianinazzi and H. Schüepp. Birkhäuser Verlag, Basel, Switzerland. pp. 191–200.

Ibekwe, A.M., and Kennedy, A.C. 1998. Fatty acid methyl ester (FAME) profiles as a tool to investigate community structure of two agricultural soils. Plant Soil, **206**: 151–161.

Jakobsen, I., Smith, S., and Smith, F. 2002. Function and diversity of arbuscular mycorrhizae in carbon and mineral nutrition. *In* Mycorrhizal ecology. *Edited by* M.G.A. van der Heijden and I.R. Sanders. Springer-Verlag, Berlin. pp. 75–92.

Jastrow, J.D., Miller, R.M., and Lussenhop, J. 1998. Contributions of interacting biological mechanisms to soil aggregate stabilization in restored prairie. Soil Biol. Biochem. **30**: 905–916.

Jeffries, P., and Dodd, J.C. 1996. Functional ecology of mycorrhizal fungi in sustainable soil–plant systems. *In* Mycorrhizas in integrated systems: from genes to plant development. *Edited by* C. Azcón-Aguilar and J.M. Barea. European Commission, Brussels. pp. 497–501.

Jeffries, P., Gianinazzi, S., Perotto, S., Turnau, K., and Barea, J.M. 2003. The contribution of arbuscular mycorrhizal fungi in sustainable maintenance of plant health and soil fertility. Biol. Fertil. Soils, **37**: 1–16.

Johansson, J.F., Paul, L.R., and Finlay, R.D. 2004. Microbial interactions in the mycorrhizosphere and their significance for sustainable agriculture. FEMS Microbiol. Ecol. **48**: 1–13.

Johnson, D., Leake, J.R., and Read, D.J. 2002. Transfer of recent photosynthate into mycorrhizal mycelium of an upland grassland: short-term respiratory losses and accumulation of ^{14}C. Soil Biol. Biochem. **34**: 1521–1524.

Kabir, Z., O'Halloran, I.P., Fyles, J.W., and Hamel, C. 1997. Seasonal changes of arbuscular mycorrhizal fungi as affected by tillage practices and fertilization: hyphal density and mycorrhizal root colonization. Plant Soil, **192**: 285–293.

Klironomos, J.N. 2002. Feedback with soil biota contributes to plant rarity and invasiveness in communities. Nature (Lond.), **417**: 67–70.

Klironomos, J.N., Ursic, M., Rillig, M., and Allen, M.F. 1998. Interspecific differences in the response of arbuscular mycorrhizal fungi to *Artemisia tridentata* grown under elevated atmospheric CO_2. New Phytol. **138**: 599–605.

Koide, R.T., and Kabir, Z. 2000. Extraradical hyphae of the mycorrhizal fungus *Glomus intraradices* can hydrolyse organic phosphate. New Phytol. **148**: 511–517.

Krishna, K.R., and Bagyaraj, D.J. 1983. Interaction between *Glomus fasciculatum* and *Sclerotium rolfsii* in peanut. Can. J. Bot. **61**: 2349–2351.

Labour, K., Jolicoeur, M., and St-Arnaud, M. 2003. Arbuscular mycorrhizal responsiveness of *in vitro* tomato root lines is not related to growth and nutrient uptake rates. Can. J. Bot. **81**: 645–656.

Larkin, R.P., and Fravel, D.R. 1999. Mechanisms of action and dose–response relationships governing biological control of Fusarium wilt of tomato by nonpathogenic *Fusarium* spp. Phytopathology, **89**: 1152–1161.

Larsen, J., and Bødker, L. 2001. Interactions between pea root-inhabiting fungi examined using signature fatty acids. New Phytol. **149**: 487–493.

Larsen, J., and Jakobsen, I. 1996. Interactions between a mycophagous *Collembola*, dry yeast and the external mycelium of an arbuscular mycorrhizal fungus. Mycorrhiza, **6**: 259–264.

Larsen, J., Olsson, P.A., and Jakobsen, I. 1998. The use of fatty acid signatures to study mycelial interactions between the arbuscular mycorrhizal fungus *Glomus intraradices* and the saprotrophic fungus *Fusarium culmorum* in root-free soil. Mycol. Res. **102**: 1491–1496.

Larsen, J., Ravnskov, S., and Jakobsen, I. 2003. Combined effect of an arbuscular mycorrhizal fungus and a biocontrol bacterium against *Pythium ultimum* in soil. Folia Geobot. **38**: 145–154.

Leake, J., Johnson, D., Donnelly, D., Muckle, G., Boddy, L., and Read, D. 2004. Networks of power and influence: the role of mycorrhizal mycelium in controlling plant communities and agroecosystem functioning. Can. J. Bot. **82**: 1016–1045.

Lee, D.H., Zo, Y.G., and Kim, S.J. 1996. Nonradioactive method to study genetic profiles of natural bacterial communities by PCR – single-strand-conformation polymorphism. Appl. Environ. Microbiol. **62**: 3112–3120.

Linderman, R.G. 1992. Vesicular–arbuscular mycorrhizae and soil microbial interactions. *In* Mycorrhizae in sustainable agriculture. *Edited by* G.J. Bethlenfalvay and R.G. Linderman. American Society of Agronomy, Madison, Wisc. ASA Spec. Publ. 54. pp. 45–70.

Linderman, R.G. 1994. Role of VAM fungi in biocontrol. *In* Mycorrhizae and plant health. *Edited by* F.L. Pfleger and R.G. Linderman. American Phytopathological Society Press,

St. Paul, Minn. pp. 1–25.

Linderman, R.G. 2000. Effects of mycorrhizas on plant tolerance to diseases. *In* Arbuscular mycorrhizas: physiology and functions. *Edited by* Y. Kapulnik and D. Douds. Kluwer Academic Publishers, Dordrecht, the Netherlands. pp. 345–365.

Linderman, R.G. 2001. Mycorrhizae and their effects on diseases. *In* Diseases of woody ornamentals and trees in nurseries. *Edited by* R.K. Jones and D.M. Benson. American Phytopathological Society Press, St Paul, Minn. pp. 433–434.

Linderman, R.G., and Paulitz, T.C. 1990. Mycorrhizal–rhizobacterial interactions. *In* Biological control of soil-borne plant pathogens. *Edited by* D. Hornby, R.J. Cook, Y. Henis, W.H. Ko, A.D. Rovira, B. Schippers, and P.R. Scott. CAB International, Wallingford, UK. pp. 261–283.

Lioussanne, L. 2007. Rôles des modifications de la microflore bactérienne et de l'exsudation racinaire de la tomate par la symbiose mycorhizienne dans le biocontrôle sur le *Phytophthora nicotianae*. Ph.D. thesis, Université de Montréal, Montréal, Que.

Lioussanne, L., Jolicoeur, M., and St-Arnaud, M. 2008. Mycorrhizal colonization with *Glomus intraradices* and development stage of transformed tomato roots significantly modify the chemotactic response of zoospores of the pathogen *Phytophthora nicotianae*. Soil Biol. Biochem. **40**: 2217–2224.

Ludwig-Muller, J. 2000. Hormonal balance in plants during colonization by mycorrhizal fungi. *In* Arbuscular mycorrhizas: physiology and function. *Edited by* Y. Kapulnik and D.D. Douds. Kluwer Academic Publisher, Dordrecht, the Netherlands. pp. 263–285.

Manjaiah, K.M., Voroney, R.P., and Sen, U. 2000. Soil organic carbon stocks, storage profile and microbial biomass under different crop management systems in a tropical agricultural ecosystem. Biol. Fertil. Soils, **32**: 273–278.

Mansfeld-Giese, K., Larsen, J., and Bødker, L. 2002. Bacterial populations associated with mycelium of the arbuscular mycorrhizal fungus *Glomus intraradices*. FEMS Microbiol. Ecol. **41**: 133–140.

Marschner, P., and Baumann, K. 2003. Changes in bacterial community structure induced by mycorrhizal colonisation in split-root maize. Plant Soil, **251**: 279–289.

Marschner, P., and Crowley, D.E. 1996*a*. Physiological activity of a bioluminescent *Pseudomonas fluorescens* (strain 2-79) in the rhizosphere of mycorrhizal and non-mycorrhizal pepper (*Capsicum annuum* L.). Soil Biol. Biochem. **28**: 869–876.

Marschner, P., and Crowley, D.E. 1996*b*. Root colonization of mycorrhizal and nonmycorrhizal pepper (*Capsicum annuum*) by *Pseudomonas fluorescens* 2-79RL. New Phytol. **134**: 115–122.

Marschner, P., Crowley, D.E., and Higashi, R.M. 1997. Root exudation and physiological status of a root-colonizing fluorescent pseudomonad in mycorrhizal and non-mycorrhizal pepper (*Capsicum annuum* L.). Plant Soil, **189**: 11–20.

Marschner, P., Crowley, D.E., and Lieberei, R. 2001. Arbuscular mycorrhizal infection changes the bacterial 16 S rDNA community composition in the rhizosphere of maize. Mycorrhiza, **11**: 297–302.

Martinez, A., Obertello, M., Pardo, A., Ocampo, J.A., and Godeas, A. 2004. Interactions between *Trichoderma pseudokoningii* strains and the arbuscular mycorrhizal fungi *Glomus mosseae* and *Gigaspora rosea*. Mycorrhiza, **14**: 79–84.

Mazzola, M. 2002. Mechanisms of natural soil suppressiveness to soilborne diseases. Antonie Leeuwenhoek, **81**: 557–564.

McAllister, C.B., Garcia-Romera, I., Godeas, A., and Ocampo, J.A. 1994. Interactions between *Trichoderma koningii*, *Fusarium solani* and *Glomus mosseae*: effects on plant growth, arbuscular mycorrhizas and the saprophyte inoculants. Soil Biol. Biochem. **26**: 1363–1367.

McAllister, C.B., Garcia-Romera, I., Martin, J., Godeas, A., and Ocampo, J.A. 1995. Interaction between *Aspergillus niger* van Tiegh. and *Glomus mosseae* (Nicol. & Gerd.) Gerd. & Trappe. New Phytol. **129**: 309–316.

Medina, A., Vassilev, N., Alguacil, M.M., Roldan, A., and Azcón, R. 2004*a*. Increased plant growth, nutrient uptake, and soil enzymatic activities in a desertified mediterranean soil amended with treated residues and inoculated with native mycorrhizal fungi and a plant growth-promoting yeast. Soil Sci. **169**: 260–270.

Medina, A., Vassileva, M., Caravaca, F., Roldan, A., and Azcón, R. 2004*b*. Improvement of soil characteristics and growth of *Dorycnium pentaphyllum* by amendment with agrowastes and inoculation with AM fungi and/or the yeast *Yarowia lipolytica*. Chemosphere, **56**: 449–456.

Meyer, J.R., and Linderman, R.G. 1986*a*. Response of subterranean clover to dual inoculation with vesicular–arbuscular mycorrhizal fungi and a plant growth-promoting bacterium, *Pseudomonas putida*. Soil Biol. Biochem. **18**: 185–190.

Meyer, J.R., and Linderman, R.G. 1986*b*. Selective influence on populations of rhizosphere or rhizoplane bacteria and actinomycetes by mycorrhizas formed by *Glomus fasciculatum*. Soil Biol. Biochem. **18**: 191–196.

Miethling, R., Wieland, G., Backhaus, H., and Tebbe, C.C. 2000. Variation of microbial rhizosphere communities in response to crop species, soil origin, and inoculation with *Sinorhizobium meliloti* L33. Microb. Ecol. **40**: 43–56.

M'piga, P., Bélanger, R.R., Paulitz, T.C., and Benhamou, N. 1997. Increased resistance to *Fusarium oxysporum* f. sp. *radicis-lycopersici* in tomato plants treated with the endophytic bacterium *Pseudomonas fluorescens* strain 63-28. Physiol. Mol. Plant Pathol. **50**: 301–320.

Nakano, A., Takahashi, K., and Kimura, M. 1999. The carbon origin of arbuscular mycorrhizal fungi estimated from delta ^{13}C values of individual spores. Mycorrhiza, **9**: 41–47.

Nehl, D.B., Allen, S.J., and Brown, J.F. 1997. Deleterious rhizosphere bacteria: an integrating perspective. Appl. Soil Ecol. **5**: 1–20.

Norman, J.R., and Hooker, J.E. 2000. Sporulation of *Phytophthora fragariae* shows greater stimulation by exudates of non-mycorrhizal than by mycorrhizal strawberry roots. Mycol. Res. **104**: 1069–1073.

Oehl, F., Sieverding, E., Ineichen, K., Ris, E.A., Boller, T., and Wiemken, A. 2005. Community structure of arbuscular mycorrhizal fungi at different soil depths in extensively and intensively managed agroecosystems. New Phytol. **165**: 273–283.

Olsson, P.A., Chalet, M., Bååth, E., Finlay, R.D., and Söderström, B. 1996. Ectomycorrhizal mycelia reduce bacterial activity in a sandy soil. FEMS Microbiol. Ecol. **21**: 77–86.

Olsson, P.A., Francis, R., Read, D.J., and Söderström, B. 1998. Growth of arbuscular mycorrhizal mycelium in calcareous dune sand and its interaction with other soil microorganisms as estimated by measurement of specific fatty acids. Plant Soil, **201**: 9–16.

Olsson, P.A., Thingstrup, I., Jakobsen, I., and Bååth, F. 1999. Estimation of the biomass of arbuscular mycorrhizal fungi in a linseed field. Soil Biol. Biochem. **31**: 1879–1887.

Paul, E.A., and Clark, F.E. 1996. Soil microbiology and biochemistry. Academic Press, San Diego, Calif.

Paulitz, T.C., and Linderman, R.G. 1989. Interactions between fluorescent pseudomonads and VA mycorrhizal fungi. New Phytol. **113**: 37–45.

Perrin, R. 1990. Interactions between mycorrhizae and diseases caused by soil-borne fungi. Soil Use Manage. **6**: 189–195.

Pinior, A., Wyss, U., Piché, Y., and Vierheilig, H. 1999. Plants colonized by AM fungi regulate further root colonization by AM fungi through altered root exudation. Can. J. Bot. **77**: 891–897.

Posta, K., Marschner, H., and Römheld, V. 1994. Manganese reduction in the rhizosphere of mycorrhizal and nonmycorrhizal maize. Mycorrhiza, **5**: 119–124.

Pozo, M.J., and Azcón-Aguilar, C. 2007. Unraveling mycorrhiza-induced resistance. Curr. Opin. Plant Biol. **10**: 393–398.

Pozo, M.J., Dumas-Gaudot, E., Slezack, S., Cordier, C., Asselin, A., Gianinazzi, S., Gianinazzi-Pearson, V., Azcón-Aguilar, C., and Barea, J.M. 1996. Induction of new chitinase isoforms in tomato roots during interactions with *Glomus mosseae* and/ or *Phytophthora nicotianae* var *parasitica*. Agronomie, **16**: 689–697.

Pozo, M.J., Azcón-Aguilar, C., Dumas-Gaudot, E., and Barea, J.M. 1998. Chitosanase and chitinase activities in tomato roots during interactions with arbuscular mycorrhizal fungi or *Phytophthora parasitica*. J. Exp. Bot. **49**: 1729–1739.

Pozo, M.J., Cordier, C., Dumas-Gaudot, E., Gianinazzi, S., Barea, J.M., and Azcón-Aguilar, C. 2002a. Localized versus systemic effect of arbuscular mycorrhizal fungi on defence responses to *Phytophthora* infection in tomato plants. J. Exp. Bot. **53**: 525–534.

Pozo, M.J., Slezack-Deschaumes, S., Dumas-Gaudot, E., Gianinazzi, S., and Azcón-Aguilar, C. 2002b. Plant defense responses induced by arbuscular mycorrhizal fungi. *In* Mycorrhizal technology in agriculture: from genes to bioproducts. *Edited by* S. Gianinazzi, H. Schuepp, J.M. Barea, and K. Haselwandter. Birkhauser Verlag, Basel, Switzerland. pp. 103–111.

Puppi, G., Azcón, R., and Hoflich, G. 1994. Management of positive interactions of arbuscular mycorrhizal fungi with essential groups of soil microorganisms. *In* Impact of arbuscular mycorrhizas on sustainable agriculture and natural ecosystems. *Edited by* S. Gianinazzi and H. Schuepp. Birkhauser Verlag, Basel, Switzerland. pp. 201–216.

Ravnskov, S., Nybroe, O., and Jakobsen, I. 1999. Influence of an arbuscular mycorrhizal fungus on *Pseudomonas fluorescens* DF57 in rhizosphere and hyphosphere soil. New Phytol. **142**: 113–122.

Ravnskov, S., Larsen, J., and Jakobsen, I. 2002. Phosphorus uptake of an arbuscular mycorrhizal fungus is not affected by the biocontrol bacterium *Burkholderia cepacia*. Soil Biol. Biochem. **34**: 1875–1881.

Read, D.J. 1991. Mycorrhizas in ecosystems. Cell. Mol. Life Sci. **47**: 376–391.

Remy, W., Taylor, T.N., Hass, H., and Kerp, H. 1994. Four hundred-million-year-old vesicular arbuscular mycorrhizae. Proc. Natl. Acad. Sci. U.S.A. **91**: 11 841 – 11 843.

Requena, N., Jimenez, I., Toro, M., and Barea, J.M. 1997. Interactions between plant-growth-promoting rhizobacteria (PGPR), arbuscular mycorrhizal fungi and *Rhizobium* spp. in the rhizosphere of *Anthyllis cytisoides*, a model legume for revegetation in mediterranean semi-arid ecosystems. New Phytol. **136**: 667–677.

Richardson, A.E. 1994. Soil microorganisms and phosphorus availability. *In* Soil biota management in sustainable farming systems. *Edited by* C. Pankhurst, B. Doube, V. Gupta, and P. Grace. CSIRO Publishing, Melbourne, Australia. pp. 50–62.

Richardson, A.E. 2001. Prospects for using soil microorganisms to improve the acquisition of phosphorus by plants. Aust. J. Plant Physiol. **28**: 897–906.

Rivera, R., and Fernández, K. 2003. Bases cientifico. Técnicas para el manejo de los sistemas agrícolas micorrízados eficientemente. *In* El manejo eficiente de la simbiosis micorrízica, una via hacia la agricultura sostenible. Estudio de caso: El Caribe. *Edited by* R. Rivera, F. Fernández, A. Hernandez, and J.R. Martin. Instituto Nacional de Ciencias Agricolas, Havana, Cuba. pp. 45–98.

Rousseau, A., Benhamou, N., Chet, I., and Piché, Y. 1996. Mycoparasitism of the extramatrical phase of *Glomus intraradices* by *Trichoderma harzianum*. Phytopathology, **86**: 434–443.

Saito, K., Suyama, Y., Sato, S., and Sugawara, K. 2004. Defoliation effects on the community structure of arbuscular mycorrhizal fungi based on 18S rDNA sequences. Mycorrhiza, **14**: 363–373.

Sampedro, I., Aranda, E., Scervino, J.M., Fracchia, S., Garcia-Romera, I., Ocampo, J.A., and Godeas, A. 2004. Improvement by soil yeasts of arbuscular mycorrhizal symbiosis of soybean (*Glycine max*) colonized by *Glomus mosseae*. Mycorrhiza, **14**: 229–234.

Scervino, J., Ponce, M., Erra-Bassells, R., Vierheilig, H., Ocampo, J., and Godeas, A. 2005a. Flavonoids exhibit fungal species and genus specific effects on the presymbiotic growth of *Gigaspora* and *Glomus*. Mycol. Res. **109**: 789–794.

Scervino, J.M., Ponce, M.A., Erra-Bassells, R., Vierheilig, H., Ocampo, J.A., and Godeas, A. 2005b. Arbuscular mycorrhizal colonization of tomato by *Gigaspora* and *Glomus* species in the presence of root flavonoids. J. Plant Physiol. **162**: 625–633.

Scheffknecht, S., Mammerler, R., Steinkellner, S., and Vierheilig, H. 2006. Root exudates of mycorrhizal tomato plants exhibit a different effect on microconidia germination of *Fusarium oxysporum* f. sp. *lycopersici* than root exudates from non-mycorrhizal tomato plants. Mycorrhiza, **16**: 365–370.

Scheffknecht, S., St-Arnaud, M., Khaosaad, T., Steinkellner, S., and Vierheilig, H. 2007. An altered root exudation pattern through mycorrhization affecting microconidia germination of the highly specialized tomato pathogen *Fusarium oxysporum* f. sp. *lycopersici* (*Fol*) is not tomato specific but also occurs in *Fol* non-host plants. Can. J. Bot. **85**: 347–351.

Scheublin, T.R., Ridgway, K.P., Young, J.P.W., and van der Heijden, M.G.A. 2004. Nonlegumes, legumes, and root nodules harbor different arbuscular mycorrhizal fungal communities. Appl. Environ. Microbiol. **70**: 6240–6246.

Semenov, A., van Bruggen, A., and Zelenev, V. 1999. Moving waves of bacterial populations and total organic carbon along roots of wheat. Microb. Ecol. **37**: 116–128.

Sharma, A.K., and Johri, B.N. 2002. Arbuscular-mycorrhiza and plant disease. *In* Arbuscular mycorrhizae: interactions in plants, rhizosphere and soils. *Edited by* A.K. Shaxena and B.N. Johri. Science Publishers Inc., Enfield, N.H. pp. 69–96.

Siddiqui, I.A., and Shaukat, S.S. 2002. Mixtures of plant disease suppressive bacteria enhance biological control of multiple tomato pathogens. Biol. Fertil. Soils, **36**: 260–268.

Siddiqui, Z.A., and Mahmood, I. 1998. Effect of a plant growth promoting bacterium, an AM fungus and soil types on the morphometrics and reproduction of *Meloidogyne javanica* on tomato. Appl. Soil Ecol. **8**: 77–84.

Sievers, A., and Braun, M. 1996. The root cap: structure and function. *In* Plant roots: the hidden half. *Edited by* Y. Waisel, A. Eshel, and U. Kafkafi. Marcel Dekker Inc., New York. pp. 53–74.

Six, J., Bossuyt, H., Degryze, S., and Denef, K. 2004. A history of research on the link between (micro) aggregates, soil biota, and soil organic matter dynamics. Soil Tillage Res. **79**: 7–31.

Slezack, S., Dumas-Gaudot, E., Rosendahl, S., Kjøller, R., Paynot, M., Negrel, J., and Gianinazzi, S. 1999. Endoproteolytic activities in pea roots inoculated with the arbuscular mycorrhizal fungus *Glomus mosseae* and/or *Aphanomyces euteiches* in relation to bioprotection. New Phytol. **142**: 517–529.

Slezack, S., Dumas-Gaudot, E., Paynot, M., and Gianinazzi, S. 2000. Is a fully established arbuscular mycorrhizal symbiosis required for bioprotection of *Pisum sativum* roots against *Aphanomyces euteiches?* Mol. Plant Microbe Interact. **13**: 238–241.

Smith, S.E., Dickson, S., and Smith, F.A. 2001. Nutrient transfer in arbuscular mycorrhizas: how are fungal and plant processes integrated? Aust. J. Plant Physiol. **28**: 683–694.

Smith, S.E., and Read, D.J. 1997. Mycorrhizal symbiosis. 2nd ed. Academic Press, San Diego, Calif.

Söderberg, K.H., Olsson, P.A., and Bååth, E. 2002. Structure and activity of the bacterial community in the rhizosphere of different plant species and the effect of arbuscular mycorrhizal colonisation. FEMS Microbiol. Ecol. **40**: 223–231.

Spanu, P., and Bonfante-Fasolo, P. 1988. Cell-wall-bound peroxidase activity in roots of mycorrhizal *Allium porrum*. New Phytol. **109**: 119–124.

Spanu, P., Boller, T., Ludwig, A., Wiemken, A., Faccio, A., and Bonfante-Fasolo, P. 1989. Chitinases in roots of mycorrhizal *Allium porrum*: regulation and localization. Planta, **177**: 447–455.

Srinath, J., Bagyaraj, D., and Satyanarayana, B. 2003. Enhanced growth and nutrition of micropropagated *Ficus benjamina* to *Glomus mosseae* co-inoculated with *Trichoderma harzianum* and *Bacillus coagulans*. World J. Microbiol. Biotechnol. **19**: 69–72.

St-Arnaud, M., and Elsen, A. 2005. Interaction of arbuscular mycorrhizal fungi with soil-borne pathogens and non-pathogenic rhizosphere micro-organisms. *In* In vitro culture of mycorrhizas. *Edited by* S. Declerck, D.-G. Strullu, and J.A. Fortin. Springer-Verlag, Berlin, Heidelberg. pp. 217–231.

St-Arnaud, M., and Vujanovic, V. 2007. Effect of the arbuscular mycorrhizal symbiosis on plant diseases and pests. *In* Mycorrhizae in crop production. *Edited by* C. Hamel and C. Plenchette. Haworth Press, Binghampton, N.Y. pp. 67–122.

St-Arnaud, M., Hamel, C., Caron, M., and Fortin, J.A. 1994. Inhibition of *Pythium ultimum* in roots and growth substrate of mycorrhizal *Tagetes patula* colonized with *Glomus intraradices*. Can. J. Plant Pathol. **16**: 187–194.

St-Arnaud, M., Hamel, C., Caron, M., and Fortin, J.A. 1995*a*. Endomycorhizes VA et sensibilité aux maladies: synthèse de la littérature et mécanismes d'interaction potentiels. *In* La symbiose mycorhizienne—état des connaissances. *Edited by* J.A. Fortin, C. Charest, and Y. Piché. Orbis Publishing, Frelighsburg, Que. pp. 51–87.

St-Arnaud, M., Hamel, C., Vimard, B., Caron, M., and Fortin, J.A. 1995*b*. Altered growth of *Fusarium oxysporum* f. sp. *chrysanthemi* in an *in vitro* dual culture system with the vesicular arbuscular mycorrhizal fungus *Glomus intraradices* growing on *Daucus carota* transformed roots. Mycorrhiza, **5**: 431–438.

St-Arnaud, M., Hamel, C., Vimard, B., Caron, M., and Fortin, J.A. 1997. Inhibition of *Fusarium oxysporum* f. sp. *dianthi* in the non-VAM species *Dianthus caryophyllus* by co-culture with *Tagetes patula* companion plants colonized by *Glomus intraradices*. Can. J. Bot. **75**: 998–1005.

Steinberg, P.D., and Rillig, M.C. 2003. Differential decomposition of arbuscular mycorrhizal fungal hyphae and glomalin. Soil Biol. Biochem. **35**: 191–194.

Stevenson, F.J., and Cole, M.A. 1999. Cycles of soil: carbon, nitrogen, phosphorus, sulphur, micronutrients. John Wiley & Sons, Inc., New York.

Stewart, L.I., Hamel, C., Hogue, R., and Moutoglis, P. 2005. Response of strawberry to inoculation with arbuscular mycorrhizal fungi under very high soil phosphorus conditions. Mycorrhiza, **15**: 612–619.

Subramanian, K.S., and Charest, C. 1998. Arbuscular mycorrhizae and nitrogen assimilation in maize after drought and recovery. Physiol. Plant. **102**: 285–296.

Subramanian, K.S., and Charest, C. 1999. Acquisition of N by external hyphae of an arbuscular mycorrhizal fungus and its impact on physiological responses in maize under drought-stressed and well-watered conditions. Mycorrhiza, **9**: 69–75.

Talavera, M., Itou, K., and Mizukubo, T. 2001. Reduction of nematode damage by root colonization with arbuscular mycorrhiza (*Glomus* spp.) in tomato – *Meloidogyne incognita* (Tylenchida: Meloidognidae) and carrot – *Pratylenchus penetrans* (Tylenchida: Pratylenchidae) pathosystems. Appl. Entomol. Zool. (Jpn.), **36**: 387–392.

Tanaka, Y., and Yano, K. 2005. Nitrogen delivery to maize via mycorrhizal hyphae depends on the form of N supplied. Plant Cell Environ. **28**: 1247–1254.

Tawaraya, K., Naito, M., and Wagatsuma, T. 2006. Solubilization of insoluble inorganic phosphate by hyphal exudates of arbuscular mycorrhizal fungi. J. Plant Nutr. **29**: 657–665.

Thygesen, K., Larsen, J., and Bodker, L. 2004. Arbuscular mycorrhizal fungi reduce development of pea root-rot caused by *Aphanomyces euteiches* using oospores as pathogen inoculum. Eur. J. Plant Pathol. **110**: 411–419.

Tiedje, J.M., Asuming-Brempong, S., Nusslein, K., Marsh, T.L., and Flynn, S.J. 1999. Opening the black box of soil microbial diversity. Appl. Soil Ecol. **13**: 109–122.

Toro, M., Azcón, R., and Barea, J.M. 1997. Improvement of arbuscular mycorrhiza development by inoculation of soil with phosphate-solubilizing rhizobacteria to improve rock phosphate bioavailability (^{32}P) and nutrient cycling. Appl. Environ. Microbiol. **63**: 4408–4412.

Toussaint, J.P., St-Arnaud, M., and Charest, C. 2004. Nitrogen transfer and assimilation between the arbuscular mycorrhizal fungus *Glomus intraradices* Schenck & Smith and Ri T-DNA roots of *Daucus carota* L. in an *in vitro* compartmented system. Can. J. Microbiol. **50**: 251–260.

Traquair, J.A. 1995. Fungal biocontrol of root diseases: endomycorrhizal suppression of Cylindrocarpon root rot. Can. J. Bot. **73**(Suppl. 1): 89–95.

Trotta, A., Varese, G.C., Gnavi, E., Fusconi, A., Sampo, S., and Berta, G. 1996. Interactions between the soilborne root pathogen *Phytophthora nicotianae* var *parasitica* and the arbuscular mycorrhizal fungus *Glomus mosseae* in tomato plants. Plant Soil, **185**: 199–209.

Uren, N.C. 2000. Types, amounts, and possible functions of compounds released into the rhizosphere by soil-grown plants. *In* The rhizosphere: biochemistry and organic substances at the soil–plant interface. *Edited by* R. Pinton, Z. Varanini, and P. Nannipieri. Marcel Dekker, Inc., New York. pp. 19–40.

Usha, K., Saxena, A., and Singh, B. 2004. Rhizosphere dynamics influenced by arbuscular mycorrhizal fungus (*Glomus deserticola*) and related changes in leaf nutrient status and yield of Kinnow mandarin {King (*Citrus nobilis*) × Willow Leaf (*Citrus deliciosa*)}. Aust. J. Agric. Res. **55**: 571–576.

van der Heijden, M.G.A., Klironomos, J.N., Ursic, M., Moutoglis, P., Streitwolf-Engel, R., Boller, T., Wiemken, A., and Sanders, I.R. 1998. Mycorrhizal fungal diversity determines plant biodiversity, ecosystem variability and productivity. Nature (Lond.), **396**: 69–72.

Vassileva, M., Vassilev, N., and Azcón, R. 1998. Rock phosphate solubilization by *Aspergillus niger* on olive cake-based medium and its further application in a soil–plant system. World J. Microbiol. Biotechnol. **14**: 281–284.

Vazquez, M.M., Cesar, S., Azcón, R., and Barea, J.M. 2000. Interactions between arbuscular mycorrhizal fungi and other microbial inoculants (*Azospirillum, Pseudomonas, Trichoderma*) and their effects on microbial population and enzyme activities in the rhizosphere of maize plants. Appl. Soil Ecol. **15**: 261–272.

Vessey, J.K. 2003. Plant growth promoting rhizobacteria as biofertilizers. Plant Soil, **255**: 571–586.

Vestberg, A., Kukkonen, S., Saari, K., Parikka, P., Huttunen, J., Tainio, L., Devos, D., Weekers, F., Kevers, C., Thonart, P., Lemoine, M.C., Cordier, C., Alabouvette, C., and Gianinazzi, S. 2004. Microbial inoculation for improving the growth and health of micropropagated strawberry. Appl. Soil Ecol. **27**: 243–258.

Vierheilig, H. 2004. Regulatory mechanisms during the plant – arbuscular mycorrhizal fungus interaction. Can. J. Bot. **82**: 1166–1176.

Vierheilig, H., and Piché, Y. 2002. Signalling in arbuscular mycorrhiza: facts and hypotheses. *In* Flavonoids in cell function. *Edited by* B. Bushlig and J. Manthey. Kluwer Academic/Plenum Publishers, New York. pp. 23–39.

Vierheilig, H., Garcia-Garrido, J.M., Wyss, U., and Piché, Y. 2000. Systemic suppression of mycorrhizal colonization of barley roots already colonized by AM fungi. Soil Biol. Biochem. **32**: 589–595.

Vierheilig, H., Lerat, S., and Piché, Y. 2003. Systemic inhibition of arbuscular mycorrhiza development by root exudates of cucumber plants colonized by *Glomus mosseae*. Mycorrhiza, **13**: 167–170.

Vigo, C., Norman, J.R., and Hooker, J.E. 2000. Biocontrol of the pathogen *Phytophthora parasitica* by arbuscular mycorrhizal fungi is a consequence of effects on infection loci. Plant Pathol. **49**: 509–514.

Villegas, J., and Fortin, J.A. 2001. Phosphorus solubilization and pH changes as a result of the interactions between soil bacteria and arbuscular mycorrhizal fungi on a medium containing NH_4^+ as nitrogen source. Can. J. Bot. **79**: 865–870.

Villegas, J., and Fortin, J.A. 2002. Phosphorus solubilization and pH changes as a result of the interactions between soil bacteria and arbuscular mycorrhizal fungi on a medium containing NO_3^- as nitrogen source. Can. J. Bot. **80**: 571–576.

Vivas, A., Azcón, R., Biro, B., Barea, J.M., and Ruiz Lozano, J.M. 2003. Influence of bacterial strains isolated from lead-polluted soil and their interactions with arbuscular mycorrhizae on the growth of *Trifolium pratense* L. under lead toxicity. Can. J. Microbiol. **49**: 577–588.

Vujanovic, V., Hamel, C., Yergeau, E., and St-Arnaud, M. 2006. Biodiversity and biogeography of *Fusarium* species from northeastern North American asparagus fields based on microbiological and molecular approaches. Microb. Ecol. **51**: 242–255.

Wamberg, C., Christensen, S., Jakobsen, I., Muller, A.K., and Sorensen, S.J. 2003. The mycorrhizal fungus (*Glomus intraradices*) affects microbial activity in the rhizosphere of pea plants (*Pisum sativum*). Soil Biol. Biochem. **35**: 1349–1357.

Whipps, J.M. 2004. Prospects and limitations for mycorrhizas in biocontrol of root pathogens. Can. J. Bot. **82**: 1198–1227.

Whitelaw, M.A. 2000. Growth promotion of plants inoculated with phosphate-solubilizing fungi. Adv. Agron. **69**: 99–151.

Witter, E., and Kanal, A. 1998. Characteristics of the soil microbial biomass in soils from a long-term field experiment with different levels of C input. Appl. Soil Ecol. **10**: 37–39.

Yao, M.K., Desilets, H., Charles, M.T., Boulanger, R., and Tweddell, R.J. 2003. Effect of mycorrhization on the accumulation of rishitin and solavetivone in potato plantlets challenged with *Rhizoctonia solani*. Mycorrhiza, **13**: 333–336.

Yergeau, E., Filion, M., Vujanovic, V., and St-Arnaud, M. 2005. A PCR-denaturing gradient gel electrophoresis approach to assess *Fusarium* diversity in asparagus. J. Microbiol. Methods, **60**: 143–154.

Yergeau, E., Vujanovic, V., and St-Arnaud, M. 2006. Changes in communities of *Fusarium* and arbuscular-mycorrhizal fungi as related to different asparagus cultural factors. Microb. Ecol. **52**: 104–113.

Zelenev, V., van Bruggen, A., and Semenov, A. 2000. "BACWAVE," a spatial-temporal model for traveling waves of bacterial populations in response to a moving carbon source in soil. Microb. Ecol. **40**: 260–272.

Zhang, Y., Guo, L.D., and Liu, R.J. 2004. Survey of arbuscular mycorrhizal fungi in deforested and natural forest land in the subtropical region of Dujiangyan, southwest China. Plant Soil, **261**: 257–263.

Zhu, H.H., and Zao, Q. 2004. Localized and systemic increase of phenols in tomato roots induced by *Glomus versiforme* inhibits *Ralstonia solanacearum*. J. Phytopathol. **152**: 537–542.

Chapter 5
Arbuscular mycorrhiza: where nature and industry meet

Tandra Fraser, Atul Nayyar, Walid Ellouze, Juan Perez, Keith Hanson, Jim Germida, Zadok Bouzid, and Chantal Hamel

Introduction

Climate change, decreasing quality of Canada's soil and water resources, and increases in the country's greenhouse gas emissions have raised concerns for over 15 years. Better use of arbuscular mycorrhizal (AM) fungi and other soil biota in agriculture, coupled with a reduction in fertilizer applications, could reduce Canada's emissions of CO_2 and N_2O, nitrate seepage into ground water reserves, and P concentration in agricultural soils. The latter has increased to such an extent that soil P saturation has become a threat to water quality in intensively farmed areas of Canada and many other industrialized countries (Beauchemin and Simard 2000; Dorioz et al. 2006). The management of the AM symbiosis in agriculture offers an opportunity for reducing fertilizer applications (see Chapter 6). However, the utilization of AM fungi is more complicated than that of fertilizers. It is now realized that, for the AM symbiosis to function properly in managed ecosystems, certain basic ecological requirements of the mycobionts must be respected (Rivera et al. 2007).

Increasing agricultural input costs and shrinking profit margins have led to the development of highly mechanized agricultural systems in most industrialized countries. By contrast, many developing countries cannot afford to invest in fertilizers and other conventional inputs to produce the food needed to feed their increasing populations. It is now clear that the next green revolution depends on the development of technologies allowing food production to be maintained with lower levels of inputs. Arbuscular mycorrhizal-related technologies are perceived as an important component of this revolution.

A better understanding of the conditions necessary for the development of an effective AM symbiosis is the first step toward the exploitation of this symbiosis in crop production and reduction of fertilizer use. With this goal in mind, the first section of this chapter presents our current knowledge on the AM symbiosis. In the second section, the potential for using the AM symbiosis in agriculture and the constraints imposed by current intensive agricultural practices are described.

T. Fraser, A. Nayyar, and C. Hamel.[1] Environmental Health / Water and Nutrients, Agriculture and Agri-Food Canada, 1 Airport Road, P.O. Box 1030, Swift Current, SK S9H 3X2, Canada, and Department of Soil Science, University of Saskatchewan, 51 Campus Drive, Saskatoon, SK S7N 5A8, Canada.

W. Ellouze. Environmental Health / Water and Nutrients, Agriculture and Agri-Food Canada, 1 Airport Road, P.O. Box 1030, Swift Current, SK S9H 3X2, Canada, Department of Soil Science, University of Saskatchewan, 51 Campus Drive, Saskatoon, SK S7N 5A8, Canada, and Département de Biologie, Faculté des Sciences Mathématiques, Physiques et Naturelles, Université Tunis El Manar, Campus Universitaire, Tunis 1060, Tunisia.

J.C. Perez. Environmental Health / Water and Nutrients, Agriculture and Agri-Food Canada, 1 Airport Road, P.O. Box 1030, Swift Current, SK S9H 3X2, Canada, Department of Soil Science, University of Saskatchewan, 51 Campus Drive, Saskatoon, SK S7N 5A8, Canada, and Universidad Nacional de Colombia – Sede Medellin, AA 3840 Medellin, Colombia.

K. Hanson. Environmental Health / Water and Nutrients, Agriculture and Agri-Food Canada, 1 Airport Road, P.O. Box 1030, Swift Current, SK S9H 3X2, Canada.

J. Germida. Department of Soil Science, University of Saskatchewan, 51 Campus Drive, Saskatoon, SK S7N 5A8, Canada.

Z. Bouzid. Département de Biologie, Faculté des Sciences Mathématiques, Physiques et Naturelles, Université Tunis El Manar, Campus Universitaire, Tunis 1060, Tunisia.

[1]Corresponding author (e-mail: hamelc@agr.gc.ca).

The Nature of Arbuscular Mycorrhiza

Nutrient uptake by arbuscular mycorrhizal fungi

Arbuscular mycorrhizal fungi are abundant in natural terrestrial ecosystems, and they influence plant growth in a number of ways (Klironomos 2003). Although they improve the soil environment by stabilizing soil aggregates (Jastrow et al. 1998; Franzluebbers et al. 2000) and improve plant health through interactions with soil microorganisms (Andrade et al. 1998; Filion et al. 1999; Chapter 4), their most important effect on plant growth is in the enhancement of plant nutrition (Smith and Read 1997). The main physiological basis for the AM association is the bidirectional transfer of nutrients (Smith and Smith 1990), which probably occurs principally via the arbuscules formed by the fungus in the host's root cells (Saito 2000). During this exchange, plants supply AM fungi with C; in return, AM fungi supply plants with mineral nutrients, in particular those with low mobility in the soil matrix (e.g., P, Cu, and Zn). Increased nutrient uptake in mycorrhizal plants is mainly due to the large network of plant-associated extraradical hyphae. The latter allow mycorrhizal plants to explore a greater volume of soil than nonmycorrhizal plants (Hattingh et al. 1973). Furthermore, extraradical hyphae allow the uptake of nutrients of low mobility located outside the depletion zone that develops in the immediate vicinity of the roots.

Numerous studies on nutrient uptake by AM fungi have been conducted, and the topic has been widely reviewed (see Kothari et al. 1990; Smith and Read 1997; Clark and Zeto 2000; Jeffries et al. 2003; Takacs and Voros 2003). The major effect of AM fungi on plant growth is mainly attributed to improved plant P nutrition. The greater P uptake of plants colonized by AM fungi has been associated with the capacity of AM fungi to lower soil pH (Li et al. 1991), to release phosphatase (Tarafdar and Marschner 1994), and to explore a large soil volume (Hattingh et al. 1973). Soil P is absorbed by inorganic P transporters located on the plasma membrane of extraradical AM fungal hyphae (Harrison and van Buuren 1995) and translocated to the intraradical hyphae by protoplasmic streaming. The observation of reduced amounts of available P under AM plants with larger P content than noncolonized plants supports the importance of AM fungi in P uptake (Liu et al. 2003). Although plant roots can take up P, most of the P in a mycorrhizal plant is provided by AM fungi, even when the plant growth response to AM inoculation is small (Smith et al. 2003).

Although most of the P taken up by AM fungi comes from the labile soil P pool (Bolan 1991; Nurlaeny et al. 1996), it has been shown that mycorrhizal plants can also obtain P from inorganic P sources that are generally unavailable to plants (Bolan et al. 1987; Jayachandrana et al. 1989). A few reports on the uptake of P by AM fungi from organic sources have also been published. Certain AM fungi hydrolyze organic P through the release of phosphatase enzymes (Joner and Johansen 2000); the hydrolyzed P is then taken up by the extraradical hyphae of the AM fungus and transported to plant roots (Joner et al. 2000).

Although the contribution of the AM symbiosis to plant P nutrition is well documented, its impact on plant N nutrition is still unclear. Nitrogen is the nutrient that plants require in the largest amount. Inorganic N is very soluble and moves in soil by mass flow. The predominant forms of N in soils are NO_3^- and NH_4^+; however, the latter is retained on soil cation exchange sites and, thus, moves more slowly in soil than the former. Under conditions of water deficit, which limits the mass flow transportation of NO_3^-, the contribution of AM fungi to plant N nutrition can be significant (Tobar et al. 1994). Because the contribution of AM fungi to plant nutrient uptake is more important for nutrients of low mobility, most of the studies related to N uptake have concentrated on NH_4^+.

Recently, the ability of AM fungi to access N from organic sources has been demonstrated (Hodge et al. 2001). A number of studies (Azcón et al. 1992; Subramanian and Charest 1998, 1999) have also investigated the different enzymes involved in N metabolism in the AM symbiosis. Govindarajulu et al. (2005) showed that inorganic N is taken up and assimilated via nitrate reductase and glutamine synthase in the extraradical mycelium. Nitrogen is then translocated as arginine to the intraradical mycelium where arginine is broken down to give NH_4^+, which is then taken up by the plant via ammonia channels (Jin et al. 2005).

Although most research concerning nutrient uptake by AM fungi has concentrated on the role of AM fungi in the uptake of P and N, it has also been shown that AM fungi can improve plant uptake of all other essential nutrients (Clark and Zeto 2000; Liu et al. 2007). In general, AM fungi have a positive impact on plant uptake of those nutrients that are difficult to obtain from the soil matrix because of their low mobility or of the low abundance of their available forms in some soils.

Role of arbuscular mycorrhizal fungi in soil aggregation

Soil physical structure is an important aspect of soil quality. Productive agricultural soils must exhibit good water infiltration and gas exchange rates. These characteristics are determined by soil porosity, which, in turn, is determined by the extent to which soil particles are bound into stable aggregates. Because soil aggregation increases porosity, it reduces run off and soil erosion. Soil aggregation also reduces the rate of oxidation of organic matter and influences microbial community structure by creating diverse microenvironments with differing oxygen, water, and nutrient levels. The stabilization of soil aggregates is related to root length, the amount of soil organic matter (SOM), and soil microbial biomass (Tisdall and Oades 1979). Therefore, soils under pastures are more stable than those under crops (Ramsay et al. 1986), soils under grasses are more stable than those under legumes, and soils under C3 plants are more stable than those under C4 plants (Tisdall and Oades 1979; Miller and Jastrow 2000).

The effect of plants on soil structure has been reviewed by Degens (1997) and Angers and Caron (1998). Roots influence soil structure through root penetration, changes in the soil water regime, root exudation, root decomposition (Gale et al. 2000), and root entanglement (Jastrow et al. 1998; Miller and Jastrow 2000). The compression of soil by growing roots decreases soil porosity in the zone between roots; however, the reorientation of clay particles along the root surface induces the formation of microaggregates (Dorioz et al. 1993). Water uptake by plant roots also influences soil aggregation by modifying the soil water status. This causes localized drying of the soil, which promotes the binding of root exudates to clay particles (Reid and Goss 1982). The release of organic substances by plant roots into the rhizosphere affects soil aggregation both directly by sticking the soil particles together (Morel et al. 1991) and indirectly through microbial stimulation. Microorganisms present in soil play a pivotal role in soil aggregation and both microbial biomass and water extractable carbohydrates have been correlated to varying degrees with soil aggregation (Degens 1997). Various explanations have been proposed to explain the contribution of AM fungi to soil aggregation. According to Miller et al. (1995), the abundance of extraradical hyphae of AM fungi is an important factor in the formation and stability of soil aggregates. Typically, between 50 and 160 m of AM fungal hyphae are associated with 1 g of macroaggregates (Tisdall and Oades 1979) and increased hyphal length increases aggregate stability.

The AM fungal mycelium within the soil has been described as a "sticky string bag" because it entangles particles within a hyphal network and cements them together with extracellular polysaccharides (Oades and Waters 1991). This process is considered to be one of the major factors in the formation of microaggregates (Tisdall et al. 1997; Bossuyt et al. 2001). In clay soils, stability is assured by fine particles sticking firmly to hyphae with polysaccharide-based mucilage (Clough and Sutton 1978); however, the means by which the stability of aggregates in sandy soils is achieved is less clear (Degens et al. 1994). It may be due to the production of higher amounts of mucilage by those AM fungal species and their associated microbiota frequently found in sandy soil. Glomalin, a glycoprotein produced by AM fungi (Wright et al. 1996), is important in soil aggregate stability because of its slow decomposition rate and often high concentration in soils (Wright and Upadhyaya 1998). Glomalin is present on the walls of extraradical hyphae and is hydrophobic in nature. It has been reported to be produced by all the AM fungi examined, but studies suggests that species of *Gigaspora* (more frequently observed in sandy soils) produces more glomalin than species of *Glomus* (Wright and Upadhyaya 1996). However, recent studies have shown that glomalin is not necessarily secreted by living AM fungi, as was previously thought, and that its role in soil is secondary (see review by Rillig and Mummey (2006)). Furthermore, AM fungal associated microbiota are likely responsible for at least some of the soil aggregation previously attributed directly to AM fungi (Rillig et al. 2005).

Carbon uptake and cycling by arbuscular mycorrhizal fungi

Soil organic matter is the major source of C in soil and occurs in both stable and readily available pools. Various soil microorganisms are involved in the turnover of SOM. In turn, the cell walls of dead microorganisms accumulate in the soil where they constitute part of the readily available SOM pool.

Arbuscular mycorrhizal fungi are an important component of the soil biota. They are obligate biotrophs and use host plant C rather than SOM C as a source of C and energy. Because of their reliance on a C source external to the soil, they play a valuable role in increasing SOM. According to Graham (2000), up to 20% of the C fixed by plants is directed to mycorrhizal root systems. Host plants are able to compensate for the transfer of C to AM fungi by increasing their photosynthetic activity. Arbuscular mycorrhizal fungi are rich in triacylglycerides, and their major form of C is lipidic (Cox et al. 1975). Spores of AM fungi are particularly rich in lipids; depending on the species, these may account for between 45% and 95% of the spore's C reserve (Bécard et al. 1991). Other important C-containing compounds in AM fungi are glycogen, trehalose, chitin (Bonfante-Fasolo 1988), and glomalin (Wright et al. 1996; Rillig et al. 2001).

The production of extraradical hyphae, their lifespans, and their decomposition rates are some of the important factors determining the cycling of AM fungal C in soil. Because hyphal biomass production varies between species (Sanders et al. 1998), different species are likely to contribute different amounts of C to the SOM pools. The lifespan of hyphae determines the flux rate of C from live fungal biomass to the dead soil microbial biomass C pool. For example, smaller lateral hyphae have been estimated to have a lifespan of approximately one week (e.g., Allen et al. 2003). Carbon from dead tissues enters into either the readily available or the more stable SOM pool, depending on the complexity of the molecules with which it is associated (Parton et al. 1988). Triacylglycerides, carbohydrates, and other metabolites that are processed relatively quickly by decomposers enter into the available or active SOM pool, whereas chitin and glomalin enter into the pool of more stable SOM. As SOM decomposes, CO_2 is produced through soil respiration, and the C is returned to the atmosphere. The cycling of C through AM fungi influences the soil microbial community (see Chapter 4) and soil processes.

Important amounts of C are believed to enter the soil via the extraradical mycelium of AM fungi. This is supported by the fact that AM fungi acquire plant photosynthate directly (Johnson et al. 2002), by the abundance of AM fungal biomass (approximately 25% of the total soil microbial biomass; Olsson et al. 1999), and by the rapid turnover of the hyphal network (Staddon et al. 2003). However, the exact quantity of C involved is difficult to assess. Some of the C translocated from plants to AM fungi cycles rapidly. Johnson et al. (2002), using a pulse labelling experiment, showed that between 5% and 8% of the assimilated labelled C was released to the atmosphere from the fungal mycelium within 21 h. Nevertheless, part of the AM fungal C cycles slowly and accumulates in soil. Carbon dating suggests that the total glomalin pool extracted from tropical soils had a minimum residence time of 6–42 years, which supports the hypothesis that glomalin is an important component of the stable SOM pool (Rillig et al. 2001). In these soils, glomalin accounted for approximately 2.7% of the soil mass and accounted for 4% of total soil C and 5% of total soil N.

Arbuscular mycorrhizal fungi under conditions of water deficit

The impact of AM fungi on the water status of plants has become an important topic in recent years. There are conflicting reports in the literature: whereas some researchers reported no effect of the AM symbiosis on plant water relations or growth response under drought conditions (Simpson and Daft 1991; Bryla and Duniway 1997a, 1997b), others provide evidence of enhanced drought tolerance of colonized plants (see review by Augé (2001)).

It is now accepted that the contribution of AM symbiosis to plant drought tolerance is the result of a number of physical, nutritional, physiological, and cellular effects (Ruiz-Lozano 2003). This symbiotic association changes the rate of water movement into, through, and out of host plants, with consequent effects on root hydraulic conductivity (Sands and Theodorou 1978; Sweatt and Davies 1984; Nelsen 1987; Newman and Davies 1988), leaf gas exchange (Hardie 1985; Davies et al. 1993), leaf expansion (Koide 1985), osmotic adjustment (Augé et al. 1986), cell-wall elasticity (Augé et al. 1987; Sánchez-Díaz and Honrubia 1994), extraradical hyphal development (Davies et al. 1992), phytohormone production (Gogala 1991, Duan et al. 1996), and nonhydraulic root-to-shoot communication during soil drying (Augé and Duan 1991). Moreover, soil colonization by AM fungi seems to have a greater influence on the water relations of host plants than does the percentage of the root colonized by AM fungi (Augé 2004). An increase in the quantity of extraradical hyphae within a given volume of soil has been shown to improve plant water uptake by increasing the root system's absorptive surface area and the soil moisture retention capacity (Augé 2004). Each cubic centimetre of soil may contain over 100 m of AM fungal hyphae (Miller et al. 1995) that can penetrate soil pores inaccessible to root hairs (Ruiz-Lozano and Azcón 1995).

Under drought conditions when low soil moisture results in decreased nutrient diffusion rates and movement from the soil matrix to the absorbing root surface (Viets 1972), the improved drought resistance of mycorrhizal plants has usually been associated with improved nutrient (particularly P) acquisition. Furthermore, AM plants have been shown to have higher yields in dry soils than nonmycorrhizal controls (Augé 2001).

Linkage between arbuscular mycorrhizal fungi and plant communities

Arbuscular mycorrhizal fungi show generally low host specificity, and a given AM fungal species may colonize the roots of a large number of plant species if environmental conditions are conducive. However, AM fungi and host plants are selective, and certain plant species do favour the development of certain AM fungal species (Sanders 2003). Preferential plant–fungal associations are formed in different environments, and as a result, legumes and nonlegumes (Scheublin et al. 2004), different species of plants from a tall grass prairie (Eom et al. 2000), and different grass species (Gollotte et al. 2004) have been shown to develop AM fungal populations that differ in composition and diversity. Furthermore, some plant species were shown to host a greater AM fungal biodiversity than others (Liu and Wang 2003). Because of plant preference for different AM fungal associates, plant diversity may reflect AM fungal diversity (Burrows and Pfleger 2002).

The selective impact of plant species on genera and species of AM fungi may be related to the production of specific flavonoids by plants (Scervino et al. 2005; Chapter 2). In turn, the fact that host plants modify AM fungal populations could influence plant community composition, contributing to the coexistence or disappearance of certain plant species (Bever 2002). Furthermore, AM fungal species or isolates have different effects on their host's biomass production (Gange et al. 2005; Stewart et al. 2005; Youpensuk et al. 2005) and nutrient content (Sylvia et al. 2003), which influences the competitive ability of plants within a given community (Bray et al. 2003).

In addition to host specificity, AM fungal community richness may affect the coexistence of certain plants in a given area (Hart et al. 2003). A larger number of plant species are likely to find a suitable AM fungal partner in a more species-rich AM fungal community. This is supported by a study by van der Heijden et al. (1998), showing a positive correlation between AM fungal richness and plant biodiversity, nutrient uptake, and productivity (see also Cuenca et al. (2004)).

Arbuscular mycorrhizal fungi are able to form large mycelial networks in soil, and the sharing of a common AM fungal mycelial network may modify plant competitive interactions. Several species of the Glomeraceae are known to form anastomoses with different yet compatible mycelia (de la Providencia et al. 2005). In theory, the resulting mycelial network may be of indefinite size and may be connected to many plants within a community (Giovannetti et al. 2004). The sharing of a common mycelial network may result in a more equal distribution of soil resources between dominant and subordinate host plant species and, thus, allow their coexistence. By contrast, the mycelia of certain other AM fungal species, in particular that of the Gigasporaceae, have a reduced ability to form amastamoses, and their impact on soil heterogeneity may be more localized, thus creating different niches that can be occupied by plant species with different adaptations.

Arbuscular mycorrhizal fungi usually have a positive influence on their host plants when soil resources are limiting, and thus, the positive feedback of AM fungi on their host plants may have a homogenizing effect in a plant community, allowing the most efficient plant – AM fungi combination to eventually dominate (Bever 2003). However, in highly productive systems where light becomes limiting, the benefit of AM fungi may become relatively smaller than the C drain they create. Under such circumstances, the feedback from AM fungi could be negative, and AM fungi may enhance diversity (Reynolds et al. 2003).

Arbuscular mycorrhizal fungi are in dynamic equilibrium with their hosts; therefore, fungal populations are affected by biotic and abiotic environmental factors that affect plant growth. For certain species of AM fungi, spore production has been correlated with SOM, calcium carbonate (Mohammad et al. 2003), and nitrate concentrations (Burrows and Pfleger 2002). Furthermore, benefits from inoculation with different AM fungal species appear to depend on soil P availability, soil pH, soil aggregation (Hamel et al. 1997), and soil type (Rivera and Fernandez 2003). This suggests that environmental conditions play an important role in determining the nature of the interaction between the two symbionts.

Temporal and spatial variations in symbiotic function

A number of recent papers have reiterated the important role of the AM symbiosis in plant health (Jeffries et al. 2003), plant competition (Aerts 2002), and nutrition. However, there are also reports of growth depression following colonization (e.g., Ryan et al. 2005). According to Fitter (2005), about two-thirds of modern plant species form AM symbioses. The widespread occurrence of AM plants has raised several questions about AM fungal biodiversity and its potential importance in ecosystem processes.

The availability of molecular tools to study AM fungi under field conditions has stimulated the research into the role of the AM symbiosis in ecosystem functioning. However, the results are highly divergent, and there seem to be no steadfast rules. The outcome of an AM association appears to depend on the interaction between specific plants and AM fungi and varies spatially and temporally. For example, in prairies where the primary limiting factor is low annual precipitation (Tannas 2003), C3 grass plants are more active under the cooler and moister conditions of early summer. By contrast, C4 plants show higher activity in midsummer, when conditions are warmer and drier and when certain C3 plants become dormant. Although these plants show temporal differences in their phenology, they may be connected to the same AM fungal network. The high photosynthetic rates and water-use efficiency of C4 species would allow them to pay the carbon cost of maintaining the symbiosis during hot periods (Heckathorn et al. 1999) unlike C3 plants, which become dormant. Such differences in plant metabolic traits provide spatially and temporally dispersed resources. In the same vein, the size of the AM extraradical mycelial network may define the C costs to plants and C distribution within the network.

Changes along time gradients

Aboveground and belowground community links are complex because they operate on a relatively wide range of spatial and temporal scales (Bardgett et al. 2005). By integrating temporal effects into studies of microbial diversity and plant performance, we may be able to link microbial and plant ecosystem dynamics. Schadt et al. (2003) found that microbial biomass in frozen alpine soils was at its maximum in late winter and declined significantly thereafter. Furthermore, the authors recorded an almost complete change in the composition of the microbial community and its functional attributes between winter and summer. However, the general assumption is that soil microbes are inactive during the winter. In a study by Moscatelli et al. (2005), potential acid phosphatase activity peaked in the Mediterranean winter, whereas microbial respiration and the production of other enzymes tended to peak in the fall. The seasonal dynamics in microbial community composition and activity are important because they control the partitioning of nutrients between plants and soil microbes (Bardgett et al. 2005). There is no apparent consistency in the way AM fungi community structure changes over time and in different ecosystems. For example, changes in AM fungi community structure was reported to take years (Vandenkoornhuyse et al. 2002), seasons (Helgason et al. 1999; Clapp et al. 2002; Lutgen et al. 2003; Heinemeyer et al. 2004), or months (Bhadalung et al. 2005; Li et al. 2005). Yet, it is clear that the benefit of the symbiosis depends on the effective functioning of a given plant – AM fungal combination, in a given environment.

Plant species as determinants of arbuscular mycorrhizal fungal spatial variation

Under field conditions, AM fungi influence not only the plant species with which they have the greatest affinity, but also neighbouring plants via complex interactions at the community level. Using DNA analysis techniques, Helgason et al. (1999) showed that adjacent plants can affect AM fungal colonization. This is further supported by Mummey et al. (2005), who showed that AM fungi colonizing the roots of a common grass (*Dactylis glomerata* L.) were negatively affected by the presence of roots of an invasive forb (*Centaurea maculosa* Lam.). The authors' results suggest that the influence of *C. maculosa* on AM fungi in the roots of its competitors, may represent the mechanism underlying its invasiveness.

Specificity in plant – AM fungal associations influences inter- and intra-specific interaction among neighbouring plants. Individual plants received a greater benefit from the mycorrhizal association when adjacent individuals were genetically identical or from the same population. This suggests that plants from the same population share more efficient hyphal networks in nature (Ronsheim and Anderson 2001).

Oehl et al. (2005) found that, under different crop management intensities, differences occurred in the distribution of AM fungal spores of different genera within the soil profile. At a different scale, Scheublin et al. (2004) found differences in AM fungal DNA sequence types between legumes and nonlegumes. Furthermore, the authors found different sequence types in different parts of the legume roots: sequence type Acau5 was almost exclusively restricted to root nodules, whereas sequence type Glo50, which was common in roots, was never found in nodules.

Arbuscular mycorrhizal fungal diversity appears to be linked to specific niches within the soil and host roots. If we want to understand the nature and function of AM symbioses in ecosystems, it is necessary to consider the temporal and spatial variations in AM associations created by climatic variation, plant phenology, and variation in soil conditions. In turn, this, may allow us to better manage AM symbioses.

Arbuscular mycorrhiza in agriculture

The biodiversity and abundance of AM fungi in soil are influenced by a variety of environmental and plant factors (Bever et al. 2001). Therefore, AM fungi may be positively or negatively influenced by different management practices, such as crop rotation, fertilization, and tillage.

Crop rotation

Highly mycorrhizal dependent crop plants (see Chapter 6 for definition) show greater root and shoot dry masses, root length, percentage colonization, and P uptake when mycorrhizal (Shibata and Yano 2003). Crop rotations involving AM plants potentially increase the diversity of the AM fungal community (Douds et al. 1993) at a given site. By contrast, the use of nonhost plants may delay mycorrhizal colonization of a subsequent crop (Gavito and Miller 1998) in the rotation, presumably as a result of a decrease in the number of active AM fungal propagules in the soil.

Although the benefits of the AM symbiosis are widely recognized, they are rarely considered when making agricultural management decisions (Plenchette et al. 2005). The use of AM crops in a rotation under a no-till management systems would promote AM fungal diversity and increase soil inoculum levels. To be sustainable, crop rotations must be agronomically and economically feasible (Zentner et al. 2002). From an economic standpoint, the successful management of AM fungi in agriculture could decrease fertilizer use and input costs.

Fertilizer use

The results of studies investigating the effect of fertilizer applications on AM fungal colonization and nutrient uptake are contradictory. High levels of applied P may lead to dramatic decreases in root colonization and spore production (e.g., Abbott and Robson 1991). However, the level of colonization does not depend solely on soil fertility, it also depends on the AM dependency of the plant studied (Plenchette et al. 1983; McGonigle et al. 1990). Balser et al. (2005) found that AM fungal abundance was much higher in N-enriched treatments than in N-limited ones. It is possible that AM fungi have nutrient requirements of their own, and these may vary between species. After comparing fertilized and unfertilized soils, Johnson (1993) found that, in general, increases in available N increased colonization, whereas increases in available P decreased it. It appears that root colonization and dry matter yield are not always related (Khaliq and Sanders 2000), that AM development does not only depend on fertilization, and that the plant species involved is an important factor (Plenchette et al. 1983; McGonigle et al. 1990).

The fact that the development of AM fungi may be limited in nutrient-rich soils has led to the conclusion that AM fungi are not important in highly fertilized systems. However, the advantages provided by AM fungi are not limited to plant nutrient uptake. In annual crop production, the impact of AM fungi on the physical and biological quality of the soil (see Chapter 4) may be more important than nutrient uptake related benefits.

Tillage

Tillage has a negative effect on AM fungal communities (Jansa et al. 2002). In past years, it was common for Canadian prairie farmers to incorporate bare fallow into cropping rotations in an attempt to conserve water and nutrients. Conventional tillage was used for weed management, but soil tillage contributes to substantial soil erosion and disruption of AM fungal hyphal networks (Evans and Miller 1990).

Tillage may be replaced by other methods that enhance water and nutrient conservation. For example, AM fungal host species may be planted instead of nonhosts. Arbuscular mycorrhizal fungal hyphae extend into the soil and, as outlined above, increase exploration of micropores, increase soil–root contact, increase nutrient and water uptake, enhance soil structural quality, which improves soil water holding capacity (Miller and Jastrow 2000; Augé et al. 2003).

Commercial inocula

Arbuscular mycorrhizal fungi may be managed in agricultural systems by planting mycorrhizal crops, decreasing fertilizer applications and tillage intensity, and by introducing selected AM strains that are available in the form of commercial inocula (see Chapter 8). Large-scale production of pure AM inoculum is complicated by the biotrophic nature of AM fungi (Wilarso Budi et al. 1998) and by the fact that they produce relatively few spores. Because of high production costs, the use of such inocula is generally limited to high-value crops but is being attempted in organically produced agronomic crops. The production of inoculum in the field is also difficult, and the results are not always predictable (Graham et al. 1995). Currently, some companies are able to mass produce and distribute AM fungal inocula (Chapter 8). However, these companies must ensure that their products offer an effective and economically feasible option for farmers.

Crop breeding and the arbuscular mycorrhizal association

The ability of different plants to affect the AM fungal inoculum potential of soils has been outlined with regard to the use of mycorrhizal and nonmycorrhizal plants in crop rotations. However, different cultivars of the same mycorrhizal crop plant may also affect AM fungal populations differently. Breeding programs tend to focus on crop yield, quality, and pest resistance rather than on mycorrhizal responsiveness (Parke and Kaeppler 2000), and modern breeding programs may lead to reduction in the mycorrhizal responsiveness of new cultivars (Zhu et al. 2001). Parke and Kaeppler (2000) argue that breeding crop plants without taking AM formation capacity into consideration is "irresponsible" and that it may result in the permanent loss of genes related to formation of the AM symbiosis (see Chapter 6). They contend that this loss would have far reaching effects on AM fungi and the potential benefits they could provide to cropping systems. Therefore, a coordinated approach to plant breeding is needed to create symbiotic plant genotypes.

Most often, the driving force behind the use of AM fungi in agriculture is the use of soils with low available P. Under such circumstances, AM fungal colonization may be considered in breeding programs (Sanginga et al. 2000), especially in areas of the world where agricultural inputs (including P fertilizers) are limited by availability and (or) cost. In addition to P uptake, AM fungal – plant interactions may improve drought tolerance by modifying host plant water relations (Augé 2001). New molecular techniques may allow the identification of genes associated with mycorrhizal responsiveness, which could then be used in breeding programs (Reynolds et al. 2005).

Breeding for mycorrhizal responsiveness is considered relatively difficult, and studies evaluating genotypic variation within a given crop species are generally lacking (Gahoonia and Nielsen 2004). However, a number of studies have evaluated AM responsiveness in wheat (*Triticum aestivum* L.; Xavier and Germida 1998; Zhu et al. 2001), maize (*Zea mays* L.; Kaeppler et al. 2000), soybean (*Glycine max* (L.) Merr.; Nwoko and Sanginga 1999), lettuce (*Lactuca sativa* L.; Jackson et al. 2002), and clover (*Trifolium repens* L.; Eason et al. 2001). Nevertheless, these studies have rarely influenced agricultural practices. Ryan and Graham (2002) showed that, for many crops, P or micronutrient fertilization and AM fungal colonization give similar results. Therefore, the authors suggested that AM fungi might not provide enough benefit to crops to be important in plant breeding programs. However, Rengel (2002) suggested that selective breeding for AM responsiveness will lead to increased crop yields under a wide range of different environmental conditions and that this will lead, ultimately, to greater "sustainability of agricultural ecosystems in which soil–plant–microbe interactions will be better exploited."

The use of genomic technologies will lead to new information concerning the AM symbiosis; however, whether this will be used in future breeding programmes remains to be seen. A study by Guimil et al. (2005) investigated the genes affected during the formation of the AM symbiosis in rice. The authors found that several of these genes were also expressed when the roots were colonized by pathogenic fungi. This may be useful when breeding for both pest resistance and mycorrhizal responsiveness. Other studies have identified other plant genes related to the AM symbiosis (Ane et al. 2004; Levy et al. 2004; Imaizumi-Anraku et al. 2005).

The effect of pesticides on the arbuscular mycorrhizal symbiosis

Pesticide use is widespread in Canada. In 2001, 73% of farmers applied herbicides, insecticides, or fungicides (Korol 2004). A European study on winter wheat found that herbicide and fungicide applications (Girvan et al. 2004) changed the community structure of soil bacteria but did not affect AM fungi. However, other studies have shown that pesticides can influence AM fungi in a variety of ways (Trappe et al. 1984).

A greenhouse study by Bethlenfalvay et al. (1996) found that, when the herbicide bentazon was applied to resistant soybean plants and the susceptible weed cocklebur (*Xanthium strumarium* L.), the AM fungal colonization of cocklebur was reduced by 43%. Colonization was also reduced in soybean, but the results were not significant. Furthermore, as cocklebur plants succumbed to the treatment, an AM-mediated flux of nutrients occurred from the weed species towards the crop species (Bethlenfalvay et al. 1996). Mujica et al. (1998) found a similar response when chlorsulfuron was applied to soybean and a weed species. However, diclofop was found to inhibit AM fungal colonization of resistant wheat when grown with susceptible ryegrass (*Lolium perenne* L.; Rejon et al. 1997). Nevertheless, wheat growth and yield were enhanced. This was attributed to interplant AM networks in which the wheat became a stronger sink for nutrients than the ryegrass.

The effect of fungicides has been shown to vary depending on the AM species and fungicides being tested. Benomyl, PCNB, and captan were all found to reduce AM formation in pea (*Pisum sativum* L.) plants inoculated with *Glomus mosseae* (Nicol. & Gerd.) Gerd & Trappe and *Gigaspora rosea* Nicol. & Schenck (Schreiner and Bethlenfalvay 1997). Pea plants inoculated with *Glomus etunicatum* Becker & Gerdeman did not show the same reductions in root colonization. Furthermore, a mixed inoculum of the three AM fungi overcame any deleterious effects of the fungicides. Foliar applications to leek (*Allium porrum* L.) plants of fosetyl-Al (a systemic fungicide used to control pathogenic Oomycetes), was found to increase AM root colonization by *Glomus intraradices* Schenck & Smith and plant growth; however, a soil drench using the systemic fungicide metalaxyl was found to significantly decrease both parameters (Jabaji-Hare and Kendrick 1987). By contrast, fosetyl-Al was found to inhibit AM root colonization and both root and shoot growth of onion (*Allium cepa* L.) when applied as a soil drench (Sukarno et al. 1998).

The systemic fungicide benomyl severely affects the AM symbiosis (Spokes et al. 1981; Menge 1982; Verkade and Hamilton 1983; Salem et al. 2003). This is probably due to inhibition of spore germination and hyphal growth (Carr and Hinkley 1985). Benomyl disrupts the translocation of phosphate from AM fungi to their host plant (Boatman et al. 1978; Hale and Sanders 1982). In a study by Salem et al. (2003), the fungicide captan inhibited AM fungal colonization of tomato (*Solanum lycopersicum* L.) plants for several weeks and reduced AM root colonization intensity (Salem et al. 2003). When captan was used in combination with benomyl, AM fungal colonization was completely suppressed during the 15-week study.

Propiconazole, a systemic fungicide inhibiting ergosterol synthesis, was expected to have little effect on AM fungi because of the low amount of ergosterol in these fungi (Schmitz et al. 1991; Frey et al. 1994). However, deleterious effects were found on root colonization, spore production, and plant growth (Nemec 1985; von Alten et al. 1993). By contrast, in a study by Kling and Jakobsen (1997), propiconazole was found to have no influence on AM hyphal function and P transport when applied only to the AM hyphae and at recommended field rates. In the same study, carbendazim, the active ingredient in benomyl, was shown to greatly affect AM fungi. Kjøller and Rosendahl (2000) also found that propiconazole had little affect on AM fungi; however, another sterol inhibitor, fenpropimorph, reduced both root and soil hyphal length and activity.

Two strobilurin fungicides, azoxystrobin and kresoxim-methyl, were found to have no influence on AM formation in maize when applied foliarly; however, when applied as soil drenches they dramatically decreased AM activity and eventually completely inhibited the AM fungi (Diedhiou et al. 2004).

The use of selective insecticides has been found to have little influence on AM fungal P uptake (Schweiger and Jakobsen 1998) or root colonization (Wan and Rahe 1998). Interestingly, chlorpyrifos was found to increase AM fungal diversity when applied to a mixed grassland (Vandenkoornhuyse et al. 2003). The authors attributed this to a decrease in mesofaunal populations and a subsequent decrease in the grazing pressure on AM fungal tissue.

It is perhaps of little surprise that, of the pesticides used in cropping systems, fungicides have the greatest impact on AM. However, there are a range of pesticides that do not negatively affect AM fungi, and some inoculum producers (e.g., Premier Tech Biotechnologies, Rivière-du-Loup, Que.) provide a list those (including seed treatments that contain fungicides) that are compatible with their products.

Conclusion

Conventional agricultural practices place several constraints on the use of AM fungi. Nevertheless, AM fungi are present in agricultural fields and colonize the roots of most host crops in most Canadian agricultural soils. Whether or not we develop and use cropping systems and technologies that optimize the benefits to soil, crops, and the environment that can be derived from the AM symbiosis will depend on the extent to which our society values environmental quality. In industrialized countries, high-input agriculture is practiced, and the use of AM fungi will likely improve environmental quality rather than yields.

References

Abbott, L.K., and Robson, A.D. 1991. Factors influencing the occurrence of vesicular–arbuscular mycorrhizas. Agric. Ecosyst. Environ. **35**: 121–150.

Aerts, R. 2002. The role of various types of mycorrhizal fungi in nutrient cycling and plant competition. *In* Mycorrhizal ecology. *Edited by* M.G.A. van der Heijden and I.R. Sanders. Springer-Verlag, Berlin. pp. 117–133.

Allen, M.F., Swenson, W., Querejeta, J.I., Egerton-Warburton, L.M., and Treseder, K.K. 2003. Ecology of mycorrhizae: a conceptual framework for complex interactions among plants and fungi. Annu. Rev. Phytopathol. **41**: 271–303.

Andrade, G., Linderman, R.G., and Bethlenfalvay, G.J. 1998. Bacterial associations with the mycorrhizosphere and hyphosphere of the arbuscular mycorrhizal fungus *Glomus mosseae*. Plant Soil, **202**: 79–87.

Ane, J.-M., Kiss, G.B., Riely, B.K., Penmetsa, R.V., Oldroyd, G.E.D., Ayax, C., Levy, J., Debellé, F., Baek, J.-M., Kalo, P., Rosenberg, C., Roe, B.A., Long, S.R., Dénarié, J., and Cook, D.R. 2004. *Medicago truncaluta* DMI1 required for bacterial and fungal symbioses in legumes. Science (Washington, D.C.), **303**: 1364–1367.

Angers, D.A., and Caron, J. 1998. Plant-induced changes in soil structure: processes and feedbacks. Biogeochemistry, **42**: 55–72.

Augé, R.M. 2001. Water relations, drought and vesicular–arbuscular mycorrhizal symbiosis. Mycorrhiza, **11**: 3–42.

Augé, R.M. 2004. Arbuscular mycorrhizae and soil/plant water relations. Can. J. Soil Sci. **84**: 373–381.

Augé, R.M., and Duan, X. 1991. Mycorrhizal fungi and nonhydraulic root signals of soil drying. Plant Physiol. **97**: 821–824.

Augé, R.M., Schekel, K.A., and Wample, R.L. 1986. Osmotic adjustment in leaves of VA mycorrhizal and nonmycorrhizal rose plants in response to drought stress. Plant Physiol. **82**: 765–770.

Augé, R.M., Schekel, K.A., and Wample, R.L. 1987. Rose leaf elasticity changes in response to mycorrhizal colonization and drought acclimation. Physiol. Plant. **70**: 175–182.

Augé, R.M., Moore, J.L., Cho, K., Stutz, J.C., Sylvia, D.M., Al-Agely, A.K., and Saxton, A.M. 2003. Relating foliar dehydration tolerance of mycorrhizal *Phaseolus vulgaris* to soil and root colonization by hyphae. J. Plant Physiol. **160**: 1147–1156.

Azcón, R., Gomez, M., and Tobar, R. 1992. Effects of nitrogen source on growth, nutrition, photosynthetic rate and nitrogen metabolism of mycorrhizal and phosphorus-fertilized plants of *Lactuca sativa* L. New Phytol. **121**: 227–234.

Balser, T.C., Treseder, K.K., and Ekenler, M. 2005. Using lipid analysis and hyphal length to quantify AM and saprotrophic fungal abundance along a soil chronosequence. Soil Biol. Biochem. **37**: 601–604.

Bardgett, R.D., Bowman, W.D., Kaufmann, R., and Schmidt, S.K. 2005. A temporal approach to linking aboveground and belowground ecology. Trends Ecol. Evol. **20**: 634–641.

Beauchemin, S., and Simard, R.R. 2000. Phosphorus status of intensively cropped soils of the St. Lawrence lowlands. Soil Sci. Soc. Am. J. **64**: 659–670.

Bécard, G., Doner, L.W., Rolin, D.B., Douds, D.D., and Pfeffer, P.E. 1991. Identification and quantification of trehalose in vesicular–arbuscular mycorrhizal fungi by *in vivo* ^{13}C NMR and HPLC analyses. New Phytol. **118**: 547–552.

Bethlenfalvay, G.J., Mihara, K.L., Schreiner, R.P., and McDaniel, H. 1996. Mycorrhizae, biocides, and biocontrol. 1. Herbicide–mycorrhiza interactions in soybean and cocklebur treated with bentazon. Appl. Soil Ecol. **3**: 197–204.

Bever, J.D. 2002. Host-specificity of AM fungal population growth rates can generate feedback on plant growth. Plant Soil, **244**: 281–290.

Bever, J.D. 2003. Soil community feedback and the coexistence of competitors: conceptual frameworks and empirical tests. New Phytol. **157**: 465–473.

Bever, J.D., Schultz, P.A., Pringle, A., and Morton, J.B. 2001. Arbuscular mycorrhizal fungi: more diverse than meets the eye and the ecological tale of why. Bioscience, **51**: 923–931.

Bhadalung, N.N., Suwanarit, A., Dell, B., Nopamornbodi, O., Thamchaipenet, A., and Rungchuang, J. 2005. Effects of long-term NP-fertilization on abundance and diversity of arbuscular mycorrhizal fungi under a maize cropping system. Plant Soil, **270**: 371–382.

Boatman, N., Paget, D., Hayman, D.S., and Mosse, B. 1978. Effects of systemic fungicides on vesicular–arbuscular mycorrhizal infection and plant phosphate uptake. Trans. Br. Mycol. Soc. **70**: 443–450.

Bolan, N.S. 1991. A critical review on the role of mycorrhizal fungi in the uptake of phosphorus by plants. Plant Soil, **134**: 189–207.

Bolan, N.S., Robson, A.D., and Barrow, N.J. 1987. Effects of vesicular–arbuscular mycorrhizae on the availability of iron phosphates to plants. Plant Soil, **99**: 401–410.

Bonfante-Fasolo, P. 1988. The role of the cell wall as a signal in mycorrhizal associations. *In* Cell to cell signals in plant, animal and microbial symbiosis. *Edited by* S. Scannerini, D.G. Smith, P. Bonfante-Fasolo, and V. Gianinazzi-Pearson. Springer-Verlag, Berlin. pp. 219–235.

Bossuyt, H., Denef, K., Six, J., Frey, S.D., Merckx, R., and Paustian, K. 2001. Influence of microbial populations and residue quality on aggregate stability. Appl. Soil Ecol. **16**: 195–208.

Bray, S.R., Kitajima, K., and Sylvia, D.M. 2003. Mycorrhizae differentially alter growth, physiology and competitive ability of an invasive shrub. Ecol. Appl. **13**: 565–574.

Bryla, D.R., and Duniway, J.M. 1997*a*. Growth, phosphorus uptake and water relations of safflower and wheat infected with an arbuscular mycorrhizal fungus. New Phytol. **136**: 581–590.

Bryla, D.R., and Duniway, J.M. 1997*b*. Water uptake by safflower and wheat roots infected with arbuscular mycorrhizal fungi. New Phytol. **136**: 591–601.

Burrows, R.L., and Pfleger, F.L. 2002. Arbuscular mycorrhizal fungi respond to increasing plant diversity. Can. J. Bot. **80**: 120–130.

Carr, G.R., and Hinkley, M.A. 1985. Germination and hyphal growth of *Glomus caledonicum* on water agar containing benomyl. Soil Biol. Biochem. **17**: 313–316.

Clapp, J.P., Helgason, T., Daniell, T.J., and Young, J.P.W. 2002. Genetic studies of the structure and diversity of arbuscular mycorrhizal fungal communities. *In* Mycorrhizal ecology. *Edited by* M.G.A. van der Heijden and I.R. Sanders. Springer-Verlag, Berlin. pp. 201–224.

Clark, R.B., and Zeto, S.K. 2000. Mineral acquisition by arbuscular mycorrhizal plants. J. Plant Nutr. **23**: 867–902.

Clough, K.S., and Sutton, J.C. 1978. Direct observation of fungal aggregates in sand dune soil. Can. J. Microbiol. **24**: 333–335.

Cox, G., Sanders, F.E., Tinker, P.B., and Wild, J.A. 1975. Ultrastructual evidence relating to host–endophyte transfer in vesicular–arbuscular mycorrhiza. *In* Endomycorrhizas. *Edited by* F.E. Sanders, B. Mosse, and P.B. Tinker. Academic Press, London. pp. 297–312.

Cuenca, G., De Andrade, Z., Lovera, M., Fajardo, L., and Meneses, E. 2004. The effect of two arbuscular mycorrhizal inocula of contrasting richness and the same mycorrhizal potential on the growth and survival of wild plant species from La Gran Sabana, Venezuela. Can. J. Bot. **82**: 582–589.

Davies, F.T., Jr., Potter, J.R., and Linderman, R.G. 1992. Mycorrhiza and repeated drought exposure affect drought resistance and extraradical hyphae development of pepper plants independent of plant size and nutrient content. J. Plant Physiol. **139**: 289–294.

Davies, F.T., Jr., Potter, J.R., and Linderman, R.G. 1993. Drought resistance of mycorrhizal pepper plants independent of leaf P concentration—response in gas exchange and water relations. Physiol. Plant. **87**: 45–53.

Degens, B.P. 1997. Macro-aggregation of soils by biological bonding and binding mechanisms and the factors affecting these: a review. Aust. J. Soil Res. **35**: 431–459.

Degens, B.P., Sparling, G.P., and Abbott, L.K. 1994. The contribution from hyphae, roots and organic carbon constituents to the aggregation of a sandy loam under long-term clover-based and grass pastures. Eur. J. Soil Sci. **45**: 459–468.

de la Providencia, I.E., de Souza, F.A., Fernandez, F., Delmas, N.S., and Declerck, S. 2005. Arbuscular mycorrhizal fungi reveal distinct patterns of anastomosis formation and hyphal healing mechanisms between different phylogenic groups. New Phytol. **165**: 261–271.

Diedhiou, P.M., Oerke, E.C., and Dehne, H.W. 2004. Effects of the strobilurin fungicides azoxystrobin and kresoxim-methyl on arbuscular mycorrhiza. J. Plant Dis. Prot. **111**: 545–556.

Dorioz, J.M., Robert, M., and Chenu, C. 1993. The role of roots, fungi and bacteria on clay particle organization: an experimental approach. Geoderma, **56**: 179–194.

Dorioz, J.M., Wang, D., Poulenard, J., and Trevisan, D. 2006. The effect of grass buffer strips on phosphorus dynamics—a critical review and synthesis as a basis for application in agricultural landscapes in France. Agric. Ecosyst. Environ. **117**: 4–21.

Douds, D.D., Jr., Janke, R.R., and Peters, S.E. 1993. VAM fungus spore populations and colonization of roots of maize and soybean under conventional and low-input sustainable agriculture. Agric. Ecosyst. Environ. **43**: 325–335.

Duan, X., Neuman, D.S., Reiber, J.M., Green, C.D., Saxton, A.M., and Augé, R.M. 1996. Mycorrhizal influence on hydraulic and hormonal factors implicated in the control of stomatal conductance during drought. J. Exp. Bot. **47**: 1541–1550.

Eason, W.R., Webb, K.J., Michaelson-Yeates, T.P.T., Abberton, M.T., Griffith, G.W., Culshaw, C.M., Hooker, J.E., and Dhanoa, M.S. 2001. Effect of genotype of *Trifolium repens* on mycorrhizal symbiosis with *Glomus mosseae*. J. Agric. Sci. **137**: 27–36.

Eom, A.H., Hartnett, D.C., and Wilson, G.W.T. 2000. Host plant species effects on arbuscular mycorrhizal fungal communities in tallgrass prairie. Oecologia (Berl.), **122**: 435–444.

Evans, D.G., and Miller, M.H. 1990. The role of the external mycelial network in the effect of soil disturbance upon vesicular–arbuscular mycorrhizal colonization of maize. New Phytol. **114**: 65–71.

Filion, M., St-Arnaud, M., and Fortin, J.A. 1999. Direct interaction between the arbuscular mycorrhizal fungus *Glomus intraradices* and different rhizosphere microorganisms. New Phytol. **141**: 525–533.

Fitter, A.H. 2005. Darkness visible: reflections on underground ecology. J. Ecol. **93**: 231–243.

Franzluebbers, A.J., Wright, S.F., and Stuedemann, J.A. 2000. Soil aggregation and glomalin under pastures in the Southern Piedmont USA. Soil Sci. Soc. Am. J. **64**: 1018–1026.

Frey, B., Vilarino, A., Schüepp, H., and Arines, J. 1994. Chitin and ergosterol content of extraradical and intraradical mycelium of the vesicular–arbuscular mycorrhizal fungus

Glomus intraradices. Soil Biol. Biochem. **26**: 711–717.

Gahoonia, T.S., and Nielsen, N.E. 2004. Root traits as tools for creating phosphorus efficient crop varieties. Plant Soil, **260**: 47–57.

Gale, W.J., Cambardella, C.A., and Bailey, T.B. 2000. Root-derived carbon and the formation and stabilization of aggregates. Soil Sci. Soc. Am. J. **64**: 201–207.

Gange, A.C., Brown, V.K., and Aplin, D.M. 2005. Ecological specificity of arbuscular mycorrhizae: evidence from foliar- and seed-feeding insects. Ecology, **86**: 603–611.

Gavito, M.E., and Miller, M.H. 1998. Changes in mycorrhiza development in maize induced by crop management practices. Plant Soil, **198**: 185–192.

Giovannetti, M., Sbrana, C., Avio, L., and Strani, P. 2004. Patterns of below-ground plant interconnections established by means of arbuscular mycorrhizal networks. New Phytol. **164**: 175–181.

Girvan, M.S., Bullimore, J., Ball, A.S., Pretty, J.N., and Osborn, A.M. 2004. Responses of active bacterial and fungal communities in soils under winter wheat to different fertilizer and pesticide regimens. Appl. Environ. Microbiol. **70**: 2692–2701.

Gogala, N. 1991. Regulation of mycorrhizal infection by hormonal factors produced by hosts and fungi. Experientia, **47**: 331–340.

Gollotte, A., van Tuinen, D., and Atkinson, D. 2004. Diversity of arbuscular mycorrhizal fungi colonising roots of the grass species *Agrostis capillaris* and *Lolium perenne* in a field experiment. Mycorrhiza, **14**: 111–117.

Govindarajulu, M., Pfeffer, P.E., Jin, H., Abubaker, J., Douds, D.D., Jr., Allen, J.W., Bucking, H., Lammers, P.J., and Shachar-Hill, Y. 2005. Nitrogen transfer in the arbuscular mycorrhizal symbiosis. Nature (Lond.), **435**: 819–823.

Graham, J.H. 2000. Assessing costs of arbuscular mycorrhizal symbiosis in agroecosystems. *In* Current advances in mycorrhizae research. *Edited by* G.K. Podila and D.D. Douds, Jr. American Phyotpathological Society Press, St. Paul, Minn. pp. 127–140.

Graham, J.H., Hodge, N.C., and Morton, J.B. 1995. Fatty acid methyl ester profiles for characterization of Glomalean fungi and their endomycorrhizae. Appl. Environ. Microbiol. **61**: 58–64.

Guimil, S., Chang, H.-S., Zhu, T., and Sesma, A. Osbourn, A., Roux, C., Ioannidis, V., Oakeley, E.J., Docquier, M., Descombes, P., Briggs, S.P., and Paszkowski, U. 2005. Comparative transcriptomics of rice reveals an ancient pattern of response to microbial colonization. Proc. Natl. Acad. Sci. U.S.A. **102**: 8066–8070.

Hale, K.A., and Sanders, F.E.O. 1982. Effects of benomyl on vesicular–arbuscular mycorrhizal infection in red clover (*Trifolium pratense* L.) and consequences for phosphorus inflow. J. Plant Nutr. **5**: 1355–1367.

Hamel, C., Dalpé, Y., Furlan, V., and Parent, S. 1997. Indigenous populations of arbuscular mycorrhizal fungi and soil aggregate stability are major determinants of leek (*Allium porrum* L.) response to inoculation with *Glomus intraradices* Schenck & Smith or *Glomus versiforme* (Karsten) Berch. Mycorrhiza, **7**: 187–196.

Hardie, K. 1985. The effect of removal of extraradical hyphae on water uptake by vesicular–arbuscular mycorrhizal plants. New Phytol. **101**: 677–684.

Harrison, M.J., and van Buuren, M.L. 1995. A phosphate transporter from the mycorrhizal fungus *Glomus versiforme*. Nature (Lond.), **378**: 626–629.

Hart, M.M., Reader, R.J., and Klironomos, J.N. 2003. Plant coexistence mediated by arbuscular mycorrhizal fungi. Trends Ecol. Evol. **18**: 418–423.

Hattingh, M.J., Gray, L.E., and Gerdemann, J.W. 1973. Uptake and translocation of ^{32}P labelled phosphate to onion roots by endomycorrhizal fungi. Soil Sci. **116**: 383–387.

Heckathorn, S.A., McNaugthon, S.J., and Coleman, J.S. 1999. C4 plants and herbivory. *In* C4 plant biology. *Edited by* R.F. Sage and K.R. Monson. Academic Press, San Diego, Calif. pp. 285–312.

Heinemeyer, A., Ridgway, K.P., Edwards, E.J., Benham, D.G., Young, J.P.W., and Fitter, A.H. 2004. Impact of soil warming and shading on colonization and community structure of arbuscular mycorrhizal fungi in roots of a native grassland plant community. Glob. Change Biol. **10**: 52–64.

Helgason, T., Fitter, A.H., and Young, J.P.W. 1999. Molecular diversity of arbuscular mycorrhizal fungi colonising *Hyacinthoides non-scripta* (bluebell) in a seminatural woodland. Mol. Ecol. **8**: 659–666.

Hodge, A., Campbell, C.D., and Fitter, A.H. 2001. An arbuscular mycorrhizal fungus accelerates decomposition and acquires nitrogen directly from organic material. Nature (Lond.), **413**: 297–299.

Imaizumi-Anraku, H., Takeda, N., Charpentier, M., Perry, J., Miwa, H., Umehara, Y., Kouchi, H., Murakami, Y., Mulder, L., Vickers, K., Pike, J., Downie, J.A., Wang, T., Sato, S., Asamizu, E., Tabata, S., Yoshikawa, M., Murooka, Y., Wu, G.-J., Kawaguchi, M., Kawasaki, S., Parniske, M., and Hayashi, M. 2005. Plastid proteins crucial for symbiotic fungal and bacterial entry into plant roots. Nature (Lond.), **433**: 527–531.

Jabaji-Hare, S.H., and Kendrick, W.B. 1987. Response of an endomycorrhizal fungus in *Allium porrum* L. to different concentrations of the systemic fungicides, metalaxyl (Ridomil®) and fosetyl-Al (Aliette®). Soil Biol. Biochem. **19**: 95–99.

Jackson, L.E., Miller, D., and Smith, S.E. 2002. Arbuscular mycorrhizal colonization and growth of wild and cultivated lettuce in response to nitrogen and phosphorus. Sci. Hortic. (Amsterdam), **94**: 205–218.

Jansa, J., Mozafar, A., Anken, T., Ruh, R., Sanders, I.R., and Frossard, E. 2002. Diversity and structure of AMF communities as affected by tillage in a temperate soil. Mycorrhiza, **12**: 225–234.

Jastrow, J.D., Miller, R.M., and Lussenhop, J. 1998. Contributions of interacting biological mechanisms to soil aggregate stabilization in restored prairie. Soil Biol. Biochem. **30**: 905–916.

Jayachandrana, K., Schwab, A.P., and Hetrick, B.A.D. 1989. Mycorrhizal mediation of phosphorus availability: synthetic iron chelate effects on phosphorus solubilization. Soil Sci. Soc. Am. J. **53**: 1701–1706.

Jeffries, P., Gianinazzi, S., Perotto, S., Katarzyna, S.P., Turnau, K., and Barea, J.M. 2003. The contribution of arbuscular mycorrhizal fungi in sustainable maintenance of plant health and soil fertility. Biol. Fertil. Soils, **37**: 1–16.

Jin, H., Pfeffer, P.E., Douds, D.D., Jr., Piotrowski, E., Lammers, P.J., and Shachar-Hill, Y. 2005. The uptake, metabolism, transport and transfer of nitrogen in an arbuscular mycorrhizal symbiosis. New Phytol. **168**: 687–696.

Johnson, D., Leake, J.R., Ostle, N., Ineson, P., and Read, D.J. 2002. In situ $^{13}CO_2$ pulse-labelling of upland grassland demonstrates a rapid pathway of carbon flux from arbuscular mycorrhizal mycelia to the soil. New Phytol. **153**: 327–334.

Johnson, N.C. 1993. Can fertilization of soil select less mutualistic mycorrhizae? Ecol. Appl. **3**: 749–757.

Joner, E.J., and Johansen, A. 2000. Phosphatase activity of external hyphae of two arbuscular mycorrhizal fungi. Mycol. Res. **104**: 81–86.

Joner, E.J., Ravnskov, S., and Jakobsen, I. 2000. Arbuscular mycorrhizal phosphate transport under monoxenic conditions using radio-labelled inorganic and organic phosphate. Biotechnol. Lett. **22**: 1705–1708.

Kaeppler, S.M., Parke, J.L., Mueller, S.M., Senior, L., Stuber, C., and Tracy, W.F. 2000. Variation among maize inbred lines and detection of quantitative trait loci for growth at low phosphorus and responsiveness to arbuscular mycorrhizal fungi. Crop Sci. **40**: 358–364.

Khaliq, A., and Sanders, F.E. 2000. Effects of vesicular–arbuscular mycorrhizal inoculation on the yield and phosphorus uptake of field-grown barley. Soil Biol. Biochem. **32**: 1691–1696.

Kjøller, R., and Rosendahl, S. 2000. Effects of fungicides on arbuscular mycorrhizal fungi: differential responses in alkaline phosphatase activity of external and internal hyphae. Biol. Fertil. Soils, **31**: 361–365.

Kling, M., and Jakobsen, I. 1997. Direct application of carbendazim and propiconazole at field rates to the external mycelium of three arbuscular mycorrhizal fungi species: effect on ^{32}P transport and succinate dehydrogenase activity. Mycorrhiza, **7**: 33–37.

Klironomos, J.N. 2003. Variation in plant response to native and exotic arbuscular mycorrhizal fungi. Ecology, **84**: 2292–2301.

Koide, R. 1985. The effect of VA mycorrhizal infection and phosphorus status on sunflower hydraulic and stomatal properties. J. Exp. Bot. **36**: 1087–1098.

Korol, M. 2004. Fertilizer and pesticide management in Canada. In Farm environmental management in Canada. Vol. 1. No. 3. Edited by R. Koroluk. Statistics Canada, Ottawa, Ont. Catalogue No. 21-02-MIE. No. 00341. pp. 1–41.

Kothari, S.K., Marschner, H., and Römheld, V. 1990. Direct and indirect effects of VA mycorrhizal fungi and rhizosphere microorganisms on acquisition of mineral nutrients by maize (Zea mays L.) in a calcareous soil. New Phytol. **116**: 637–645.

Levy, J., Bres, C., Geurts, R., Chalhoub, B., Kulikova, O., Duc, G., Journet, E.-P., Ane, J.-M., Lauber, E., Bisseling, T., Dénarié, J., Rosenberg, C., and Debellé, F. 2004. A putative Ca^{2+} and calmodulin-dependent protein kinase required for bacterial and fungal symbioses. Science (Washington, D.C.), **303**: 1361–1364.

Li, L.F., Yang, A., and Zhao, Z.W. 2005. Seasonality of arbuscular mycorrhizal symbiosis and dark septate endophytes in a grassland site in southwest China. FEMS Microbiol. Ecol. **54**: 367–373.

Li, X.-L., George, E., and Marschner, H. 1991. Phosphorus depletion and pH decrease at the root–soil and hyphae–soil interface of VA mycorrhizal white clover fertilized with ammonium. New Phytol. **119**: 397–404.

Liu, A., Hamel, C., Begna, S.H., Ma, B.L., and Smith, D.L. 2003. Soil phosphorus depletion capacity of arbuscular mycorrhizae formed by maize hybrids. Can. J. Soil Sci. **83**: 337–342.

Liu, A., Plenchette, C., and Hamel, C. 2007. Soil nutrient and water providers: How arbuscular mycorrhizal mycelia support plant performance in a resource limited world. In Mycorrhizae in crop production. Edited by C. Hamel and C. Plenchette. Haworth Press, Binghamton, N.Y. pp. 37–66.

Liu, R.J., and Wang, F.Y. 2003. Selection of appropriate host plants used in trap culture of arbuscular mycorrhizal fungi. Mycorrhiza, **13**: 123–127.

Lutgen, E.R., Muir-Clairmont, D., Graham, J., and Rillig, M.C. 2003. Seasonality of arbuscular mycorrhizal hyphae and glomalin in a western Montana grassland. Plant Soil, **257**: 71–83.

McGonigle, T.P., Evans, D.G., and Miller, M.H. 1990. Effect of degree of soil disturbance on mycorrhizal colonization and phosphorus absorption by maize in growth chamber and field experiments. New Phytol. **116**: 629–636.

Menge, J.A. 1982. Effect of soil fumigants and fungicides on vesicular–arbuscular fungi. Phytopathology, **72**: 1125–1132.

Miller, R.M., and Jastrow, J.D. 2000. Mycorrhizal fungi influence soil structure. In Arbuscular mycorrhizas: physiology and function. Edited by Y. Kapulnik and D.D. Douds, Jr. Kluwer Academic Publishers, Boston. pp. 3–18.

Miller, R.M., Reinhardt, D.R., and Jastrow, J.D. 1995. External hyphal production of vesicular–arbuscular mycorrhizal fungi in pasture and tallgrass prairie communities. Oecologia (Berl.), **103**: 17–23.

Mohammad, M.J., Hamad, S.R., and Malkawi, H.I. 2003. Population of arbuscular mycorrhizal fungi in semi-arid environment of Jordan as influenced by biotic and abiotic factors. J. Arid Environ. **53**: 409–417.

Morel, J.L., Habib, L., and Planturuex, S. 1991. Influence of maize root mucilage on soil aggregate stability. Plant Soil, **136**: 111–119.

Moscatelli, M.C., Lagomarsino, A., De Angelis, P., and Grego, S. 2005. Seasonality of soil biological properties in a poplar plantation growing under elevated atmospheric CO_2. Appl. Soil Ecol. **30**: 162–173.

Mujica, M.T., Fracchia, S., Menendez, A., Ocampo, J.A., and Godeas, A. 1998. Influence of chlorsulfuron herbicide on arbuscular mycorrhizas and plant growth of Glycine max intercropped with the weeds Brassica campestris. In Proceedings of the 2nd International Conference on Mycorrhiza, 5–10 July 1998, Uppsala, Sweden. Sveriges Lantbruksuniversitet, Uppsala, Sweden. Available from www-icom2.slu.se/ABSTRACTS/Mujica.html [accessed 22 April 2008].

Mummey, D.L., Rillig, M.C., and Holben, W.E. 2005. Neighboring plant influences on arbuscular mycorrhizal fungal community composition as assessed by T-RFLP analysis. Plant Soil, 271: 83–90.

Nelsen, C.E. 1987. The water relations of vesicular–arbuscular mycorrhizal systems. *In* Ecophysiology of VA mycorrhizal plants. *Edited by* G.R. Safir. CRC Press, Boca Raton, Fla. pp. 71–91.

Nemec, S. 1985. Influence of selected pesticides on *Glomus* species and their infection in citrus. Plant Soil, 84: 133–137.

Newman, S.E., and Davies, F.T., Jr. 1988. High root-zone temperatures, mycorrhizal fungi, water relations and root hydraulic conductivity of container-grown woody plants. J. Am. Soc. Hortic. Sci. 113: 138–146.

Nurlaeny, N., Marschner, H., and George, E. 1996. Effects of liming and mycorrhizal colonization on soil phosphate depletion and phosphate uptake by maize (*Zea mays* L.) and soybean (*Glycine max* L.) grown in two tropical acid soils. Plant Soil, 181: 275–285.

Nwoko, H., and Sanginga, N. 1999. Dependence of promiscuous soybean and herbaceous legumes on arbuscular mycorrhizal fungi and their response to bradyrhizobial inoculation in low P soils. Appl. Soil Ecol. 13: 251–258.

Oades, J.M., and Waters, A.G. 1991. Aggregate hierarchy in soils. Aust. J. Soil Res. 29: 815–828.

Oehl, F., Sieverding, E., Ineichen, K., Ris, E.A., Boller, T., and Wiemken, A. 2005. Community structure of arbuscular mycorrhizal fungi at different soil depths in extensively and intensively managed agroecosystems. New Phytol. 165: 273–283.

Olsson, P.A., Thingstrup, I., Jakobsen, I., and Bååth, E. 1999. Estimation of the biomass of arbuscular mycorrhizal fungi in a linseed field. Soil Biol. Biochem. 31: 1879–1887.

Parke, J.L., and Kaeppler, S.W. 2000. Effects of genetic differences among crop species and cultivars upon the arbuscular mycorrhizal symbiosis. *In* Arbuscular mycorrhizas: physiology and function. *Edited by* Y. Kapulnik and D.D. Douds, Jr.. Kluwer Academic Press, Boston. pp. 131–146.

Parton, W.J., Stewart, J.W.B., and Cole, C.V. 1988. Dynamics of C, N, P and S in grassland soils: a model. Biogeochemistry, 5: 109–131.

Plenchette, C., Fortin, J.A., and Furlan, V. 1983. Growth responses of several plant species to mycorrhizae in a soil of moderate P fertility. I. Mycorrhizal dependency under field conditions. Plant Soil, 70: 199–209.

Plenchette, C., Clermont-Dauphin, C., Meynard, J.M., and Fortin, J.A. 2005. Managing arbuscular mycorrhizal fungi in cropping systems. Can. J. Plant Sci. 85: 31–40.

Ramsay, A.J., Stannard, R.E., and Churchman, G.J. 1986. Effect of conversion from ryegrass pasture to wheat cropping on aggregation and bacterial populations in a silt loam soil in New Zealand. Aust. J. Soil Res. 24: 253–264.

Reid, J.B., and Goss, M.J. 1982. Interactions between soil drying due to plant water use and decrease in aggregate stability caused by maize roots. J. Soil Sci. 33: 47–53.

Rejon, A., Garcia-Romera, I., Ocampo, J.A., and Bethlenfalvay, G.J. 1997. Mycorrhizal fungi influence competition in a wheat–ryegrass association treated with the herbicide diclofop. Appl. Soil Ecol. 7: 51–57.

Rengel, Z. 2002. Breeding for better symbiosis. Plant Soil, 245: 147–162.

Reynolds, H.L., Packer, A., Bever, J.D., and Clay, K. 2003. Grassroots ecology: plant–microbe–soil interactions as drivers of plant community structure and dynamics. Ecology, 84: 2281–2291.

Reynolds, M.P., Mujeeb-Kazi, A., and Sawkins, M. 2005. Prospects for utilising plant-adaptive mechanisms to improve wheat and other crops in drought- and salinity-prone environments. Ann. Appl. Biol. 146: 239–259.

Rillig, M.C., and Mummey, D.L. 2006. Mycorrhizas and soil structure. New Phytol. 171: 41–53.

Rillig, M.C., Wright, S.F., Nichols, K.A., Schmidt, W.F., and Torn, M.S. 2001. Large contribution of arbuscular mycorrhizal fungi to soil carbon pools in tropical forest soils. Plant Soil, 233: 167–177.

Rillig, M.C., Lutgen, E.R., Ramsey, P.W., Klironomos, J.N., and Gannon, J.E. 2005. Microbiota accompanying different arbuscular mycorrhizal fungal isolates influence soil aggregation. Pedobiologia, 49: 251–259.

Rivera, R., and Fernandez, K. 2003. Bases cientifico—téchnicas para el manejo de los sistemas agricolas micorrizados eficientemente. *In* El manejo efectivo de la simbiosis micorrizica, una via hacia la agricultura sostenible. Estudio de caso: El Caribe. *Edited by* R. Rivera, F. Fernandez, A. Hernandez, J.R. Martin, and K. Fernandez. INCA, Havana, Cuba. pp. 45–98.

Rivera, R., Fernández, F., Fernández, K., Ruiz, L., Sánchez, C., and Riera, M. 2007. Advances in the management of effective arbuscular mycorrhizal symbiosis in tropical ecosystems. *In* Mycorrhizae in crop production. *Edited by* C. Hamel and C. Plenchette. Haworth Press, Binghamton, N.Y. pp. 151–196.

Ronsheim, M.L., and Anderson, S.E. 2001. Population-level specificity in the plant–mycorrhizae association alters intraspecific interactions among neighboring plants. Oecologia (Berl.), 128: 77–84.

Ruiz-Lozano, J.M. 2003. Arbuscular mycorrhizal symbiosis and alleviation of osmotic stress. New perspectives for molecular studies. Mycorrhiza, 13: 309–317.

Ruiz-Lozano, J.M., and Azcón, R. 1995. Hyphal contribution to water uptake in mycorrhizal plants as affected by the fungal species and water status. Physiol. Plant. 95: 472–478.

Ryan, M.H., and Graham, J.H. 2002. Is there a role for arbuscular mycorrhizal fungi in production agriculture? Plant Soil, 244: 263–271.

Ryan, M.H., van Herwaarden, A.F., Angus, J.F., and Kirkegaard, J.A. 2005. Reduced growth of autumn-sown wheat in a low-P soil is associated with high colonization by arbuscular mycorrhizal fungi. Plant Soil, 270: 275–286.

Saito, M. 2000. Symbiotic exchange of nutrients in arbuscular mycorrhizas: transport and transfer of phosphorus. *In* Arbuscular mycorrhizas: physiology and function. *Edited by* Y. Kapulnik and D.D. Douds, Jr. Kluwer Academic Press, Boston. pp. 85–106.

Salem, S.F., Dobolyi, C., Helyes, L., Pék, Z., and Dimény, J. 2003. Side-effect of benomyl and captan on arbuscular mycorrhiza formation in tomato plant. *In* ISHS Acta Horticulturae, 613: VIII International Symposium on the

Processing Tomato. Available from www.actahort.org/members/showpdf?booknrarnr=613_37 [accessed 22 December 2005].

Sánchez-Díaz, M., and Honrubia, M. 1994. Water relations and alleviation of drought stress in mycorrhizal plants. *In* Impact of arbuscular mycorrhizas on sustainable agriculture and natural ecosystems. Edited by S. Gianinazzi and H. Schüepp. Birkhauser Verlag AG, Basel, Switzerland. pp. 167–178.

Sanders, I.R. 2003. Preference, specificity and cheating in the arbuscular mycorrhizal symbiosis. Trends Plant Sci. **8**: 143–145.

Sanders, I.R., Streitwolf-Engel, R., van der Heijden, M.G.A., Boller, T., and Wiemken, A. 1998. Increased allocation to external hyphae of arbuscular mycorrhizal fungi under CO_2 enrichment. Oecologia (Berl.), **117**: 496–503.

Sands, R., and Theodorou, C. 1978. Water uptake by mycorrhizal roots of radiata pine seedlings. Aust. J. Plant Physiol. **5**: 301–309.

Sanginga, N., Lyasse, O., and Singh, B.B. 2000. Phosphorus use efficiency and nitrogen balance of cowpea breeding lines in a low P soil of the derived savanna zone in West Africa. Plant Soil, **220**: 119–128.

Scervino, J.M., Ponce, M.A., Erra-Bassells, R., Vierheilig, H., Ocampo, J.A., and Godeas, A. 2005. Flavonoids exhibit fungal species and genus specific effects on the presymbiotic growth of *Gigaspora* and *Glomus*. Mycol. Res. **109**: 789–794.

Schadt, C.W., Martin, A.P., Lipson, D.A., and Schmidt, S.K. 2003. Seasonal dynamics of previously unknown fungal lineages in tundra soils. Science (Washington, D.C.), **301**: 1359–1361.

Scheublin, T.R., Ridgway, K.P., Young, J.P.W., and van der Heijden, M.G.A. 2004. Nonlegumes, legumes and root nodules harbor different arbuscular mycorrhizal fungal communities. Appl. Environ. Microbiol. **70**: 6240–6246.

Schmitz, O., Danneberg, G., Hundeshagen, B., Klingner, A., and Bothe, H. 1991. Quantification of vesicular–arbuscular mycorrhiza by biochemical parameters. J. Plant Physiol. **139**: 106–114.

Schreiner, R.P., and Bethlenfalvay, G.J. 1997. Plant and soil response to single and mixed species of arbuscular mycorrhizal fungi under fungicide stress. Appl. Soil Ecol. **7**: 93–102.

Schweiger, P.F., and Jakobsen, I. 1998. Dose–response relationships between four pesticides and phosphorus uptake by hyphae of arbuscular mycorrhizas. Soil Biol. Biochem. **30**: 1415–1422.

Shibata, R., and Yano, K. 2003. Phosphorus acquisition from non-labile sources in peanut and pigeonpea with mycorrhizal interaction. Appl. Soil Ecol. **24**: 133–141.

Simpson, D., and Daft, M.J. 1991. Effects of *Glomus clarum* and water stress on growth and nitrogen fixation in two genotypes of groundnut. Agric. Ecosyst. Environ. **35**: 47–54.

Smith, S.E., and Read, D.J. 1997. Mycorrhizal symbiosis. 2nd ed. Academic Press, San Diego, Calif.

Smith, S.E., and Smith, F.A. 1990. Structure and function of the interfaces in biotrophic symbioses as they relate to nutrient transport. New Phytol. **114**: 1–38.

Smith, S.E., Smith, F.A., and Jakobsen, I. 2003. Mycorrhizal fungi can dominate phosphate supply to plants irrespective of growth responses. Plant Physiol. **133**: 16–20.

Spokes, J.R., Macdonald, R.M., and Hayman, S. 1981. Effects of plant protection chemicals on vesicular–arbuscular mycorrhizas. Pestic. Sci. **12**: 346–350.

Staddon, P.L., Ramsey, C.B., Ostle, N., Ineson, P., and Fitter, A.H. 2003. Rapid turnover of hyphae of mycorrhizal fungi determined by AMS microanalysis of ^{14}C. Science (Washington, D.C.), **300**: 1138–1140.

Stewart, L.I., Hamel, C., Hogue, R., and Moutoglis, P. 2005. Response of strawberry to inoculation with arbuscular mycorrhizal fungi under very high soil phosphorus conditions. Mycorrhiza, **15**: 612–619.

Subramanian, K.S., and Charest, C. 1998. Arbuscular mycorrhizae and nitrogen assimilation in maize after drought and recovery. Physiol. Plant. **102**: 285–296.

Subramanian, K.S., and Charest, C. 1999. Acquisition of N by external hyphae of an arbuscular mycorrhizal fungus and its impact on physiological responses in maize under drought-stressed and well-watered conditions. Mycorrhiza, **9**: 69–75.

Sukarno, N., Smith, F.A., Scott, E.S., Jones, G.P., and Smith, S.E. 1998. The effect of fungicides on the vesicular–arbuscular mycorrhizal symbiosis. III. The influence of VA mycorrhiza on phytotoxic effects following application of fosetyl-Al and phosphonate. New Phytol. **139**: 321–330.

Sweatt, M.R., and Davies, F.T., Jr. 1984. Mycorrhizas, water relations, growth and nutrient uptake of geranium grown under moderately high phosphorus regimes. J. Am. Soc. Hortic. Sci. **109**: 210–213.

Sylvia, D.M., Alagely, A.K., Kane, M.E., and Philman, N.L. 2003. Compatible host/mycorrhizal fungus combinations for micropropagated sea oats. I. Field sampling and greenhouse evaluations. Mycorrhiza, **13**: 177–183.

Takacs, T., and Voros, I. 2003. Role of arbuscular mycorrhizal fungi in the water and nutrient supplies of the host plant. Novenytermeles, **52**: 583–593.

Tannas, K. 2003. Common plants of the western rangelands. Vol. 1. Grasses and grass like species. Alberta Agriculture Food and Rural Development, Edmonton, Alta.

Tarafdar, J.C., and Marschner, H. 1994. Phosphatase activity in the rhizosphere and hyphosphere of VA mycorrhizal wheat supplied with inorganic and organic phosphorus. Soil Biol. Biochem. **26**: 387–395.

Tisdall, J.M., and Oades, J.M. 1979. Stabilization of soil aggregates by the root systems of ryegrass. Aust. J. Soil Res. **17**: 429–441.

Tisdall, J.M., Smith, S.E., and Rengasamy, P. 1997. Aggregation of soil by fungal hyphae. Aust. J. Soil Res. **35**: 55–60.

Tobar, R., Azcón, R., and Barea, J.M. 1994. Improved nitrogen uptake and transport from ^{15}N-labelled nitrate by external hyphae of arbuscular mycorrhiza under water-stressed conditions. New Phytol. **126**: 119–122.

Trappe, J.M., Molina, R., and Castellano, M. 1984. Reactions of mycorrhizal fungi and mycorrhiza formation to pesticides. Annu. Rev. Phytopathol. **22**: 331–359.

Vandenkoornhuyse, P., Husband, R., Daniell, T.J., Watson, I.J., Duck, .M., Fitter, A.H., and Young, J.P.W. 2002. Arbuscular mycorrhizal community composition associated with two plant species in a grassland ecosystem. Mol. Ecol. **11**: 1555–1564.

Vandenkoornhuyse, P., Ridgway, K.P., Watson, I.J., Fitter, A.H., and Young, J.P.W. 2003. Co-existing grass species have distinctive arbuscular mycorrhizal communities. Mol. Ecol. **12**: 3085–3095.

van der Heijden, M.G.A., Klironomos, J.N., Ursic, M., Moutoglis, P., Streitwolf-Engel, R., Boller, T., Wiemken, A., and Sanders, I.R. 1998. Mycorrhizal fungal diversity determines plant biodiversity, ecosystem variability and productivity. Nature (Lond.), **396**: 69–72.

Verkade, S.D., and Hamilton, D.F. 1983. Effects of benomyl on growth of *Liriodendron tulipifera* L. seedlings inoculated with the vesicular–arbuscular fungus, *Glomus fasciculatus*. Sci. Hortic. (Amsterdam), **21**: 253–260.

Viets, F.G.J. 1972. Water deficits and nutrient availability. *In* Water deficits and plant growth, Vol. II. *Edited by* T.T. Kozlowski. Academic Press, New York. pp. 217–239.

von Alten, H., Lindermann, A., and Schonbeck, F. 1993. Stimulation of vesicular–arbuscular mycorrhiza by fungicides or rhizosphere bacteria. Mycorrhiza, **2**: 167–173.

Wan, M.T., and Rahe, J.E. 1998. Impact of azadirachtin on *Glomus intraradices* and vesicular–arbuscular mycorrhiza in root inducing transferred DNA transformed roots of *Daucus carota*. Environ. Toxicol. Chem. **17**: 2041–2050.

Wilarso Budi, S., Caussanel, J.-P., Trouvelot, A., and Gianinazzi, S. 1998. The biotechnology of mycorrhizas. *In* Microbial interactions in agriculture and forestry. Vol. 1. *Edited by* N. Subba Rao and Y.R. Dommergues. Science Publishers, Inc. Enfield, N.H. pp. 149–162.

Wright, S.F., and Upadhyaya, A. 1996. Extraction of an abundant and unusual protein from soil and comparison with hyphal protein of arbuscular mycorrhizal fungi. Soil Sci. **161**: 575–586.

Wright, S.F., and Upadhyaya, A. 1998. A survey of soils for aggregate stability and glomalin, a glycoprotein produced by hyphae of arbuscular mycorrhizal fungi. Plant Soil, **198**: 97–107.

Wright, S.F., Franke-Snyder, M., Morton, J.B., and Upadhyaya, A. 1996. Time-course study and partial characterization of a protein on hyphae of arbuscular mycorrhizal fungi during active colonization of roots. Plant Soil, **181**: 193–203.

Xavier, L.J.C., and Germida, J.J. 1998. Response of spring wheat cultivars to *Glomus clarum* NT4 in a P-deficient soil containing arbuscular mycorrhizal fungi. Can. J. Soil Sci. **78**: 481–484.

Youpensuk, S., Rerkasem, B., Dell, B., and Lumyong, S. 2005. Effects of arbuscular mycorrhizal fungi on a fallow enriching tree (*Macaranga denticulata*). Fungal Divers. **18**: 189–199.

Zentner, R.P., Lafond, G.P., Derksen, D.A., and Campbell, C.A. 2002. Tillage method and crop diversification: effect on economic returns and riskiness of cropping systems in a Thin Black Chernozem of the Canadian Prairies. Soil Tillage Res. **67**: 9–21.

Zhu, Y.-G., Smith, S.E., Barritt, A.R., and Smith, F.A. 2001. Phosphorus (P) efficiencies and mycorrhizal responsiveness of old and modern wheat cultivars. Plant Soil, **237**: 249–255.

Chapter 6
The relative field mycorrhizal dependency concept and its usefulness in agronomy

Christian Plenchette and J. André Fortin

Introduction

Arbuscular mycorrhizal (AM) fungi belong to the order Glomeromycota (Schüßler et al. 2001). They are the most universal and fundamental of the mycorrhizal fungi, forming symbiotic associations with over 200 000 vascular plant species, including the majority of cultivated crops. Paradoxically, <200 species of AM fungi are recognized. This mycorrhizal symbiosis is the most ancient, and its origin dates back at least 450 million years (Simon et al. 1993). This timeframe also coincides with the appearance and early diversification of land plants. Four-hundred-million-year-old fossils of the plant *Rhynia* show characteristic arbuscules in the cortical cells of their rhizomes, lending support to the theory that colonization of dry land by plants was made possible through a symbiotic association with soilborne AM fungi (Malloch et al. 1980; Pirozynski 1981). Such an association would have allowed plants to acquire sufficient nutrients and water to survive in the much drier environment encountered. Furthermore, it seems reasonable to assume that the AM association has played an essential role in the evolution and diversification of land plants ever since.

The ability to form AM associations is considered to be an ancestral trait (Selosse and Le Tacon 1997). This implies that mycorrhizal dependency (MD) has also existed since plants first appeared on dry land. Mycorrhizal dependency is genetically controlled (Janos 1993); if different plant species are grown under low-nutrient conditions, some will barely grow if they are not colonized by AM fungi, whereas others will only show a slight growth reduction. By contrast, during evolutionary selection, certain plant families, such as the Brassicaceae and Chenopodiaceae, have emerged that are typical of disturbed, nutrient-rich sites and that lack the capacity to form mycorrhizal associations. These are referred to as non-mycorrhizal plants, and thus, they have no MD. Similarly, crop plant selection can also result in the development of strains that have lower MD levels (e.g., Azcón and Ocampo 1981) and higher available phosphorus (P) requirements.

During the co-evolution of plants and mycorrhizal fungi, natural selection has led to the development of a wide diversity of plants from the early land plants with their simple rhizomatous 'root' systems. These more recent species exhibit a wide variety of root systems that range from coarsely to finely branched. Furthermore, these may or may not possess root hairs (Baylis 1975). These morphological alterations to the root systems are closely linked with MD (Baylis 1975). However, the MD of a given plant also varies according to the plant–fungal interactions involved (Plenchette et al. 1982) and the availability of P in the soil (Menge et al. 1978; Habte and Manjunath 1987).

If coarsely branched or magnoloid-type roots lacking root hairs are compared with the finely branched or graminoid-type roots, which possess numerous root hairs, it is easy to understand that the latter are better equipped to explore soil and take up low-labile ions Thus, for a given soil, the MD threshold of a host plant such as wheat (*Triticum aestivum* L.), with a fine, highly branched root system with abundant root hairs, occurs at a much lower level of available P than for a plant such as carrot (*Daucus*

C. Plenchette.[1] INRA, UMR BGA, 17, rue Sully, 21065 Dijon, CEDEX, France.
J.A. Fortin. Centre d'étude de la forêt, Pavillon C.-E.-Marchand, Université Laval, Québec, QC G1V 0A6, Canada.
[1]Corresponding author (e-mail: plenchet@dijon.inra.fr).

carota L.), which has a coarse root system (see also Granger et al. 1983). The latter has a greater dependence on AM fungi for P uptake and, therefore, has a high MD.

Intensive farming based on selected strains of crop species and high P application levels contrasts strongly with sustainable agriculture practices involving low fertilizer application. The latter system has the potential to allow the development of crop plants that are less nutrient demanding and that have a much higher MD than crops with high yield but low MD.

In this chapter, we revisit the concept of MD. We expand on previous publications, and in particular, we consider the relative field mycorrhizal dependency (RFMD) under different levels of plant available P (RFMD^P). Furthermore, we underline the importance of using RFMD^P for the development of more sustainable agricultural practices.

Effect of soil phosphorus content on mycorrhizal dependency

The effect of AM fungi on plant P nutrition is more important in nutrient-poor soils (Mosse 1973*a*) and in lightly fertilized or unfertilized cultivated soils. The bioavailability of P in a soil is influenced by the quantity of labile P (Hayman and Mosse 1971; Mosse 1973*b*; Owusu-Bennoah and Mosse 1979; Bloss and Pfeiffer 1981; Clarke and Mosse 1981) and (or) the soil's P retention (fixation) capacity (Plenchette et al. 1981; Plenchette and Fardeau 1988).

Plants only absorb ions in solution; although soils naturally contain P in a number of forms, most of these are strongly adsorbed on the surface of soil particles. The concentration of P within the soil solution, and so available for plant uptake, varies from <0.1 μg/g in poorly fertile soils to >1 μg/g in heavily fertilized agricultural soils. Generally, the concentration of available P in a soil is negatively correlated with the level of root colonization by AM fungi (e.g., Habte and Manjunath 1987; but see Ortas and Akpinar 2006). In nonmycorrhizal plants, P uptake essentially occurs via root hairs from a zone extending 1–2 mm out from the root. However, in mycorrhizal plants, P uptake occurs principally via the external mycelium (Smith et al. 2003), and the zone exploited may extend several centimetres from the root surface. When fertilizers containing P are applied, the soil P reserve and, more importantly, the quantity of P in the soil solution, increase. Under certain conditions, the P concentration in the rhizosphere becomes high enough for a plant to bypass the mycorrhizal mycelium and obtain its P directly from the soil solution that bathes the root. As a result, the MD of the plant is reduced.

To obtain a maximum yield, a crop plant requires a certain level of P in the soil solution. This is referred to as the external phosphorus requirement and is plant specific (Fox 1981). The amount of P fertilizer necessary to reach the external phosphorus requirement is soil specific and is referred to as the phosphorus fertilizer requirement. The situation is complicated by the fact that, to supply the same amount of P to a given plant, the concentration of the soil solution needs to be higher in a sandy soil than a clay soil. This is due to two factors influencing P bioavailability: the P diffusion rate and the P retention (fixation) capacity of the soil. The P diffusion rate is lower in sandy soils than clay soils (Olsen and Watanabe 1963), and the soil P fixation capacity of a sandy soil is also lower than that of a clay soil. Thus, although the MD of a plant decreases when the soil solution P concentration increases, the relationship between these two factors is affected by the fact that mycorrhizal fungi decrease the diffusion distances of P ions to roots and also increase P desorption from the solid phase of the soil (Plenchette and Fardeau 1988).

It must be borne in mind that it is not necessary for a mycorrhizal plant to be able to absorb P at lower soil solution P levels than a nonmycorrhizal plant. This is due to the fact that the P depletion zone observed around the root and root hairs (Barber et al. 1963) does not occur around hyphae. The P depletion zone around a root occurs when P uptake is faster than the rate of diffusion of P to the root. In this common situation, a P concentration gradient is established in the depletion zone, and P ions diffuse along this gradient. Because of the smaller radius of hyphae (0.005 mm) than of root hairs (0.025 mm) and roots (0.15 mm), no gradient is created because the distance of diffusion is shorter, and therefore, the P concentration of the soil solution is higher at the hyphal surface than at the root surface (Barber 1984).

Plants can be placed in one of many different MD categories, ranging from highly to marginally dependent with regards to the soil solution P concentration (Habte and Manjunath 1991). According to Janos (1993), "Dependence can be experimentally measured along an axis of soil fertility, especially one of phosphorus availability. Either the threshold level of soil fertility below which an individual plant without mycorrhizas fails to grow, or the level of soil fertility above which an asymptotic, maximum growth rate is reached measures dependence." To take into account these different aspects, Plenchette et al. (1983) proposed the concept of RFMD.

Relative field mycorrhizal dependency and its agricultural application

Although several approaches have been used to express MD (Gerdemann 1975; Menge et al. 1978), Plenchette et al. (1983) considered the RFMD to be the most inclusive and, in practical terms, the most useful:

$$RFMD = \frac{\text{dry wt. of mycorrhizal plant} - \text{dry wt. of non-mycorrhizal}}{\text{dry wt. of mycorrhizal plant}} \times 100$$

If the soil P concentration at a given site is taken into account, it is possible to class species (Table 6.1) and cultivars on a relative scale of dependency. For this reason, we propose that the acronym RFMD should be replaced by $RFMD^P$, where the quantity of available P in the soil is expressed in micrograms per gram soil. Because several different methods for determining P bioavailability in soil exist and because these give slightly different results, it is important that a reference be made to the method used in each study.

The $RFMD^P$ allows plants to be ranked from 0% (for nonmycorrhizal species or plants growing in the presence of a soil solution that meets their external P requirements) to 100% dependent (for plants completely dependent on the mycorrhizal symbiosis for growth). Although Gerdemann (1975) defined the MD of plants as "the degree to which a plant species is dependent on the mycorrhizal condition to produce its maximum growth at a given soil fertility," he proposed an index that was calculated using the percentage of the ratio of dry masses of mycorrhizal and nonmycorrhizal plants. This index, which was also used by Menge et al. (1978) and has been considered to be a measure of plant receptiveness (Janos 1993) to mycorrhizal fungi, has no limits and very high values. This reduces its usefulness, especially when making comparisons among different plant species and growing conditions. The $RFMD^P$ index is based on the concept that a given plant may be completely dependent on the mycorrhizal symbiosis or not at all (e.g., for members the Chenopodiaceae).

Furthermore, the $RFMD^P$ index can be applied to other situations such as greenhouse experiment, where it becomes the RMD^P index.

The study by Plenchette et al. (1983) was conducted under field conditions. The authors proposed to qualify MD as RFMD because the results obtained were relative to edaphic conditions, particularly the level of bioavailable P, which was 100 ppm P_2O_5 (determined using the Bray II method; Bray and Kurtz 1945). In this experiment, carrot, which develops a taproot, exhibited the highest RFMD. Leek (*Allium ampeloprasum* var. *porrum* (L.) J.Gay), which has a coarse root systems, and legumes also exhibited high MD values, whereas wheat exhibited no MD (Table 6.1). The lack of a MD response for the latter species may be explained by the level of plant available P, which was considered sufficient for cereal growth according to published norms at the time (Conseil des Productions Végétales du Québec 1980).

Use of relative field mycorrhizal dependency in cropping systems

Cropping systems, which consist of the plant species and the agricultural practices used, vary greatly between different countries, climates, and soil types. Coupled with this, MD varies greatly among plant species (Saif 1987) and with soil fertility. Together, these two concepts create an almost infinite number of different situations under which MD may be considered. The concept of $RFMD^P$ can be used under all of these and is fundamental to the understanding of the role of AM fungi in cropping systems. Furthermore, it offers a means of demonstrating the benefits of AM fungi in a particular situation or of highlighting favourable or unfavourable agricultural practices.

Benefits of mycorrhiza are undeniable, but they are still underestimated in agriculture. It has been shown that cropping systems that are based on high fertilizer application rates can be detrimental to AM fungi (Plenchette et al. 2005). Furthermore, cropping systems include a range of nonmycotrophic (e.g., oil seed rape, *Brassica napus* L.; sugar beet, *Beta vulgaris* L.; and mustard, *Brassica juncea* (L.) Czern.)) and low (e.g., wheat and barley, *Hordeum vulgare* L.), moderately (e.g., maize, *Zea mays* L.; and sunflower, *Helianthus annuus* L.), and highly mycotrophic crops (e.g., potato, *Solanum tuberosum* L.; and pea, *Pisum sativum* L.), which directly influence the mycorrhizal inoculation potential of a

Table 6.1. Influence of plant species on mycorrhizal dependency under field conditions (adapted from Plenchette et al. 1983).

Species	Dry mass (g) M[†]	Dry mass (g) NM[‡]	RFMD%*
Carrot (*Daucus carota* L.)	9.2	0.07	99.2
Pea (*Pisum sativum* L.)	40.3	1.3	96.7
Leek (*Allium ampeloprasum* var. *porrum* (L.) J.Gay)	11.9	0.5	95.7
Green bean (*Phaseolus vulgaris* L.)	13.3	0.7	94.7
Broad bean (*Vicia faba* L.)	21.8	1.4	93.5
Sweet corn (*Zea mays* L.)	166.5	45.5	72.7
Green pepper (*Capsicum annuum* L.)	12.1	4.1	66.1
Tomato (*Solanum lycopersicum* L.)	174.6	71.2	59.2
Potato (*Solanum tuberosum* L.)	185.3	107.5	41.9
Wheat (*Triticum aestivum* L.)	155.5	155.6	0

*Relative field mycorrhizal dependency.
[†]M, arbuscular mycorrhizal plants.
[‡]NM, nonmycorrhizal plants obtained by soil fumigation.

given soil for subsequent crops. Knowledge of this, coupled with the concept of RFMD[P], provides the basis for a better use of AM fungi in agriculture and the conception of innovative cropping systems.

Conclusion

Without the development of mycorrhizal associations, the colonization of dry land by plants would have probably been impossible. However, some recently evolved plant species have lost the need to form mycorrhizal associations, and a number of artificially selected cultivars of crop plants show a reduced MD compared with wild types. Cropping systems that artificially increase mineral nutrient availability also reduce MD. Because MD is affected by phosphorus availability, which in turn is modulated by soil physical properties, we propose that the RFMD[P] index is the only fully integrative expression of mycorrhizal responsiveness. This index could allow better use of AM fungi in cropping systems and may explain the structure of natural plant communities in which biodiversity reflects the MD of various dominant and subordinate plant species (Urcelay and Diaz 2003).

References

Azcón, R., and Ocampo, J.A. 1981. Factors affecting the vesicular–arbuscular infection and mycorrhizal dependency of thirteen wheat cultivars. New Phytol. **87**: 677–685.

Barber, S.A. 1984. Soil nutrient bioavailability: a mechanistic approach. John Wiley & Sons, Inc., New York.

Barber, S.A., Walker, J.M., and Vasey, E.H. 1963. Mechanisms for the movement of plant nutrients from soil and fertilizer to the plant root. J. Agric. Food Chem. **11**: 204–207.

Baylis, G.T.S. 1975. The magnolioid mycorrhiza and mycotrophy in root systems derived from it. *In* Endomycorrhizas. *Edited by* F.E. Sanders, B. Mosse, and P.B. Tinker. Academic Press, London. pp. 373–389.

Bloss, H.E., and Pfeiffer, C.M. 1981. Growth and nutrition of mycorrhizal guayule plants. Ann. Appl. Biol. **99**: 267–274.

Bray, R.H., and Kurtz, L.T. 1945. Determination of total, organic, and available forms of phosphorus in soil. Soil Sci. **59**: 39–45.

Clarke, C., and Mosse, B. 1981. Plant growth responses to vesicular–arbuscular mycorrhiza. XII. Field inoculation responses of barley at two soil P levels. New Phytol. **87**: 695–703.

Fox, R.L. 1981. External phosphorus requirement of crops. *In* Chemistry in the soil environment. American Society of Agronomy, Madison, Wisc. pp. 223–239.

Gerdemann, J.W. 1975. Vesicular–arbuscular mycorrhizae. *In* The development and function of roots. *Edited by* J.G. Torrey and D.T. Clarkson. Academic Press, London. pp. 575–591.

Granger, R.L., Plenchette, C., and Fortin, J.A. 1983. Effect of vesicular arbuscular (VA) mycorrhizal fungus (*Glomus epigaeus*) on the growth and leaf mineral content of two apple clones propagated *in vitro*. Can. J. Plant Sci. **63**: 551–555.

Habte, M., and Manjunath, A. 1987. Soil solution phosphorus status and mycorrhizal dependency in *Leucaena leucocephala*. Appl. Environ. Microbiol. **53**: 797–801.

Habte, M., and Manjunath, A. 1991. Categories of vesicular–arbuscular mycorrhizal dependency of host species. Mycorrhiza, **1**: 3–12.

Hayman, D.S., and Mosse, B. 1971. Plant growth responses to vesicular–arbuscular mycorrhiza. I. Growth of *Endogone* inoculated plants in phosphate deficient soils. New Phytol. **70**: 19–27.

Janos, D.P. 1993. Vesicular–arbuscular mycorrhizae of epiphytes. Mycorrhiza, **4**: 1–4.

Malloch, D.W., Pirozynski, K.A., and Raven, P.H. 1980. Ecological and evolutionary significance of mycorrhizal symbioses in vascular plants (a review). Proc. Natl. Acad. Sci. U.S.A. **77**: 2113–2118.

Menge, J.A., Johnson, E.L.V., and Platt, R.G. 1978. Mycorrhizal dependency of several citrus cultivars under three nutrient regimes. New Phytol. **81**: 553–559.

Mosse, B. 1973*a*. Advances in the study of vesicular–arbuscular mycorrhiza. Annu. Rev. Phytopathol. **11**: 171–196.

Mosse, B. 1973*b*. Plant growth responses to vesicular–arbuscular mycorrhiza. IV. In soil given additional phosphate. New Phytol. **72**: 127–136.

Olsen, S.R., and Watanabe, F.S. 1963. Diffusion of phosphorus as related to soil texture and plant uptake. Soil Sci. Soc. Am. Proc. **27**: 648–653.

Ortas, I., and Akpinar, C. 2006. Response of kidney bean to arbuscular mycorrhizal inoculation and mycorrhizal dependency in P and Zn deficient soils. Acta Agric. Scand. Sect. B Soil Plant Sci. **56**: 101–109.

Owusu-Bennoah, E., and Mosse, B. 1979. Plant growth responses to vesicular–arbuscular mycorrhiza. XI. Field inoculation responses in barley, lucerne and onion. New Phytol. **83**: 671–679.

Pirozynski, K.A. 1981. Interactions between fungi and plants through the ages. Can. J. Bot. **59**: 1824–1827.

Plenchette, C., and Fardeau, J.C. 1988. Effet du pouvoir fixateur du sol sur le prélèvement de phosphore par les racines et les mycorhizes. C. R. Acad. Sci. Paris, **306**: 201–206.

Plenchette, C., Furlan, V., and Fortin, J.A. 1981. Growth stimulation in apple trees in unsterilized soil under field conditions by endomycorrhizal inoculation. Can. J. Bot. **59**: 2003–2008.

Plenchette, C., Furlan, V., and Fortin, J.A. 1982. Comparative effects of different endomycorrhizal fungi on five host plants grown on calcined montmorillonite clay. J. Am. Soc. Hortic. Sci. **107**: 535–538.

Plenchette, C., Fortin, J.A., and Furlan, V. 1983. Growth responses of several plant species to mycorrhizae in a soil of moderate P-fertility. I. Mycorrhizal dependency under field conditions. Plant Soil, **70**: 199–209.

Plenchette, C., Clermont-Dauphin, C., Meynard, J.M., and Fortin, J.A. 2005. Management of AM fungi in cropping systems. Can. J. Plant Sci. **85**: 31–40.

Saif, S.R. 1987. Growth responses of tropical forage plant species to vesicular–arbuscular mycorrhizae. I. Growth, mineral uptake and mycorrhizal dependency. Plant Soil, **97**: 25–35.

Schüßler, A., Schwarzott, D., and Walker, C. 2001. A new fungal phylum, the Glomeromycota: phylogeny and evolution. Mycol. Res. **105**: 1413–1421.

Selosse, M.-A., and Le Tacon, F. 1997. Des mycorhizes à l'origine de la flore terrestre. J. Bot.. Soc. Bot. Fr. **3**: 21–25.

Simon, L., Bousquet, J., Levesque, R.C., and Lalonde, M. 1993. Origin and diversification of endomycorrhizal fungi and coincidence with vascular land plants. Nature (Lond.) , **363**: 67–69.

Smith, S.E., Smith, F.A., and Jakobsen, I. 2003. Mycorrhizal fungi can dominate phosphate supply to plants irrespective of growth responses. Plant Physiol. **133**: 16–20.

Urcelay, C., and Diaz, S. 2003. The mycorrhizal dependence of subordinates determines the effect of arbuscular mycorrhizal fungi on plant diversity. Ecol. Lett. **6**: 388–391.

Chapter 7

Extraction, propagation, and conservation of arbuscular mycorrhizal fungi

Yolande Dalpé

Introduction

The obligate symbiotic nature of arbuscular mycorrhizal (AM) fungi has hampered their use in agriculture and agroforestry, and the commercialization of inocula (see Chapter 8). To overcome this biological constraint, new methodologies have been developed that allow investigation of the different aspects of the life cycle of these fungi. Scientists working on AM fungal identification, biodiversity, ecology, and the maintenance of culture collections have also had to overcome problems linked with the fact that these organisms inhabit the soil matrix, are microscopic, have a limited range of morphological features for identification purposes, and have a coenocytic mycelium. Because of the inability of AM fungi to complete their life cycle without colonizing a host plant, the new technologies that have been elaborated have had to take into account the host, and the influences of the substrate and diverse environmental parameters.

Arbuscular mycorrhizal fungi have a complex mode of growth and development (Gianinazzi-Pearson et al. 1995) that includes survival mechanisms, which allow persistence in the absence of a host plant root. During the evolution of these fungi, special metabolisms have been selected for, allowing them to colonize a wide variety of hosts, to grow under a wide range of environmental conditions, and to function synergistically with other AM fungal species with similar requirements and soil niches. In spite of this, AM fungi exhibit a relatively simple life cycle, especially when compared with obligate pathogens (Dalpé et al. 2005). The colonies grow simultaneously within host plant roots and the soil matrix. The extraradical mycelium produces unicellular, usually thick-walled, spores and the intraradical phase differentiates intracellular arbuscules and, in certain fungal families, inter- or intra-cellular vesicles. The intra- and extraradical phases are complementary, supporting plant growth and plant protection through nutrient and metabolite exchange (Bütehorn et al. 1999).

Elaboration of the methods used to isolate AM fungal propagules (e.g., spores, colonized roots, intra- and extra-radical hyphae, and root-borne vesicles) from soil, to establish cultures, to propagate, and finally, to store the fungal material have required a thorough understanding of the AM fungal life cycle and the growth pattern of the different AM fungal propagules. Over the last two decades, a number of technical manuals (Schenck 1982; Norris et al. 1992; Brundrett et al. 1994, 1996), book chapters, and review articles (Clapp et al. 1996; Verma and Adholeya 1996; Sahay et al. 1998; Singh 2003; Dalpé and Hamel 2007) outlining advances in the field of AM fungal collection, propagation and storage have been published. The present chapter aims to provide a review of the available information and to give an overview of the different steps required for the isolation, the propagation, and the conservation of AM fungi. Furthermore, it focuses on the advantages and disadvantages of some of the most frequently used techniques and provides an extensive and useful list of references to help scientists, students, and industry select appropriate methodologies for use in their different fields of activity.

Extraction of arbuscular mycorrhizal propagules

Arbuscular mycorrhizal fungi generally exhibit low host specificity. However, different species or dif-

---

Y. Dalpé. Eastern Cereal and Oilseed Research Centre, Agriculture and Agri-Food Canada, 960 Carling Avenue, Ottawa, ON K1A 0C6, Canada (e-mail: yolande.dalpe@agr.gc.ca).

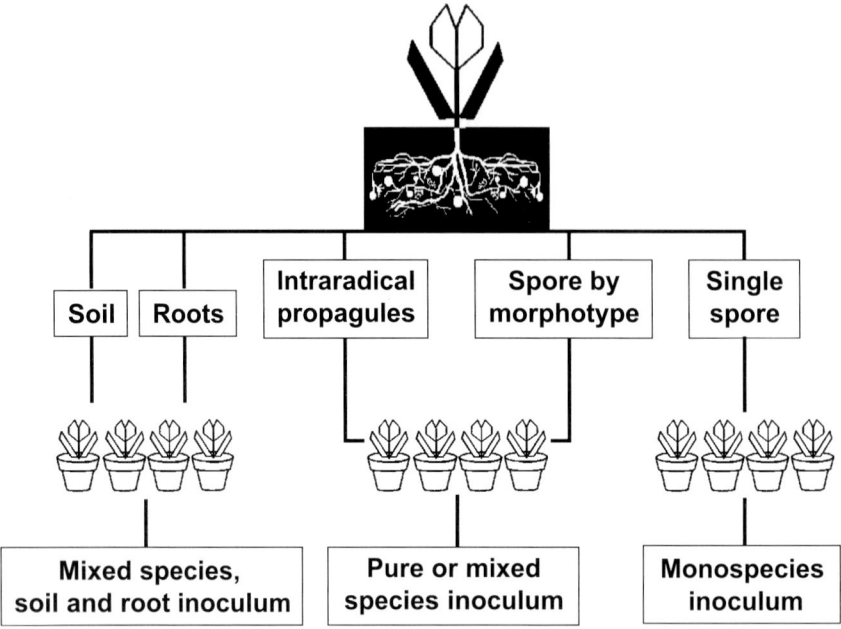

Fig. 7.1. The use of different of arbuscular mycorrhizal fungal propagules for the propagation of arbuscular mycorrhizal fungi in vivo and the respective inocula that can be produced.

ferent strains of the same species may have different capacities to colonize roots and different functional attributes. Furthermore, several AM fungal species and strains may share the same plant rhizosphere. Given the potential diversity of AM fungi within a particular soil, the choices of the extraction method and isolation technique are important and depend on the study's objectives (Fig. 7.1, Table 7.1). For example, the evaluation of a soil's inoculum potential does not require the identification of AM fungal species and (or) strains. It can be assessed directly by the capacity of the propagules it contains to colonize a host plant or, indirectly, by quantitative molecular analyses. By contrast, a study investigating AM fungal biodiversity requires the trapping and (or) isolation of the whole AM fungal population, including sporulating and nonsporulating species and aggressive and less aggressive species.

Isolation of intraradical propagules

The isolation of intraradical propagules of AM fungi (i.e., vesicles and hyphae) can be done by mechanical and enzymatic digestion of root tissues (Jabaji-Hare et al. 1984; Strullu et al. 1991). Although time consuming, this single-generation trap culture procedure allows the isolation of pure strains of those species most likely to colonize a specific plant (Table 7.1). Similar results can be obtained by first using a dissecting microscope fitted with an episcopic lighting system to detect intraradical AM fungal vesicles in unstained root fragments (Arias et al. 1987). These segments of root can then be surface sterilized and plated on a suitable medium.

Because of technical restrictions, very few attempts have been made to extract extraradical mycelium from soil. However, the sucrose-gradient centrifugation technique (Schubert et al. 1987), the use of a glass bead substrate (Chen et al. 2001), and the mechanical stirring of soil samples using ultrathin wires (Vilarino et al. 1993) allow separation of extraradical mycelium from the substrate and are the most common techniques used to achieve this. Furthermore, the latter approach may result in the recovery of sufficient material for viability tests.

In situ trapping of arbuscular mycorrhizal fungi

Arbuscular mycorrhizal fungi may also be isolated from soil using in situ trapping, whereby host plant seedlings are outplanted directly on the study site. Once colonized, their root systems serve as AM fungal propagules (Muthukumar and Udaiyan 2002). The AM fungi trapped using this method probably represent only a fraction of the species present; however, they are probably the most aggressive and those best adapted to the environment conditions and the bait plant used.

Table 7.1. Comparison of different extraction methods for arbuscular mycorrhizal fungal propagule extraction.

	AMF propagule extraction procedure			
	Intraradical propagules	Wet-sieving decanting	Flotation bubbling	Wet sieving + density gradient
Efficiency	Intemediate	Intemediate	Low	High
Purity of inoculum	High	Low	Low	Low
Labour cost	High	Low	Low	Low
Equipment required	Medium	Low	Low	Intemediate
Expertise required	High	Low	Low	Intemediate

Extraction of arbuscular mycorrhizal fungal spores

Various techniques have been developed for extracting AM fungal spores from soil (Table 7.1): wet-sieving and decanting, air-streamed fractionation of dry soils, flotation-bubbling and flotation-adhesion systems, sucrose-gradient centrifugation, and water–sucrose centrifugation (see Ianson and Allen (1986) and Kumar and Bajwa (2004) for an evaluation). A number of modifications to these methods have been proposed, including the addition of various sedimentation agents, such as gelatin, glycerol, renographin, and sucrose (Mosse and Jones 1968; Furlan et al. 1980; Vilarino and Arines 1990; Horn et al. 1992; Boddington et al. 1999; Johnson et al. 1999). However, soil texture remains the principal factor that determines the best method to use (Ianson and Allen 1986). For example, water–sucrose centrifugation should not be used when extracting spores from heavy clay soils, and simple decanting with or without wet-sieving can be used when extracting spores from sandy soil.

The most commonly used and easiest method is probably the soil wet-sieving and decanting technique (Gerdemann and Nicolson 1963). This is often coupled with density-gradient centrifugation (Khan 1999). The spores are recovered manually under a dissecting microscope by micropipetting them from the supernatant or by individually collecting them with fine tweezers from the surface of a filter paper disc following vacuum filtration of the supernatant (Dalpé and Hamel 2007). Although time consuming, this technique allows species characterization and identification, and the collection of the material required for developing mixed or pure species inocula (Brundrett et al. 1999). The use of a single spore to start a culture is highly time consuming but remains the best method to obtain pure cultures (Brundrett and Juniper 1995; Fracchia et al. 2001).

Several AM fungal strains can occur in the same rhizosphere. However, these are unlikely to sporulate at the same time; therefore, single-step spore isolation from soil will not reflect the AM fungal diversity present. Moreover, in some cases, many of the spores within a soil sample are dead or moribund and are not suitable for propagation work. For this reason, host (bait) plants are frequently grown in the collected soil prior to assessing its AM fungal population. This technique results in the multiplication of many of the fungi present and provides younger and healthier fungal propagules.

Propagation of arbuscular mycorrhizal fungi

The obligate symbiotic status of AM fungi has entailed the use of a variety of propagation methods that were originally developed for plant production. These techniques have been adapted to allow the multiplication of AM fungi and the production of inoculum (Jarstfer et al. 1992; Wood and Cummings 1992; Verma and Adholeya 1996; Sahay et al. 1998; Harikumar and Potty 2002; Singh 2002; Dalpé and Monreal 2004; Gianinazzi and Vosátka 2004) (Table 7.2).

Pot-culture inoculum

The most widely used pot-culture technique consists of growing bait plants in field-collected soil. It allows the sporulation of AM fungal strains contained in the collected soil sample and multiplies available AM fungal propagules (i.e., spores and colonized roots). The collected soil is frequently diluted with a variety of inert substrates (Guzman-Plazola et al. 1990; Feldmann and Idczak 1994; Lilly and Santhanakrishnan 1998), sterilized sand (Bragaloni et al. 1998; Gaur and Adholeya 2002), vermiculite, perlite, Turface, or Terragreen (Baltruschat 1987; Aboul-Nasr 1997; Dalpé and Hamel 2007). This propagation method usually generates

Table 7.2. Comparison of different pot and field propagation systems for arbuscular mycorrhizal fungi.

	Pot culture		Field culture
	Mutiple spores	Single spores	
Purity of culture	Low	High	Low
Controlled conditions	Low	Low	Intermediate
Equipment required	High	High	Low
Labour cost	Intermediate	High	Low
Expertise required	Low	Intermediate	Low
Commercial use	Low	Intermediate	Low

Note: Pot cultures can be maintained in either greenhouses or growth chambers.

multiple-species inocula of unknown composition. If pure cultures are required, a host plant may be potted together with an isolated spore (monospore culture), or several spores of the same morphotype, in a sterilized substrate. In all pot-culture propagation approaches, care should be taken to prevent the transfer of spores to adjacent pots, especially during watering and pot manipulation. This can be achieved by maximizing the distance between pots or by placing them in plastic bags that allow air circulation (Walker and Vestberg 1994).

Trapped species inoculum

The trap culture method consists of growing bait plants in soil over several successive generations. Because the species of host plant used may influence the AM fungal population in the soil (Bever 2002; Johnson et al. 2004), a number of different host plants may be employed, including the original plant species from under which the soil was collected. Depending on the context and objectives of the study, different culture conditions may also be used to optimize the collection of AM fungi with specific nutrient uptake capacities or that are tolerant to certain abiotic factors (Sylvia and Schenck 1983; Klironomos et al. 1993). Once sporulating on the roots of a bait plant, the fungus can be isolated and subcultured. This may be done using single spores, colonized root segments, or spores of similar morphotypes, depending on whether pure cultures or mixed-species cultures are required (Fig. 7.1). Successive generations of trap cultures using the same soil sample often allow the recovery of initially non-sporulating or dormant species (Stutz and Morton 1996). Nevertheless, the recorded diversity registered using the trap culture method rarely reaches the level of diversity found when using molecular approaches (Vandenkoornhuyse et al. 2002). However, the major advantage of trap cultures is the obtention of spores and root inoculum that probably represents the most aggressive species inhabiting the studied soil under the culture conditions used.

Inoculum production in the field

Field-based methods for the propagation of AM fungi can be used for large-scale inoculum production (Sieverding 1987; Furlan 1993; Douds et al. 2000, 2005, 2006; Gaur et al. 2000; Singh 2002). Large outdoor containers or beds filled with sterilized or nonsterilized substrates are inoculated with a starter inoculum (e.g., indigenous soil, isolated propagules, commercial inoculum, or pot-culture propagated material) and seeded with host plants having a high mycorrhizal potential. The latter are left to grow until they are well established and their root systems well colonized. A single mycorrhizal plant species, a mix of plant species, or successive generations of different plant species may also be used. Whichever method is employed, the final product (i.e., chopped colonized roots and rhizosphere soil) offers a concentrated inoculum suitable for direct field application. Using this approach, farmers can, with a minimum of investment in time and money, produce their own inoculum. Although this method is inexpensive and easy, there is little control on product quality, and the inoculum may contain a variety of soil contaminants including detrimental microorganisms. To limit the establishment of contaminants in the beds and to improve the quality of field inoculum, it is preferable to sterilize the soil prior to inoculation, to regularly weed the beds, to seed the beds with disease-free host plants, and to inoculate with a multiple species starter inoculum.

Soilless propagation

Sophisticated soilless plant production techniques have also been adapted for the production of plants colonized by AM fungi. These include (*i*) hydroponic systems, in which the plant roots are bathed in

Table 7.3. Comparison of different soilless propagation systems used for arbuscular mycorrhizal fungal inoculum production.

	Soilless propagation system			Vitroculture	
	Hydropony	Aeropony	Nutrient film	Root creep	Plant
Spore production	Intemediate	Intemediate	Intemediate	High	High
Root colonization	Low	Low	Intermediate	Low	Low
Purity of culture	Low	Low	Low	High	High
Control of culture conditions	Intermediate	Intermediate	Intermediate	High	High
Equipment required	Intermediate	Intermediate	Intermediate	Intemediate	Intemediate
Labour cost	High	High	High	High	High
Expertise required	Intermediate	Intermediate	Intermediate	High	High
Commercial use	Low	Low	Low	Intemediate	Intemediate

a liquid fertilizer solution or grown in a well-drained inert substrate (e.g., silicium and glass-wool) (Macdonald 1981; Mosse and Thompson 1984; Dehne and Backhaus 1986; Utkhede 2006); (*ii*) aeroponic systems, in which roots are surrounded by a constant or intermittent nutrient solution mist (Sylvia and Hubbell 1986; Hung and Sylvia 1988; Jarstfer and Sylvia 1994; de Souza et al. 1996; Mergulhao 2001; Paiva et al. 2003; Weber et al. 2005); and (*iii*) nutrient film systems that couple hydro- and aeroponic approaches (Elmes et al. 1984; Mosse and Thompson 1984; Verma and Adholeya 1996). Of these, the aeroponic system generally results in the highest levels of root colonization and spore production and a much higher control of inoculum quality (Table 7.3).

In vitro propagation

Some in vitro culture systems have been adapted for AM fungal propagation using either root-organ cultures (ROC) (Fortin et al. 2002; Diop 2003) or in vitro propagated plantlets (Azcón-Aguilar and Barea 1997; Hernandez-Sebastia et al. 1999; Louche-Tessandier et al. 1999). The ROC system provides single-species inocula suitable for diverse research and development activities, including large-scale bioreactor propagation, pot cultures, and field applications. The ROC method has also proved suitable for small-scale inoculum production and the development of germplasm collections (Declerck et al. 2005), such as the Glomeromycetes in vitro collection (GINCO) (Declerck and Dalpé 2001).

The propagation of in vitro colonized whole plants requires the establishment of dual cultures (AM fungi and whole plants) in a closed unit. The symbionts may or may not be subjected to the same growing conditions and nutrient media. The system must remain monoxenic and provide both partners with optimal growth conditions for root colonization, mycelial development, and sporulation. Such systems usually use either artificial gelled substrates (Elmeskaoui et al. 1995; Dupré de Boulois et al. 2006) or a mix of gelled and inert granular substrates (Fracchia et al. 2001). Nevertheless, monoxenic AM whole-plant cultures are difficult to establish; this approach remains at the experimental stage, and its use is largely restricted to fundamental research.

Each of the above-mentioned cultivation systems has its strengths and weaknesses. These are linked with the growing conditions (e.g., root aeration, nutrient availability, and the maintenance of pure cultures) and the time and costs involved in maintaining the cultures (Table 7.3). Furthermore, it should be kept in mind that AM fungal growth and sporulation varies between strains and that different AM fungi react differently to different environmental conditions, resulting in responses that range from increased sporulation to total growth inhibition (Talukdar and Germida 1993).

Finally, the intended use of the AM fungal inoculum (e.g., small- or large-scale production of pure or mixed-species inocula, field trials, assessing biodiversity, and physiological or molecular studies) dictates the most appropriate inoculum production method to use.

Conservation of arbuscular mycorrhizal fungal inocula

The long-term preservation of AM fungal inocula is

important for scientific research and inoculum production (Gagné and Moutoglis 2006) (see also Chapter 8); the primary objective being to assure the purity, viability, and infectivity of the inoculum. However, although numerous preservation methods have been tested, there is still no ideal method for storing AM fungal propagules or inocula (Table 7.4).

Preservation of arbuscular mycorrhizal fungal propagules in soil

Field-collected soil samples should be rapidly stored under cool (4 °C) conditions to prolong their inoculum potential and to reduce the damage to AM fungal propagules by soil microorganisms. The high percentage of moribund spores in field-collected soil greatly reduces its suitability for long-term conservation (Harris et al. 1987). By contrast, the rhizosphere soil of trap cultures contains younger and healthier propagules, which are usually able to survive and maintain their mycorrhizal infection potential over much longer storage periods. However, the spores of some AM fungal species (especially those of the Gigasporales and those members of the Glomerales that produce large spores; Tommerup 1983) tend to rapidly become dormant. This may considerably delay the resumption of fungal growth and, thus, reduce their mycorrhizal inoculum potential. A cold stratification period (4–10 °C) prior to inoculation, may, in some cases, break this dormancy (Camprubi et al. 1990; Juge et al. 2002).

Encapsulation

The encapsulation of AM fungal propagules in Ca-alginate gel has been done to test its potential as a long-term preservation method. Although not widely used, this method has proved successful for use with both in vitro and pot-culture propagated fungi of a number of species and with different types of fungal propagules (Declerck et al. 1996; Plenchette and Strullu 2003). Arbuscular mycorrhizal fungal propagules embedded using the method described by Strullu and Plenchette (1991) remain viable for up to six years (Sylvia and Jarstfer 1992; Plenchette and Strullu 2003). The inoculum resulting from this easy-to-use, effective, and inexpensive method has been used to develop reference collections (Declerck et al. 1996), to test infectivity (Calvet et al. 1996), to conduct plant nutrition studies (Vassilev et al. 2001a), and to act as an inoculum in micropropagation systems (Jaizme-Vega et al. 2003). Propagation tests using aeroponic systems also benefit from the vitality of embedded inoculum (Weber et al. 2005), as do field-trials using single-species inocula or mixed inocula that also contains P-solubilizing fungi (Vassilev et al. 2001b), species of *Rhizobium* (Ganry et al. 1985), or plant growth promoting bacteria (Marulanda et al. 2002).

Storage at low temperatures

The viability of AM fungal inoculum is negatively correlated with length of storage (Douds and Schenck 1990; Habte and Byappanhalli 1998); however, it can usually be maintained longer if stored at low temperatures (1–5 °C) rather than at room temperature (Ferguson and Woodhead 1982; Nemec 1987; Camprubi et al. 1990; Talukdar and Germida 1993; Wagner et al. 2001). Viability is also affected by moisture (Tommerup 1988; Douds and Schenck 1990; Wagner et al. 2001). Therefore, drying procedures coupled with adequate storage temperatures may preserve the viability of some AM fungal species for short to medium lengths of time.

The AM fungi maintained on root-organ cultures in gelled nutrient media in sealed Petri dishes, can be grown for six months before replication is required. It is also possible to replace certain nutrients and, thus, increase the life span of these cultures (Douds 2002). To reduce replication rates and maintain inoculum availability, plugs of growth medium containing AM fungal spores and mycelium can be cut and stored in vials of Ringer's solution at 4 °C for at least 18 months for large-spored species and for up to five years for strains of *Glomus intraradices* Schenck & Smith. However, fungal regrowth from these plugs may be affected by mechanical damage to the mycelium, root receptivity, spore dormancy, and the growth capacity of the germinative hypha. Nevertheless, this simple and inexpensive storage method allows the maintenance of the long-term viability of pure-cultured AM fungal species and strains.

Cryopreservation

Storage methods involving freezing have generally been less successful. The initial dessication step is crucial for the maintenance of inoculum viability. Successful AM fungal inoculum storage has been obtained using a two-step freezing method (−40 °C and −196 °C; Tommerup and Bett 1985), slow freezing of extraradical mycelium in steps down to −12 °C (Addy et al. 1998), and slow cooling to −4 °C and fast freezing to −100 °C of alginate embedded spores (Declerck and van Coppenolle 2000). The ultralow

Table 7.4. Comparison and evaluation of different of methods for the long-term storage of arbuscular mycorrhizal fungal inoculum.

	Temperature			Cryogeny		
	Room temperature	4 °C	–80 °C	Encapsulation	Freeze-drying	Liquid nitrogen
Efficiency	Low	Low	Low	High	Intemediate	Low
Long-term preservation	Low	Intemediate	Low	High	Intemediate	Low
Equipment required	Low	Low	Intemediate	Intemediate	High	High
Labour cost	Low	Low	Low	Intemediate	Intemediate	Intemediate

temperature freezing of spores (i.e., at –60 °C and –80 °C) (Douds and Schenck 1990; Kuszala et al. 2001), the freeze-drying of spores and of colonized roots (Dalpé and Mitrow 1990), and liquid nitrogen cryopreservation (–196 °C) (Kuszala et al. 2001) have all given acceptable success rates. Data on the survival of isolated freeze-dried spores from a variety of species indicated that up to 70% of the preserved spores retained their viability and mycorrhizal potential. By contrast, ultralow temperature and liquid nitrogen preserved spores of the same species did not recover (Dalpé and Mitrow 1990). When preserved at ultralow temperatures, only *Glomus mosseae* (Nicol. & Gerd.) Gerd. & Trappe peridium inoculum, comprising a dense mat of hyphae surrounding hyphal swellings, survived the cryogenic treatment. These results indicate the potential of freeze-drying treatment for the long-term preservation of AM fungal inoculum. According to Hwang et al. (1976), the reaction of an individual cell to cryogenic treatments varies and, therefore, is unpredictable regardless of the approach taken. For AM fungi, this may be linked to spore maturity, the fatty acid saturation of membranes, and general physiological status (Kockova and Hubalek 1983). Colonized roots reacted to cryopreservation in a similar way to AM fungal spores. This is not surprising given the germination potential of intraradical vesicles and the possible additional protection provided by the surrounding freeze-dried root tissues.

General conclusion

During the last decade or so, little emphasis has been placed on the development of new, more efficient extraction procedures, adapted cultivation systems, and optimized storage conditions for AM fungi. This is supported by the fact that, with the exception of in vitro growth and manipulation techniques, the vast majority of patents related to AM fungal growth stimulation, inoculum production, and propagation systems were issued prior to 1993. There is a need for improved methods for AM fungal manipulation that will provide innovative and useful information for inoculum producers, thereby allowing the widespread use of these fungi by farmers, horticulturalists, and the general public. It should be borne in mind that fundamental and applied research on AM fungi in a given scientific field relies on, and benefits from, technological breakthroughs in all the other associated fields.

References

Aboul-Nasr, A. 1997. Inoculum production for vesicular arbuscular mycorrhizal fungi on expanded clay. Alex. J. Agric. Res. **42**: 169–176.

Addy, H.D., Boswell, E.P., and Koide, R.T. 1998. Low temperature acclimation and freezing resistance of extraradical VA mycorrhizal hyphae. Mycol. Res. **102**: 582–586.

Arias, I., Sainz, M.J., Grace, C.A., and Hayman, D.S. 1987. Direct observation of vesicular–arbuscular mycorrhizal infection in fresh unstained roots. Trans. Br. Mycol. Soc. **89**: 128–131.

Azcón-Aguilar, C., and Barea, J.M. 1997. Applying mycorrhizal biotechnology to horticulture: significance and potentials. Sci. Hortic. (Amst.), **68**: 1–24.

Baltruschat, H. 1987. Field inoculation of maize with vesicular arbuscular mycorrhizal fungi by using expanded clay as carrier material for mycorrhiza. Z. Pflanzenkr. Pflanzenschutz, **94**: 419–430.

Bever, J.D. 2002. Host-specificity of AM fungal population growth rates can generate feedback on plant growth. Plant Soil, **244**: 281–290.

Boddington, C.L., Bassett, E.E., Jakobsen, I., and Dodd, J.C. 1999. Comparison of techniques for the extraction and quantification of extra-radical mycelium of arbuscular mycorrhizal fungi in soils. Soil Biol. Biochem. **31**: 479–482.

Bragaloni, M., Rea, E., and Pirazzi, R. 1998. Problems and perspectives in the inoculum production of vesicular–arbuscular fungi isolated from sand dunes. Micol. Ital. **27**: 61–67.

Brundrett, M., and Juniper, S. 1995. Non-destructive assessment of spore germination of VAM fungi and production of pot cultures from single spores. Soil Biol. Biochem. **27**: 85–91.

Brundrett, M.C., Melville, L., and Peterson, R.L. 1994. Practical methods in mycorrhiza research. Mycologue Publications, Waterloo, Ont.

Brundrett, M.C., Bougher, N., Dell, B., Grove, T., and Malajczuk, N. 1996. Working with mycorrhizas in forestry and agriculture. Australian Centre for International Agricultural Research, Canberra, Australia. ACIAR Monogr. 32.

Brundrett, M.C., Jasper, D.A., and Ashwath, N. 1999. Glomalean mycorrhizal fungi from tropical Australia. II. The effect of nutrient levels and host species on the isolation of fungi. Mycorrhiza, **8**: 315–321.

Bütehorn, B., Gianinazzi-Pearson, V., and Franken, P. 1999. Quantification of β-tubulin RNA expression during asymbiotic and symbiotic development of the arbuscular mycorrhizal fungus *Glomus mosseae*. Mycol. Res. **103**: 360–364.

Calvet, C., Camprubi, A., and Rodriguez-Kábana, R. 1996. Inclusion of arbuscular mycorrhizal fungi in alginate films for experimental studies and plant inoculation. HortScience, **31**: 285.

Camprubi, A., Calvet, C., and Estaun, V. 1990. Effect of cold storage on spore germination and inoculum infectivity of *Glomus mosseae*. Investig. Agrar. Prod. Prot. Veg. **5**: 337–343.

Chen, B.D., Christie, P., and Li, X.L. 2001. A modified glass bead compartment cultivation system for studies on nutrient and trace metal uptake by arbuscular mycorrhiza. Chemosphere, **42**: 185–192.

Clapp, J.P., Fitter, A.H., and Merryweather, J.W. 1996. Arbuscular mycorrhizas. *In* Methods for the examination of organismal diversity in soils and sediments. *Edited by* G.S. Hall. UNESCO, IUBS, CAB International, Wallingford, Oxford, UK. pp. 145–161.

Dalpé, Y., and Hamel, C. 2007. Arbuscular mycorrhizae. *In* Manual of soil sampling and methods of analysis 3rd ed. Canadian Society of Soil Science. Lewis Publishers of CRC Press, Boca Raton, Fla. pp. 355–377.

Dalpé, Y., and Mitrow, G. 1990. Long-term preservation of *Glomus mosseae*. *In* Abstracts of the 8th North American Conference on Mycorrhizae, 5–8 Sept. 1990, Jackson, Wyo. University of Wyoming, Jackson. p. 66. [Abstr.]

Dalpé, Y., and Monreal, M. 2004. Arbuscular mycorrhiza inoculum to support sustainable cropping systems. Crop Manage. 2004: 1–12. Available from www.plantmanagementnetwork.org/pub/cm/review/2004/amfungi/.

Dalpé, Y., de Souza, F.A., and Declerck, S. 2005. Life cycle of *Glomus* species in monoxenic culture. *In In vitro* culture of mycorrhizas. *Edited by* S. Declerck, D.G. Strullu, and J.A. Fortin. Springer-Verlag, Berlin. pp. 49–65.

Declerck, S., and Dalpé, Y. 2001. GINCO. Mycorrhiza, **11**: 263.

Declerck, S., and van Coppenolle, A. 2000. Cryopreservation of entrapped monoxenically produced spores of an arbuscular mycorrhizal fungus. New Phytol. **148**: 169–176.

Declerck, S., Strullu, D.G., Plenchette, C., and Guillemette, T. 1996. Entrapment of *in vitro* produced spores of *Glomus versiforme* in alginate beads: *in vitro* and *in vivo* inoculum potentials. J. Biotechnol. **48**: 51–57.

Declerck, S., Séguin, S., and Dalpé, Y. 2005. The monoxenic culture of arbuscular mycorrhizal fungi as a tool for germplasm collections. *In In vitro* culture of mycorrhizas. *Edited by* S. Declerck, D.G. Strullu, and J.A. Fortin. Springer-Verlag, Berlin. pp. 17–29.

Dehne, H.W., and Backhaus, G.F. 1986. The use of vesicular–arbuscular mycorrhizal fungi in plant production. Z. Pflanzenkr. Pflanzenschutz, **93**: 415–424.

de Souza, E.S., Burity, H.A., do Espirito Santo, A.C., and da Silva, M.L.R.B. 1996. Alternative for arbuscular mycorrhizal fungi inoculum production in aeroponic culture. Pesquisa Agropecu. Bras. **31**: 153–158.

Diop, T.A. 2003. *In vitro* culture of arbuscular mycorrhizal fungi: advances and future prospects. Afr. J. Biotechnol. **2**: 692–697.

Douds, D.D., Jr. 2002. Increased spore production by *Glomus intraradices* in the split-plate monoxenic culture system by repeated harvest, gel replacements, and resupply of glucose to the mycorrhiza. Mycorrhiza, **12**: 163–167.

Douds, D.D., Jr., and Schenck, N.C. 1990. Cryopreservation of spores of vesicular–arbuscular mycorrhizal fungi. New Phytol. **115**: 667–674.

Douds, D.D., Jr., Adholeya, A., and Gadkar, V. 2000. Mass production of VAM fungus biofertilizer. *In* Mycorrhizal biology. *Edited by* K.G. Mukerji, B.P. Chamola, and J. Singh. Kluwer Academic Press, New York. pp. 197–215.

Douds, D.D., Jr., Nagahashi, G., Pfeiffer, P.E., Kaiser, W.M., and Reider, C. 2005. On-farm production and utilization of arbuscular mycorrhizal fungus inoculum. Can. J. Plant Sci. **85**: 15–21.

Douds, D.D., Jr., Nagahashi, G., Pfeiffer, P.E., Reider, C., and Kaiser, W.M. 2006. On-farm production of AM inoculum in mixed compost and vermiculite. Bioresour. Technol. **97**: 809–818.

Dupré de Boulois, H., Voets, L., and Declerck, S. 2006. From ROC to AM-P *in vitro* culture systems for arbuscular mycorrhizal fungal transport studies. *In* Abstracts of the 5th International Conference on Mycorrhiza, 23–27 July 2006, Granada, Spain. p. 248. [Abstr.]

Elmes, R.P., Hepper, C.M., Hayman, D.S., and O'Shea, J. 1984. The use of vesicular–arbuscular mycorrhizal roots grown by the nutrient film technique as inoculum for field sites. Ann. Appl. Biol. **104**: 437–442.

Elmeskaoui, A., Damont, J.P., Poulin, M.-J., Piché, Y., and Desjardins, Y. 1995. A tripartite culture system for endomycorrhizal inoculation of micropropagated strawberry plantlets *in vitro*. Mycorrhiza, **5**: 313–319.

Feldmann, F., and Idczak, E. 1994. Inoculum production of VA-mycorrhizal fungi. *In* Techniques for mycorrhizal research. *Edited by* J.R. Norris, D.J. Read, and A.K. Varma. Academic Press, San Diego, Calif. pp. 799–817.

Ferguson, J.J., and Woodhead, S.H. 1982. Production of endomycorrhizal inoculum. Part A: increase and maintenance of vesicular–arbuscular mycorrhizal fungi. *In* Methods and principles of mycorrhizal research. *Edited by* N.C. Schenck. American Phytopathological Society, St. Paul, Minn. pp. 47–54.

Fortin, J.A., Bécard, G., Declerck, S., Dalpé, Y., St-Arnaud, M., Coughlan, A.P., and Piché, Y. 2002. Arbuscular mycorrhiza on root-organ cultures: a review. Can. J. Bot. **80**: 1–20.

Fracchia, S., Menendez, A., Godeas, A., and Ocampo, J.A. 2001. A method to obtain monosporic cultures of arbuscular mycorrhizal fungi. Soil Biol. Biochem. **33**: 1283–1285.

Furlan, V. 1993. Large scale application of endomycorrhizal

fungi and technology transfer to the farmer. *In* Proceedings of the 9th North American Conference on Mycorrhizae (NACOM9), 8–12 Aug. 1993, Guelph, Ont.. *Edited by* R.L. Peterson and M. Schelkle. p. 77. [Abstr.]

Furlan, V., Bartschi, H., and Fortin, J.A. 1980. Media for density gradient extraction of endomycorrhizal spores. Trans. Br. Mycol. Soc. **75**: 336–338.

Gagné, S., and Moutoglis, P. 2006. Challenges for development of mycorrhizal inoculants adapted for specific markets. *In* Abstracts of the 5th International Conference on Mycorrhiza, 23–27 July 2006, Granada, Spain. p. 228. [Abstr.]

Ganry, F., Diem, H.G., Wey, J., and Dommergues, Y.R. 1985. Inoculation with *Glomus mosseae* improves N_2 fixation by field grown soybeans. Biol. Fertil. Soils, **1**: 15–23.

Gaur, A., and Adholeya, A. 2002. Effects of the particle size of soil-less substrates upon AM fungus inoculum production. Biol. Fertil. Soils, **35**: 214–218.

Gaur, A., Adholeya, A., and Mukerji, K.G. 2000. On-farm production of VAM inoculum and vegetable crops in marginal soil amended with organic matter. Trop. Agric. **77**: 21–26.

Gerdemann, J.W., and Nicolson, T.H. 1963. Spores of mycorrhizal *Endogone* species extracted from soil by wet sieving and decanting. Trans. Br. Mycol. Soc. **46**: 235–244.

Gianinazzi, S., and Vosátka, M. 2004. Inoculum of arbuscular mycorrhizal fungi for production systems: science meets business. Can. J. Bot. **82**: 1264–1271.

Gianinazzi-Pearson, V., Gollotte, A., Tisserant, B., Franken, P., Dumas-Gaudot, E., Lemoine, M.-C., van Tuinen, D., Gianinazzi, S., and Lherminier, J. 1995. Cellular and molecular approaches in the characterization of symbiotic events in functional arbuscular mycorrhizal associations. Can. J. Bot. **73**: S526–S532.

Guzman-Plazola, R.A., Ferrera-Cerrato, R., Etchevers-Barra, J.D., and Volke-Haller, V. 1990. Biotecnologia de la produccion de inoculo micorrizico VA. Agroci. Ser. Agua-Suelo-Clima, **1**: 155–181. [In Spanish with English abstract.]

Habte, M., and Byappanhalli, B.N. 1998. Influence of pre-storage drying conditions and duration of storage on the effectiveness of root inoculum of *Glomus aggregatum*. J. Plant Nutr. **21**: 1375–1389.

Harikumar, V.S., and Potty, V.P. 2002. Technology for mass-multiplication of arbuscular mycorrhizal (AM) fungi for field inoculation to sweet potato. Mycorrhiza News, **14**: 11–12.

Harris, J.A., Hunter, D., and Birch, P. 1987. Vesicular–arbuscular mycorrhizal populations in stored topsoil. Trans. Br. Mycol. Soc. **89**: 600–603.

Hernandez-Sebastia, C., Piché, Y., and Desjardins, Y. 1999. Water relations of whole strawberry plantlets *in vitro* inoculated with *Glomus intraradices* in a tripartite culture system. Plant Sci. **143**: 81–91.

Horn, K., Hana, A., Pausch, F., and Hock, B. 1992. Isolation of pure spore and hyphal fractions from vesicular–arbuscular mycorrhizal fungi. J. Plant Physiol. **141**: 28–32.

Hung, L.L., and Sylvia, D.M. 1988. Production of vesicular-arbuscular mycorrhizal fungus inoculum in aeroponic culture. Appl. Environ. Microbiol. **54**: 353–357.

Hwang, S.W., Kwolek, W.F., and Haynes, W.C. 1976. Investigation of ultra low temperature for fungal cultures. III. Viability and growth rate of mycelial cultures following cryogenic storage. Mycologia, **68**: 377–387.

Ianson, D.C., and Allen, M.F. 1986. The effects of soil texture on extraction of vesicular–arbuscular mycorrhizal fungal spores from arid sites. Mycologia, **78**: 164–168.

Jabaji-Hare, S., Sridhara, S.I., and Kendrick, B. 1984. A technique for the isolation of intramatrical vesicles from vesicular–arbuscular mycorrhizae. Can. J. Bot. **62**: 1466–1468.

Jaizme-Vega, M.C., Rodriguez-Romero, A.S., Marín Hermoso, C., and Declerck, S. 2003. Growth of micropropagated bananas colonized by root-organ culture produced arbuscular mycorrhizal fungi entrapped in Ca-alginate beads. Plant Soil, **254**: 329–335.

Jarstfer, A.G., and Sylvia, D.M. 1994. Aeroponic culture of VAM fungi. *In* Mycorrhiza: structure, function, molecular biology and biotechnology. *Edited by* A.K. Varma and B. Hock, Springer-Verlag, Berlin. pp. 427–441.

Jarstfer, A.G., Sylvia, D.M., and Metting, F.B., Jr. 1992. Inoculum production and inoculation strategies for vesicular–arbuscular mycorrhizal fungi. *In* Soil microbial ecology: applications in agricultural and environmental management. *Edited by* F.B. Metting. M. Dekker, New York. pp. 349–377.

Johnson, D., Vandenkoornhuyse, P.J., Leake, J.R., Gilbert, L., Booth, R.E., Grime, J.P., Young, J.P.W., and Read, D.J. 2004. Plant communities affect arbuscular mycorrhizal fungal diversity and community composition in grassland microcosms. New Phytol. **161**: 503–515.

Johnson, N.C., Dell, T.E., and Bledsoe, C.S. 1999. Methods for ecological studies of mycorrhizae. *In* Standard soil methods for long term ecological research. Oxford University Press, New York. pp. 378–412.

Juge, C., Samson, J., Bastien, C., Vierheilig, H., Coughlan, A., and Piché, Y. 2002. Breaking dormancy in spores of the arbuscular mycorrhizal fungus *Glomus intraradices*: a critical cold-storage period. Mycorrhiza, **12**: 37–42.

Khan, Q.Z. 1999. Evaluation of techniques for isolation of vesicular arbuscular mycorrhizal fungi. Ann. Plant Prot. Sci. **7**: 239–241.

Klironomos, J.N., Moutoglis, P., Kendrick, B., and Widden, P. 1993. A comparison of spatial heterogeneity of vesicular–arbuscular mycorrhizal fungi in two maple-forest soils. Can. J. Bot. **71**: 1472–1480.

Kockova, K., and Hubalek, Z. 1983. Liquid nitrogen storage of yeast cultures. 2. Stability of characteristics of stored strains. J. Microbiol. **49**: 571–578.

Kumar, R.H., and Bajwa, R. 2004. Vesicular arbuscular mycorrhizal fungi. *In* Research methods in plant sciences: allelopathy. Vol. 3. Plant pathogens. *Edited by* S.S. Narvall and V. Pal. Scientific Publishers, Jodhpur, India. pp. 47–66.

Kuszala, C., Gianinazzi, S., and Gianinazzi-Pearson, V. 2001. Storage conditions for the long-term survival of AM fungal propagules in wet sieved soil fractions. Symbiosis, **30**: 287–299.

Lilly, S.S., and Santhanakrishnan, P. 1998. Coirpith compost a suitable medium for mass multiplication of *Glomus fasciculatum*. Madras Agric. J. **85**: 240–243.

Louche-Tessandier, D., Samson, G., Hernandez-Sebastia, C., Chagvardieff, P., and Desjardins, Y. 1999. Importance of light and CO_2 on the effects of endomycorrhizal colonization on growth and photosynthesis of potato plantlets (*Solanum tuberosum*) in an *in vitro* tripartite system. New Phytol. **142**: 539–550.

Macdonald, R.M. 1981. Routine production of axenic vesicular–arbuscular mycorrhizas. New Phytol. **89**: 87–93.

Marulanda, A., Vivas, A., Vassileva, M., and Azcón, R. 2002. Interactive effect of suspension or encapsulated inoculum of *Bacillus thuringiensis* associated with arbuscular mycorrhizal fungus on plant growth responses and mycorrhizal inoculum potential. Symbiosis, **33**: 23–38.

Mergulhao, A.C.E.S. 2001. Effect of inoculation with arbuscular mycorrhizal fungus (*Entrophospora colombiana*) on cassava seedlings propagated through an aeroponic system. Ecosistema, **26**: 125–128. [In Portugese with English abstract.]

Mosse, B., and Jones, G.W. 1968. Separation of Endogone spores from organic soil debris by differential sedimentation on gelatin columns. Trans. Br. Mycol. Soc. **51**: 604–608.

Mosse, B., and Thompson, J.P. 1984. Vesicular–arbuscular endomycorrhizal inoculum production. Exploratory experiments with beans (*Phaseolus vulgaris*) in nutrient flow culture. Can. J. Bot. **62**: 1523–1530.

Muthukumar, T., and Udaiyan, K. 2002. Seasonality of vesicular–arbuscular mycorrhizae in sedges in a semi-arid tropical grassland. Acta Oecol. **23**: 337–347.

Nemec, S. 1987. Effect of storage temperature and moisture on *Glomus* species and their subsequent effect on citrus rootstock seedling growth and mycorrhiza development. Trans. Br. Mycol. Soc. **89**: 205–212.

Norris, J.R., Read, D.J., and Varma, A.K. (*Editors*). 1992. Methods in microbiology. Vol. 24. Academic Press, New York.

Paiva, L.M., Silva, M.A., Silva, P.C., and Maia, L.C. 2003. *Glomus clarum* Nicol. & Schenck and *G. etunicatum* Becker & Gerd.: cultivated in soil and aeroponic culture. Rev. Bras. Bot. **26**: 257–262.

Plenchette, C., and Strullu, D.G. 2003. Long-term viability and infectivity of intraradical forms of *Glomus intraradices* vesicles encapsulated in alginate beads. Mycol. Res. **107**: 614–616.

Sahay, N.S., Sudha, S.A., and Varma, A. 1998. Trends in endomycorrhizal research. Indian J. Exp. Biol. **36**: 1069–1086.

Schenck, N.C. 1982. Methods and principles of mycorrhizal research. American Phytopathological Society, St. Paul, Minn.

Schubert, A., Marzachí, C., Mazzitelli, M., Cravero, M.C., and Bonfante-Fasolo, P. 1987. Development of total and viable extraradical mycelium in the vesicular–arbuscular mycorrhizal fungus *Glomus clarum* Nicol. & Schenck. New Phytol. **107**: 183–190.

Sieverding, E. 1987. On-farm production of VAM inoculum. *In* Proceedings of the 7th North American Conference on Mycorrhizae, 3–8 May 1987, Gainesville, Fla. *Edited by* D.M. Sylvia, L.L. Hung, and J.H. Graham. Institute of Food and Agricultural Sciences, University of Florida, Gainesvillle,Fla. p. 284. [Abstr.]

Singh, S. 2002. Mass production of AM fungi. Part 1. Mycorrhiza News, **14**: 2–10.

Singh, S. 2003. Mass multiplication of AM fungi. Part 2. Carrier system, storage, and field application. Mycorrhiza News, **14**: 2–10.

Strullu, G.D., and Plenchette, C. 1991. The entrapment of *Glomus* sp. in alginate beads and their use as root inoculum. Mycol. Res. **95**: 1194–1196.

Strullu, G.D., Romand, C., and Plenchette, C. 1991. Axenic culture and encapsulation of the intraradical forms of *Glomus* spp. World J. Microbiol. Biotechnol. **7**: 292–297.

Stutz, J.C., and Morton, J.B. 1996. Successive pot cultures reveal high species richness of arbuscular endomycorrhizal fungi in arid ecosystems. Can. J. Bot. **74**: 1883–1889.

Sylvia, D.M., and Hubbell, D.H. 1986. Growth and sporulation of vesicular–arbuscular mycorrhizal fungi in aeroponic and membrane systems. Symbiosis, **1**: 259–267.

Sylvia, D.M., and Jarstfer, A.G. 1992. Sheared root inocula of vesicular arbuscular mycorrhizal fungi. Appl. Environ. Microbiol. **58**: 229–232.

Sylvia, D.M., and Schenck, N.C. 1983. Application of superphosphate to mycorrhizal plants stimulates sporulation of phosphorus-tolerant vesicular–arbuscular mycorrhizal fungi. New Phytol. **95**: 655–661.

Talukdar, N.C., and Germida, J.J. 1993. Propagation and storage of vesicular–arbuscular mycorrhizal fungi isolated from Saskatchewan agricultural soils. Can. J. Bot. **71**: 1328–1335.

Tommerup, I.C. 1983. Spore dormancy in vesicular–arbuscular mycorrhizal fungi. Trans. Br. Mycol. Soc. **81**: 37–45.

Tommerup, I.C. 1988. Long-term preservation by L-drying and storage of vesicular arbuscular mycorrhizal fungi. Trans. Br. Mycol. Soc. **90**: 585–591.

Tommerup, I.C., and Bett, K.B. 1985. Cryopreservation of genotypes of VA mycorrhizal fungi. *In* Proceedings of the 6th North American Conference on Mycorrhizae, 25–29 June 1984, Corvallis, Ore. *Edited by* R. Molina. Forest Research Laboratory, Oregon State University, Corvallis, Ore. p. 379. [Abstr.]

Utkhede, R. 2006. Increased growth and yield of hydroponically grown greenhouse tomato plants inoculated with arbuscular mycorrhizal fungi and *Fusarium oxysporum* f. sp *radicis-lycopersici*. BioControl, **51**: 393–400.

Vandenkoornhuyse, P., Baldauf, S.L., Leyval, C., Straczed, J., and Young, J.P.W. 2002. Extensive fungal diversity in plant roots. Science (Washington, D.C.), **295**: 2051.

Vassilev, N., Vassileva, M., Azcón, R., and Medina, A. 2001*a*. Application of free and Ca-alginate-entrapped *Glomus deserticola* and *Yarowia lipolytica* in a soil–plant system. J. Biotechnol. **91**: 237–242.

Vassilev, N., Vassileva, M., Azcón, R., and Medina, A. 2001*b*. Preparation of gel-entrapped mycorrhizal inoculum in the presence or absence of *Yarowia lipolytica*. Biotechnol. Lett. **23**: 907–909.

Verma, A., and Adholeya, A. 1996. Cost-economics of existing methodologies for inoculum production of vesicular–arbuscular mycorrhizal fungi. *In* Concepts in mycorrhizal research. *Edited by* K.G. Mukerji. Kluwer Academic Publishers, Amsterdam. pp. 179–194.

Vilarino, A., and Arines, J. 1990. An instrumental modification of Gerdemann and Nicolson's method for extracting VAM fungal spores from soil samples. Plant Soil, **121**: 211–215.

Vilarino, A., Arines, J., and Schuepp, H. 1993. Extraction of vesicular–arbuscular mycorrhizal mycelium from sand samples. Soil Biol. Biochem. **25**: 99–100.

Wagner, S.C., Skipper, H.D., Walley, F., and Bridges, W.B., Jr. 2001. Long-term survival of *Glomus claroideum* propagules from soil pot cultures under simulated conditions. Mycologia, **93**: 815–820.

Walker, C., and Vestberg, M. 1994. A simple and inexpensive method for producing and maintaining closed pot cultures of arbuscular mycorrhizal fungi. Agric. Sci. Finl. **3**: 233–240.

Weber, J., Ducousso, M., Tham, M.Y., Nourissier-Mountou, S., Galiana, A., Prin, Y., and Lee, S.K. 2005. Co-inoculation of *Acacia mangium* with *Glomus intraradices* and *Bradyrhizobium* sp. in aeroponic culture. Biol. Fertil. Soils, **41**: 233–239.

Wood, T., and Cummings, B. 1992. Biotechnology and the future of VAM commercialization. *In* Mycorrhizal functioning: an integrative plant–fungal process. *Edited by* M.F. Allen. pp. 468–487.

Chapter 8
Industrial perspective of applied mycorrhizal research in Canada

Susan Parent and Peter Moutoglis

Introduction

Intact natural terrestrial ecosystems are typically rich in mycorrhizal fungi, and soil samples collected from them usually exhibit a high mycorrhizal inoculum potential. By contrast, disturbed sites often show a marked decline in the number of fungal propagules capable of colonizing plant roots (e.g., Lovera and Cuenca 1996; see Chapter 10, but see Wiseman and Wells 2005). This reduction in the number of active mycorrhizal fungal propagules can reduce the survival, establishment, and growth of plants. Furthermore, modern agricultural practices based on fertilizer and pesticide application and involving mechanical soil preparation can also reduce mycorrhizal inoculum potential (Kurle and Pfleger 1994; Ezawa et al. 2000; for a review of arbuscular mycorrhizal fungi in agriculture, see Ryan and Graham 2002) (see also Chapter 5). In certain cases, the application of a mycorrhizal fungal inoculum can rectify the situation (Plenchette et al. 1983, Furlan and Bernier-Cardou 1989).

Both the commercial development and the use of mycorrhizal fungi have grown considerably over the last three decades, with mycorrhizal fungi being seen as a natural means of improving plant production. Most of the initial research in the 1970s and 1980s investigated the possible use of ectomycorrhizal (ECM) fungi in the forestry sector, particularly for increasing tree seedling health (e.g., Marx 1971) and growth, and to enhance establishment following outplanting (see review by Marx and Kenney 1982).

Studies done over the last 30 years have provided a greater understanding of mycorrhizal fungal nutritional requirements and life cycles. Recently, particularly important breakthroughs have been made with regards to the growth of arbuscular mycorrhizal (AM) fungi under in vitro conditions (e.g., St-Arnaud et al. 1996; Fortin et al. 2002). These advances triggered a shift from the small-scale production of AM fungal inocula in pot cultures that started in the 1980s (Dalpé and Monreal 2004) toward the development of large-scale inoculum production and the production of a number of mycorrhizal fungal based products for use in agriculture, horticulture, forestry, and the revegetation and restoration of disturbed or polluted sites.

The large-scale use of mycorrhizal fungal inocula for plant production is still in its infancy and high production costs, variable inoculum quality, and official registration must all be overcome before a product can be marketed (Dalpé and Monreal 2004). Currently, more than 20 companies are producing mycorrhizal inoculum worldwide (Gianinazzi and Vosátka 2004) (Table 8.1). In this chapter, we outline the history of the development and use of mycorrhizal fungal inoculants in Canada, where major advances have been made through collaboration among government bodies, universities, and industry. Although ECM fungal inoculum is considered, emphasis is placed on AM fungal inoculum production (further details concerning ECM fungi can be found in Chapters 9, 10, and 11).

S. Parent.[1] Premier Tech, 1, Avenue Premier, Rivière-du-Loup, QC G5R 6C1, Canada.

P. Moutoglis. BioSynetterra Solutions, Carrefour industriel et expérimental de Lanaudière, 801, Route 344, Boîte postale 3158, L'Assomption, QC J5W 4M9, Canada.

[1]Corresponding author (e-mail: pars@premiertech.com).

The integration of governmental-, university-, and industry-based research

Eastern provinces

Mycorrhizal research in Québec initially focused on applications within the forestry sector, which led to the industrial-scale development of ECM inocula (e.g., Fortin et al. 1988). The first successful liquid and solid inocula were produced using strains of *Hebeloma cylindrosporum* Romagnesi and *Laccaria bicolor* (Maire) Orton (Langlois and Gagnon 1988). *Laccaria bicolor* extensively colonized the tree seedling used and particularly promising growth responses were observed on black spruce (*Picea mariana* (Mill.) BSP) subjected to low-nitrogen fertilizer regimes under greenhouse conditions (Gagnon et al. 1988). Additional trials were done in provincial government nurseries to evaluate the cost–benefit of a large-scale inoculation program. Subsequently, methods for scaling up inoculum production were perfected, and the large-scale inoculation of tree seedlings produced for outplanting on clear-cut sites throughout the province was done. The inoculum used was a mycelial suspension produced in bioreactors by the company Rhizotec Laboratories Inc. (St-Jean-Chrysostôme, Québec). In all, over 1×10^6 seedlings were inoculated, outplanted, and monitored. The results indicated that the ECM fungi used were not always competitive when indigenous species were present in nondisturbed field sites (McAfee and Fortin 1986). This led to the conclusion that the greatest benefits associated with the use of ECM fungi were likely to be on sites that were difficult to regenerate due to a lack of natural soil- or wind-borne inoculum. Because of the lack of conclusive results and funding, much of the research into large-scale ECM fungal inoculum production and inoculation in Québec was stopped. However, most of the initial work was done using *L. bicolor*, which is easy to culture and multiply. Nevertheless, there are many hundreds of other native ECM fungal species for which we have no information concerning their potential value for use in inoculation programs. Furthermore, the potential importance of using site-indigenous fungi should not be ignored (see Chapter 10).

With regard to AM fungi, early experiments showed increased survival, growth, and productivity of a number of agricultural crops (Plenchette et al. 1981, 1983); however, the biotrophic nature of these fungi severely limited large-scale axenic inoculum production (Hepper 1983). Until recently, moderately large-scale inoculum production has been achieved using pot cultures and adaptations of this approach (Dalpé and Monreal 2004).

In 1981, preliminary results of experiments done as part of a collaboration between Université Laval and the ministère de l'Agriculture, des Pêches et de l'Alimentation du Québec demonstrated the potential for the large-scale use of AM fungal inoculum for enhancing crop yield, in much the same way as *Rhizobium* spp. enhance legume growth. This generated much interest from the agricultural community and rapidly led to the setting up of a technological transfer program, supported by the National Research Council of Canada and Agriculture Canada. The objective of this program was to enable the peat harvesting company Les Tourbières Premier (now Premier Horticulture) to develop and supply the horticulture industry with a peat-based growing medium containing an AM fungal inoculum. This joint venture generated a large-scale inoculum production unit, based on the pot-culture technique, using host plants grown under greenhouse conditions. The result of this program was the first generation of commercially available inocula. Other companies in the United States (e.g., Native Plant Industries) also started using similar methods for large-scale inoculum production. However, although large amounts of inoculum could be generated using this method, the fact that the cultures were not aseptic meant that the production technique potentially exposed the horticultural industry to certain plant pathogens and harmful insects.

In vitro studies done by Mugnier and Mosse (1987) stimulated the development of an alternative method of propagating AM fungi that involved the use of root-organ cultures (Bécard and Fortin 1988). Bécard and Fortin (1988) showed that these monoxenic cultures could be successfully used to inoculate whole plants.

The in vitro culture technique was further adapted (Bécard and Piché 1992; Chabot et al. 1992; Jolicoeur 1998; Vimard et al. 1999; Fortin et al. 2002) for use with *Glomus intraradices* Schenck & Smith. The potential of the technique for use in large-scale bioreactors was also investigated conjointly by BIOPRO, the École Polytechnique of Montréal, the Institut de Recherche en Biologie Végétale, the Université de Montréal, and Premier Tech. This joint research project aimed to develop an aseptic inoculum production technique that could replace the conventional production methods using whole plants grown under glass in open systems using solid substrates, aeroponic, hydroponic, or nutrient-film techniques. In 2000 as a result of this collaboration, Premier Tech became the first company to introduce an aseptic commercial inoculum to the market (MYCORISE® ASP). Although the exact production process is protected, it is based on a large-scale application of the root-organ culture system. The end product is a liquid suspension of mycorrhizal fungal spores that is available in different forms. One of these forms, which is unique, comprises AM fungal spores and hyphae that are completely free of any root material (including root DNA). The purity of this product has led to its being used in the ongoing genome-sequencing project for *G. intraradices* (Lammers et al. 2006). In addition, the inoculum production technique produces other propagules, including colonized root fragments and hyphal suspensions, that can be harvested aseptically and incorporated into compatible carriers to produce low-density (e.g., perlite, vermiculite, and peat) and high-density (e.g., clays and zeolites) granules or powdered formulations. Figure 8.1 shows the evolution of AM fungal inoculum production over time.

In Ontario, many of the applied research projects that have been conducted have led to applications in the agriculture and forestry sectors. In the 1990s, industry-based research was conducted to investigate the potential for using ECM fungal inoculum in the forestry sector. The company Mikro-Tek (Timmins, Ontario) was involved in land reclamation, and their approach to the application of ECM fungi differed from that used in Québec. Their research involved the selection of fungal species adapted to specific stresses at particular reclamation sites (see also Chapter 10). They developed and tested a liquid inoculum using species of *Hebeloma* and obtained consistent and positive results with a number of tree species (Smith and Mohammed 1997). Figure 8.2 illustrates the ECM isolation, selection, production, and application processes.

In 1999, Mikro-Tek set up a carbon sequestration research project under Industry Canada's Climate Change Technology Early Action Measures (TEAM). This project was based in Chile and used Chilean EM fungi. Nursery inoculation of 1×10^6 seedlings resulted in enhanced growth following outplanting and, therefore, greater carbon sequestration (Industry Canada 2007).

In addition, with support from Sustainable Development Technology Canada (SDTC), Mikro-Tek and its partners engaged in a three-year project to test the effectiveness of AM technology for land reclamation, agriculture, and carbon sequestration. The trials were carried out in different sites across Canada and generally resulted in an increase in host plant growth, especially with inoculated grass on gold-mine tailings and golf courses (Timmins, Ontario) and with soybean (*Glycine max* (L.) Merr.) crops (New Liskeard, Ontario). Mikro-Tek was also the first company to register a turf-specific inoculant for use in Canada under the brand name MIKRO-VAM®.

Potential agriculture applications of AM fungi were also investigated at the University of Guelph (McGonigle et al. 1999) and at Agriculture Canada (S. Nelson and Y. Dalpé, unpublished data[2]). To do further large-scale studies, the latter formulated an inoculum that was designed to be compatible with existing agricultural machinery. However, the project did not reach the commercial application stage primarily because of the costs involved in the development of the field-application techniques.

Western provinces

In Saskatchewan, applied research by Talukdar and Germida (1993) showed that reinoculation of agricultural soils with AM fungi had certain advantages for crop production. However, high production costs meant that industry was less willing to investigate methods of producing inocula. The Saskatchewan-based, British-owned company Agriculture Genetics Company (which later became Micro Bio Ltd. and was finally acquired by Becker Underwood Inc.) investigated the possibility of an industrial production technique for AM fungi. However, the company ceased work after a number of years because of the difficulties associated with the formulation of AM fungal inocula that farmers could easily use.

In Alberta, a certain amount of research has been done into the use of mycorrhizal fungi by the forestry and agriculture sectors. However, the initial industrial development was linked with the use of these fungi for reclamation purposes. In the late 1990s, the company Symbiotech (see Chapter 9) produced ECM inocula for reclamation work on overburden and tailing sands produced during the extraction of oil from oil sands in northern Alberta

---

[2] Nelson, S., and Dalpé, Y. 1994. Management of soil mycorrhizal flora in agricultural systems. Poster, 10th Annual Corn Research Day. Plant Research Centre, Ottawa, Ont.

Fig. 8.1. Evolution of the arbuscular mycorrhizal inoculum production bioprocess at Premier Tech. Examples of the first-generation process are (*a*) open greenhouse production system with whole plants, aseptic starter inoculum, and biological controls, (*b*) harvest and characterization of entire colonized root system and substrate, and (*c*) a low-density granular inoculum formulation. Examples of the second-generation process are (*d*) manipulation under aseptic conditions, (*e*) characterization of liquid phase of the inoculum growth system, (*f*) industrial scale production of liquid inoculum, and (*g*) in vitro mycorrhizal root system with hyphal network and spores (approx. root diameter: 1.5 mm).

(see Chapter 10). The inoculum consisted of a *Hebeloma* sp., which had proven to be very tolerant of high salt levels and generated good growth on the nutrient-poor alkaline substrates encountered in the field (Kernaghan et al. 2002).

Further west, in British Columbia, industrial-scale mycorrhizal inoculum for use in forest tree nurseries began in the late 1980s, but agriculture applications have remained largely confined to university research groups and government agencies. Research involving ECM and forest trees has been conducted by the University of British Columbia and the Ministry of Forests (Berch and Hunt 1988); however, the initial results were inconsistent. Recently, studies using ECM inocula available in the United States (Berch and Roth 1993) showed that inoculation of Douglas-fir (*Pseudotsuga menziesii* (Mirb.) Franco) with *Rhizopogon vinicolor* A.H. Smith, enhances seedling growth in the nursery, but the advantage of inoculating was not maintained following outplanting. As a result, the forestry industry in British Columbia, tends only to use ECM tree seedling on sites that have been subjected to severe disturbances (e.g., after topsoil stripping for mining activities).

Recent advances in research and field applications across Canada

National and international partnerships among research groups has resulted in the advancement of both fundamental mycorrhizal research and field applications of mycorrhizal biotechnologies. In recent years, research groups at Université Laval, the

Fig. 8.2. Ectomycorrhizal isolation, selection, production, and application process at Mikro-Tek: (*a*) field collection of fungi associated with target host; (*b*) fungal isolation and identification; (*c*) strain selection based on growth and nutrient uptake; (*d*) strain selection based on drought stress resistance criteria; (*e*) liquid inoculum production; and (*f*) industrial-scale application of inoculum.

Université de Montréal, McGill University, the University of Guelph, and Agriculture Canada (Ottawa, Ontario; Brandon, Manitoba; Swift Current, Saskatchewan; and Summerland, British Columbia), have used inocula produced by Canadian companies. Many of the research projects have focused on applications in the agricultural, horticultural, and sport (e.g., golf) sectors. In contrast, the Ontario company Mikro-Tek has been specifically targeting the forestry sector in Canada and abroad and has been using the carbon sequestering capacity of ECM fungi, via increased host tree biomass, to generate carbon credits that can be sold to companies to offset atmospheric emissions of greenhouse gases in accordance to the Kyoto Protocol. Stemming from this work, other sectors, such as agriculture, are currently being modeled as a potential means of generating additional carbon credits.

Mycorrhizal inoculum formulation, from the laboratory to the field

The large-scale application of mycorrhizal fungi is

strongly dependant on the quality, availability and cost of inocula. Correctly formulating inocula to suit the user is vitally important and has been as big a challenge as the large-scale production of viable propagules.

Ectomycorrhizal fungi

The work done with ECM fungi led to the formulation of liquid inocula that, if used under the right conditions, has been successful in inducing beneficial survival and growth responses in host plants. Generally speaking, if the natural microflora of the soil at the outplanting site contains a large number of efficacious mycorrhizal fungi, inoculation is not recommended. Nevertheless, use of liquid ECM fungal inocula on sites where the soil has been severely altered, or in peat moss that is largely free of ECM fungal propagules, has resulted in positive survival and growth responses of host tree species (Smith and Mohammed 1997).

The long-term storage of viable liquid ECM inocula has yet to be achieved, although certain advances have been made using encapsulated mycelium in alginate beads (Jung et al. 1985); however, this formulation must still be kept refrigerated. Therefore, once produced, liquid inocula must be used within a relatively short period. Certain tree nurseries are successfully using, and even producing, liquid-based inocula (e.g., Pépinière Robin, France; Table 8.1). The market for ECM inocula shows a great degree of interannual variations (often linked to government contracts associated with forestry tree planting projects), which makes it difficult for inoculum producers to predict needs. Therefore, the inocula must be produced just prior to massive inoculation projects.

In an attempt to overcome this, certain companies in the United States have started developing spore-based inocula (see also Chapter 11). These companies organize the collection of ectomycorrhizal sporocarps from the field. These are then dried, and the spores are extracted using specific methods and formulated into dry inocula. These have a better storage capacity than liquid inocula; however, because of the high variability of the material used, significant quantities of sporocarps are required. The field-collected, spore-based inocula also have the potential to transmit contaminants to production units. Recently, companies marketing ECM inocula have diversified their range of products by including AM fungal inocula for application to ornamental trees at outplanting (Table 8.1).

Arbuscular mycorrhiza

The major challenge facing the development of a commercially viable AM fungal inoculum has been, and still is, associated with the fact that AM fungi are strict biotrophs (i.e., they must be associated with a host plant to grow and complete their life cycle) and that there is no means of producing AM fungi in axenic culture. However, the use of root-organ cultures has led to a second generation of high-quality inocula that can be combined with different carriers. Unlike mycelial-based ectomycorrhizal inoculants, root organ culture derived inocula have a longer shelf life. Initial tests using AM fungi for greenhouse crop production showed limited cost–benefits; however, these plants where generally grown under optimum conditions in terms of irrigation and fertilizer application. Subsequent research with greenhouse crops receiving lower levels of P suggest that it is possible to use these symbionts for culture under glass. However, many greenhouse crops are grown on very short production cycles, and the plants may not have enough time to be show the benefit from inoculation. Nevertheless, nurseries are required to supply plants that will survive once they leave the optimal greenhouse environment. In recent years, Premier Horticulture (formerly Les Tourbières Premier) has started to add the AM fungus *G. intraradices* to all their commercial growing media as a standard additive to enhance plant performance once they leave the producer. In the United States, Monrovia, a woody ornamental nursery that exports all over North America, claims that inoculation with mycorrhizal fungi is the main factor behind the successful performance of their plants. This has increased the visibility of the benefits associated with the use of AM fungi.

For all sectors of the horticultural market (e.g., nurseries, landscape gardeners, and golf courses), the new and stricter legislation concerning leaching, fertilizer runoff, and pesticide use should encourage growers to use AM fungi. In the agricultural sector, the soil microflora is an important component of crop production; however, the cost to farmers of adding an AM fungal inoculum at seeding is still high, and the plant responses are variable (see Chapter 6). The development of an inoculum for use in agriculture is still in the developmental stage.

Product registration

In Canada, the federal government regulates mycorrhizal fungal inocula under the Fertilizers Act, and the government mandates the Canadian Food

Table 8.1. Nonexhaustive list of companies producing arbuscular mycorrhizal (AM) and ectomycorrhizal (EM) fungal inocula.

Company name	Country	Products
Amykor	Germany	AM and EM inocula
Becker Underwood	USA	AM inocula
BioMyc	Germany	AM inocula
BioOrganics	USA	AM inocula
Biorise	France	AM inocula
BioScientific	USA	AM and EM inocula
Biotisa	France	AM and EM inocula
Buckman Laboratories	Mexico	AM inocula
If Tech	France	AM and EM inocula
Inoq	Germany	AM and EM inocula
Mikro Tek	Canada	AM and EM inocula
Mybatec	Italy	AM and EM inocula
Mycorrhizal Applications	USA	AM and EM inocula
Mycosym	Switzerland	AM inocula
MykoFlor	Poland	EM inocula
Mycovitro	Spain	AM inocula
Pépinière Robin	France	EM inocula
Plant Health Care	USA	AM and EM inocula
PlantWorks	UK.	AM inocula
Premier Tech	Canada	AM inocula
Reforestation Technologies	USA	AM inocula
Symbio M	Czech Republic	AM and EM inocula
The Energy & Resources Institute	India	AM inocula
Technologias Naturales Internacional	Mexico	AM and EM inocula

Inspection Agency to regulate all products that are sold as plant growth enhancers, supplements, or fertilizers. Registering an inoculum is difficult, time consuming, and costly. Briefly, products that are sold in Canada must undergo an important safety and efficacy assessment. The latter involves a series of experiments that allow the products' benefits to be established, together with the conditions (e.g., storage temperatures, application rates, soil types, geographic locations, and target plants) under which these benefits may be obtained.

Registration of beneficial organisms for growth stimulation will hopefully be facilitated in the future and allow the possibility of combining different biotechnologies to produce mixed inocula of AM and ECM fungi, and of combining mycorrhizal fungi with other types of inoculants, such as species of *Rhizobium* and other plant growth promoting rhizobacteria (e.g., *Bacillus subtilis* (Ehrenberg) Cohn).

Recent reports have highlighted the potential role of AM and ECM fungi in plant disease control (e.g., Whipps 2004); however, because of the extensive trials required, no mycorrhizal fungi have been successfully registered with federal agencies, such as the Pest Management Regulatory Agency, for such use within Canada. Because of this, Canadian and other western countries are starting to lag behind some of the newly emerging powers in terms of the use of biotechnological advances for sustainable agriculture.

Conclusions

Without mycorrhizal fungi, many plants show poor growth. However, most soils contain an active mycorrhizal fungal inoculum, and most plants in the field are adequately colonized by these mycobionts. Nevertheless, under certain conditions, the mycorrhizal inoculum potential of a soil is reduced, and artificial inoculation can help enhance plant survival, establishment, and growth. In the future, further research involving AM and ECM fungi will be needed if we are to lower the production cost and enhance the benefit of using these symbionts in agriculture and forestry. Presently, the benefits of using mycorrhizal fungi can be clearly measured when plants are grown under suboptimal conditions. However, this is not always a meaningful benefit to growers, and the mass screening of mycorrhizal fungal strains should be done to find strains capable of giving a more rapid and consistent benefit in terms of plant growth.

In this period of enhanced environmental awareness, there is an increasing demand for more environmentally friendly and sustainable plant-production technologies. This will hopefully lead to a quicker, but equally efficient, testing to allow the companies and government agencies developing emerging green technologies to market products more rapidly and, in turn, pave the way for a healthier environment. Existing registration regulations are driving some Canadian companies to register their products abroad, resulting in a loss in Canadian investment and innovation, which will affect the agriculture industry as a whole.

In the future, continued collaboration between the scientific community and industry will be crucial if we are to find means of maximizing plant productivity, while respecting the planet. The development of multiaction inocula may well provide one of the most exciting solutions for this. The development of multiorganism-based products coupled, or not, with signal molecules (see Chapter 2) to enhance product performance and reduce environmental variability will be a challenge for industry and the regulatory bodies alike.

Finally, we are convinced that growing concerns for the environment, novel commercial approaches, and conducive pricing will bring new mycorrhizal technologies into more general and widespread use in the near future.

References

Bécard, G., and Fortin, J.A. 1988. Early events of vesicular–arbuscular mycorrhiza formation on Ri T-DNA transformed roots. New Phytol. **108**: 211–218.

Bécard, G., and Piché, Y. 1992 Establishment of vesicular–arbuscular mycorrhiza in root organ culture: review and proposed methodology. *In* Methods in microbiology. Vol. 24. Techniques for the study of mycorrhiza. *Edited by* J.R. Norris, D.J. Read, and A.K. Varma. Academic Press, London. pp. 89–108.

Berch, S., and Hunt, G. 1988. The state of mycorrhiza research in B.C. forestry. *In* Canadian Workshop on Mycorrhizae in Forestry. *Edited by* M. Lalonde and Y. Piché. Centre de recherche en biologie forestière, Faculté de Foresterie et de Géodésie, Université Laval, Québec, Que. pp. 7–8.

Berch, S.M., and Roth, A.L. 1993. Ectomycorrhizae and growth of Douglas-fir seedlings preinoculated with *Rhizopogon vinicolor* and outplanted on eastern Vancouver Island. Can. J. For. Res. **23**: 1711–1715.

Chabot, S., Bécard, G., and Piché, Y. 1992. Life cycle of *Glomus intraradix* in root organ culture. Mycologia, **84**: 315–321.

Dalpé, Y., and Monreal, M. 2004. Arbuscular mycorrhizal inoculum to support sustainable cropping systems [online]. Crop Manage. 2004. Available from www.plantmanagementnetwork.org/pub/cm/review/2004/amfungi/.

Ezawa, T., Yamamoto, K., and Yoshida, S. 2000. Species composition and spore density of indigenous vesicular–arbuscular mycorrhizal fungi under different conditions of P-fertility as revealed by soybean trap culture. Soil Sci. Plant Nutr. **46**: 291–297.

Fortin, C., Fortin, J.A., Gaulin, G., Jomphe, N., and Lemay, S. 1988. Large-scale inoculation of containerized-grown seedlings. *In* Canadian Workshop on Mycorrhizae in Forestry. *Edited by* M. Lalonde and Y. Piché. Centre de recherche en biologie forestière, Faculté de Foresterie et de Géodésie, Université Laval, Québec, Que. pp. 115–118.

Fortin, J.A., Bécard, G., Declerck, S., Dalpé, Y., St-Arnaud, M., Coughlan, A.P., and Piché, Y. 2002. Arbuscular mycorrhiza on root-organ cultures. Can. J. Bot. **80**: 1–20.

Furlan, V., and Bernier-Cardou, M. 1989. Effects of N, P, and K on formation of vesicular–arbuscular mycorrhizae, growth and mineral content of onion. Plant Soil, **113**: 167–174.

Gagnon, J., Langlois, C.-G., and Fortin, J.A. 1988. Growth and ectomycorrhizal formation of containerized black spruce seedlings as affected by nitrogen fertilization, inoculum type, and symbiont. Can. J. For. Res. **18**: 922–929.

Gianinazzi, S., and Vosátka, M. 2004. Inoculum of arbuscular mycorrhizal fungi for production systems: science meets business. Can. J. Bot. **82**: 1264–1271.

Hepper, C.M. 1983. Limited independent growth of a vesicular–arbuscular mycorrhizal fungus. New Phytol. **93**: 537–542.

Industry Canada. 2007. Custom profile: Mikro-tek Inc. Industry Canada, Ottawa, Ont. Available from strategis.ic.gc.ca/app/ccc/srch/nvgt.do?lang=eng&app=1&prtl=1&sbPrtl=&estblmntNo=123456215563&profile=cstmPrfl&tile=tile.browseSelectCustomProfile [accessed 18 April 2008].

Jolicoeur, M. 1998. Optimisation d'un procédé de production de champignons endomycorhiziens en bioréacteur. Ph.D. thesis, École Polytechnique de Montréal, Montréal, Que.

Jung, G., Mugnier, J., Le Tacon, F., Michelot, P., and Mauperin, C. 1985. Efficiency in a forest nursery of an ectomycorrhizal fungus inoculum produced in a fermentor and entrapped in polymeric gels. Can. J. Bot. **63**: 1664–1668.

Kernaghan, G., Hambling, B., Fung, M., and Khasa, D. 2002. *In vitro* selection of boreal ectomycorrhizal fungi for use in reclamation of saline–alkaline habitats. Restor. Ecol. **10**: 43–51.

Kurle, J.E., and Pfleger, F.L. 1994. The effects of cultural practices and pesticides on VAM fungi. *In* Mycorrhizae and plant health. *Edited by* F.L. Pfleger and R.G. Linderman. American Phytopathological Society Press, St. Paul, Minn. pp. 101–131

Lammers, P.J., Koul, R., Colonne, P., Shapiro, H.J., Detter, C., Linquist, E., Richardson, P., and Moutoglis, P. 2006. *Glomus* Genome Consortium I. The genome project for *Glomus intraradices*, a model arbuscular mycorrhizal fungus. *In* Abstracts of the 5th International Conference on Mycorrhiza, 23–27 July 2006, Granada, Spain. p. 45. [Abstr.]

Langlois, C.G., and Gagnon, J. 1988. The production of mycorrhizal conifer seedlings in Québec: the progression of the project. *In* Canadian Workshop on Mycorrhizae in Forestry. *Edited by* M. Lalonde and Y. Piché. Centre de recherche en biologie forestière, Faculté de Foresterie et de Géodésie, Université Laval, Québec, Que. pp. 9–13.

Lovera, M., and Cuenca, G. 1996. Arbuscular mycorrhizal infection in Cyperaceae and Gramineae from natural, disturbed and restored savannas in La Gran Sabana, Venezuela. Mycorrhiza, **6**: 111–118.

Marx, D.H. 1971. Ectomycorrhizae as biological deterrents to pathogenic root infections. *In* Mycorrhizae. *Edited by* E. Hacskaylo. U.S. Government Printing Office, Washington, D.C. pp. 81–96.

Marx, D.H., and Kenney, D.S. 1982. Production of ectomycorrhizal fungus inoculum. *In* Methods and principles of mycorrhizal research. *Edited by* N.C. Schenck. American Phytopathologcial Society Press, St. Paul, Minn. pp. 131–146.

McAfee, B.J., and Fortin, J.A. 1986. Competitive interactions of ectomycorrhizal mycobionts under field conditions. Can. J. Bot. **64**: 848–852.

McGonigle, T.P., Miller, M.H., and Young, D. 1999. Mycorrhizae, crop growth, and crop phosphorus nutrition in maize–soybean rotations given various tillage treatments. Plant Soil, **210**: 33–42.

Mugnier, J., and Mosse, B. 1987. Vesicular–arbuscular mycorrhizal infection in transformed root-inducing T-DNA roots grown axenically. Phytopathology, **77**: 1045–1050.

Plenchette, C., Furlan, V., and Fortin, J.A. 1981. Growth stimulation of apple tree in unsterilized soil under field conditions with VA mycorrhiza inoculation. Can. J. Bot. **59**: 2003–2008.

Plenchette, C., Fortin, J.A., and Furlan, V. 1983. Growth responses of several plant species to mycorrhizae in a soil of moderate P-fertility. 1. Mycorrhizal dependency under field conditions. Plant Soil, **70**: 199–209.

Ryan, M.H., and Graham, J.H. 2002. Is there a role for arbuscular mycorrhizal fungi in production agriculture? Plant Soil, **244**: 263–271.

Smith, W., and Mohammed, G.H. 1997. Inoculation with mycorrhizal fungi (*Hebeloma* spp.) can increase drought stress resistance and improve field performance of jack pine, black spruce, and white spruce. Ontario Ministry of Natural Resources, Ontario Forest Research Institute, Sault Ste. Marie, Ont. For. Res. Rep. 145. pp. 1–10.

St-Arnaud, M., Hamel, C., Vimard, B., Caron, M., and Fortin, J.A. 1996. Enhanced hyphal growth and spore production of the arbuscular mycorrhizal fungus *Glomus intraradices* in an *in vitro* system in absence of host roots. Mycol. Res. **100**: 328–332.

Talukdar, N.C., and Germida, J.J. 1993. Propagation and storage of vesicular–arbuscular mycorrhizal fungi isolated from Saskatchewan agricultural soils. Can. J. Bot. **71**: 1328–1335.

Vimard, B., St-Arnaud, M., Furlan, V., and Fortin, J.A. 1999. Colonization potential of *in vitro*-produced arbuscular mycorrhizal fungus spores compared with a root-segment inoculum from open pot culture. Mycorrhiza, **8**: 335–338.

Whipps, J.M. 2004. Prospects and limitations for mycorrhizas in biocontrol of root pathogens. Can. J. Bot. **82**: 1198–1227.

Wiseman, P.E., and Wells, C. 2005. Soil inoculum potential and arbuscular mycorrhizal colonization of *Acer rubrum* in forested and developed landscapes. J. Arboric. **31**: 296–301.

Chapter 9
Mycorrhizal fungi in Canadian forest nurseries and field performance of inoculated seedlings

Ali M. Quoreshi, Gavin Kernaghan, and Gary A. Hunt

Introduction

The success of outplanted nursery-grown tree seedlings depends on their ability to rapidly access nutrients and water held within the soil matrix (Dunabeitia et al. 2004). In nature, this process is enhanced by the formation of symbiotic mycorrhizal associations. However, on many disturbed sites (e.g., mine spoils, roadsides, log landings, and abandoned agricultural land), suitable mycorrhizal fungi are lacking, and this can limit seedling establishment and growth (Ortega et al. 2004).

Nursery managers are aware of the potential importance of the development of mycorrhizas on the roots of plants that they produce (Trappe and Strand 1969). However, state-of-the-art techniques with rapid throughput allow seedlings of outplanting size to be produced in a single growing season. The roots of such seedlings are typically poorly colonized by mycorrhizal fungi or colonized by a limited number of nursery-adapted species. This situation can be ameliorated by artificial inoculation with selected mycorrhizal fungi using one of several existing techniques (Trappe 1977; Molina and Trappe 1982, 1984). Many studies have shown that inoculation of nursery-grown seedlings with selected mycorrhizal fungi enhances plant survival, growth, and establishment following outplanting (Marx et al. 1977, 1985; Ruehle 1982; Molina and Trappe 1984; Castellano and Trappe 1985; Valdes 1986; Kropp and Fortin 1988; Browning and Whitney 1992; Le Tacon et al. 1992; Castellano 1996; Pera et al. 1999; Quoreshi and Timmer 2000; Gagné et al. 2006). This is particularly true for plants used in the reclamation of disturbed sites (Danielson and Visser 1989) (see also Chapter 10), for the successful establishment of introduced forest tree species (Marx 1991), and for the restoration of natural ecosystems (Miller and Jastrow 1992).

Although the importance of suitable mycorrhizal associations on the roots of plants used for land reclamation and reforestation work is well recognized, the production of inoculated seedlings in tree nurseries is less widespread in Canada than in other parts of the world. In the first part of this chapter, we briefly describe the indigenous mycorrhizal fungi typically present in northern temperate and boreal forest tree nurseries. We then present some of the techniques used for seedling inoculation and the subsequent performance of inoculated seedlings. Examples are drawn largely from experiments and field trials conducted in Canada.

Potential benefits of mycorrhizas to nursery-grown tree seedlings

The majority of tree seedlings grown in Canadian forest nurseries are of species that form ectomycorrhizal (ECM) associations. The colonization of these seedlings by suitable mycorrhizal fungi can improve their vigour and, ultimately, their quality (Allen 1991; Browning and Whitney 1992, 1993; Le Tacon et al. 1992; Smith and Read 1997). The

---

A.M. Quoreshi.[1] Symbiotech Research Inc. Unit 201, 509 11th Avenue, Nisku, AB T9E 7N5, Canada and Centre d'étude de la forêt, Université Laval, Québec, QC G1V 0A6, Canada.
G. Kernaghan. Biology Department, Mount Saint Vincent University, 166 Bedford Highway, Halifax, NS B3M 2J6, Canada.
G.A. Hunt. Department of Natural Resource Sciences, Thompson Rivers University, 900 McGill Road, Kamloops, BC V2C 5N3, Canada.

[1]Corresponding author (e-mail: symbiotech.ali@2020seedlabs.ca, ali@2020seedlabs.ca).

benefits gained from the ECM association vary greatly, depending on the host–fungus combination and the conditions prevalent on the outplanting site. Although increased nutrient uptake is the single most significant benefit associated with mycorrhizal formation, the symbiosis offers numerous other benefits to the host plant. These include enhanced water absorption efficiency (Duddridge et al. 1980; Molina et al. 1992), increased resistance to pathogens (Duchesne et al. 1988; Schelkle and Peterson 1997) (see also Chapter 4), and protection against damage from heavy metals and other pollutants (Colpaert and van Assche 1987; Jones and Hutchinson 1988; Rivera-Becerril et al. 2002). Furthermore, mycorrhizal fungi improve soil structure and play an active role in nutrient-cycling processes and carbon sequestration (Treseder and Allen 2000; Staddon 2005; Treseder et al. 2007). Together, these reduce the need for fertilizer application and irrigation in the nursery, reduce the negative effects of a number of plant stresses, improve seedling survival and growth, and enhance transplanting success and seedling performance following outplanting.

Indigenous mycorrhizal fungi in forest nurseries

Bare-root and container-grown tree seedlings produced in nurseries may be colonized by a diverse array of mycorrhizal fungi. However, forest tree nurseries represent a highly specialized habitat, typically characterized by high fertilizer application rates and water regimes. For container-grown seedlings, the latter results in relatively poor aeration of the growth substrate (Hunt 1992; Kernaghan et al. 2003; El Karkouri et al. 2005; Trocha et al. 2006). Therefore, the mycorrhizal fungal communities present in tree nurseries are necessarily dominated by species capable of taking advantage of these conditions. Furthermore, host specificity may also play a role in determining the mycorrhizal fungal species present (Massicotte et al. 1999).

Mycorrhizal associations found in forest tree nurseries

Three of the seven morphologically different mycorrhizal types typically recognized (Smith and Read 1997) (see also Chapter 1) commonly occur in tree nurseries (i.e., ectomycorrhizas, arbuscular mycorrhizas, and ectendomycorrhizas). Ectomycorrhizal associations are predominant on most coniferous species in the Pinaceae and, therefore, are of primary importance to the forest tree nursery industry in Canada. However, the endomycorrhizal fungi forming arbuscular mycorrhizas are found in the roots of bare-root produced yew (*Taxus*), cedar (*Thuja* and *Xanthocyparis*), maple (*Acer*), ash (*Fraxinus*), and cherry (*Prunus*). Moreover, some tree species, including alder (*Alnus*) and poplar (*Populus*), may support species of both ECM and arbuscular mycorrhizal (AM) fungi. In such cases, colonization by fungi from these two groups may be simultaneous or sequential. In the latter case, AM fungi tend to colonize young seedlings but are eventually replaced by ECM fungi. In other cases, the colonization may shift between the two mycorrhizal types depending on environmental conditions, such as water availability (Lodge 2000). Coniferous species may also support cocolonization by AM and ECM fungi (Cázares and Smith 1995; Wagg et al. 2008); however, the extent to which this occurs under nursery conditions is unknown.

Ectendomycorrhizal, or "E-strain," associations were originally described from nursery-grown *Pinus* and *Larix* seedlings (Laiho 1965; Mikola 1965), and they appear to be restricted to these two genera (Yu et al. 2001) (see also Chapter 1). They are also found on the roots of seedlings growing on disturbed sites (Mikola 1988; Danielson 1991) (see Chapter 10). Ectendomycorrhizas exhibit distinctive morphologies, sharing some of the structural attributes of both ectomycorrhizas and arbuscular mycorrhizas (i.e., Hartig net formation, thin or nonexistent mantles, and intracellular root cells penetration). The fungi most commonly involved in this association are ascomycetes within the genus *Wilcoxina* (e.g., *Wilcoxina mikolae* var. *mikolae* Yang & Korf, *Wilcoxina mikolae* var. *tetraspora* Wilcox, Yang & Korf, and *Wilcoxina rehmii* Yang & Korf; Egger 1996). However, other ectendomycorrhizal ascomycetes, such as *Sphaerosporella brunnea* (Alb. & Schwein) Svrcek & Kubicka, *Phialophora finlandia* Wang & Wilcox, and *Chloridium paucisporum* Wang & Wilcox have also been identified (Yu et al. 2001).

The roots of nursery-grown forest tree seedlings are also often colonized by other endophytic fungi. However, the function of these remains largely unknown. Some are considered latent pathogens that may produce disease symptoms under suitable conditions, whereas many others remain asymptomatic. Although there are no visible mycorrhizal structures formed (e.g., Hartig net and fungal mantle; but see Peterson et al. (2008) and references therein), there is a growing body of evidence that suggests that some may form mutualistic associations with host plants, similar to those formed by mycorrhizal fungi (Jumpponen and Trappe 1998; Barrow and Osuna 2002). Use of DNA-based molecular techniques (e.g., cloning and sequencing) has highlighted the ubiquitous nature of fungal endophytes in the roots of nursery seedlings in recent years. These methods potentially allow the detection of all fungi colonizing the roots of a seedling and not just those species that are easily cultured or that form structures that are obvious mycorrhizal. The fungi forming these endophytic associations include "dark septate" species within the genus *Phialocephala* (Menkis et al. 2005, 2006) and members of the ascomycetous order Leotiales. Some of the latter appear to be taxonomically similar to species of ericoid mycorrhizal fungi (Kernaghan et al. 2003).

Because ECM fungi are by far the most common mycorrhizal fungi in tree nurseries in the Northern Hemisphere, they have been the focus of the majority of studies investigating both common nursery species and the possible advantages of inoculating trees with selected species or strains.

Influence of host tree species on indigenous fungi

Northern forest nurseries typically focus on the production of seedlings of species of *Picea* and *Pinus*. Nurseries in Canada are no exception, and the bulk of the production is of *Picea glauca* (Moench) Voss (white spruce) and various species of *Pinus*, together with smaller quantities of *Abies*, *Pseudotsuga*, and miscellaneous softwoods (National Forestry Database Program; nfdp.ccfm.org). Because of a certain degree of host specificity, the tree species being propagated has a major influence on the species of ECM fungi found in the nursery. In a study in Lithuania using DNA sequencing, Menkis et al. (2005) compared the ectomycorrhizas of *Pinus sylvestris* L. (Scots pine) and *Picea abies* (L.) Karst (Norway spruce) grown either as bare-root stock or in containers. The authors found clear differences between the ECM fungi colonizing the two host species: *Suillus luteus* L. ex Fr. was more common on *Pinus* seedlings, whereas *Amphinema byssoides* (Pers. ex Fr.) Erikss. dominated *Picea* roots. This apparent preference of *Amphinema* for *Picea* and of members of the Boletales (e.g., *Suillus* and *Rhizopogon*) for *Pinus* has also been observed in forest nurseries in Poland (El Karkouri et al. 2005; Iwanski et al. 2006) and Canada (Danielson and Visser 1990; Hunt 1992; Kernaghan et al. 2003). By contrast, other common forest tree nursery fungi, such as *Thelephora americana* Lloyd (synonymous with *Thelephora terrestris* Ehrh.) and species of *Wilcoxina*, exhibit far less host specificity and can form an important part of the mycorrhizal community on either host species (Kernaghan et al. 2003; Menkis et al. 2005; Iwanski et al. 2006).

Abiotic influences on indigenous fungi

The effect of the host tree species on the ECM fungal community is likely attenuated by other important factors that influence these fungi in forest nurseries. Hunt (1992), studying ECM fungal communities in conifer nurseries in British Columbia, found that *T. americana* dominated under conditions of poor aeration (normally caused by waterlogging) and high fertilizer application rates. However, the fungal community became more diverse following the amelioration of these conditions. The capacity of *Thelephora* species to tolerate high fertilizer application rates (Hunt 1992) makes them typical nursery mycobionts, and this may also explain why they are rapidly replaced by other ECM fungi following outplanting (Danielson 1991; Jones et al. 2002).

Kernaghan et al. (2003) also found that tree nursery ECM fungal community structure was related to fertilization levels. Nurseries producing *P. glauca* in Alberta (Canada) under high N application regimes only had a small subset of the fungi found in nurseries using lower N application rates. Furthermore, roots from nurseries with high levels of applied N were dominated by *T. americana*, whereas those from nurseries where lower amounts of N were applied were dominated by *A. byssoides*. The type of fertilizer may also have an important influence on tree nursery ECM fungal communities. In *P. abies* nurseries in Poland, the use of organic fertilizer (compost) appeared to be linked to the dominance of ascomycetes over basidiomycetes in the symbiotic fungal community (Trocha et al. 2006).

Many of the ECM fungi found on the roots of bare-root tree seedlings show a significant overlap with the types found on the roots of container-grown seedlings (i.e., species of *Amphinema*, *Thelephora*, and *Wilcoxina*). However, Menkis et al. (2005) compared the two systems directly and found that some ECM fungal species common on bare-root seedlings were absent in containerized systems and vice versa. Moreover, the studies by Ursic and Peterson (1997), Rudawska et al. (2006) and Trocha et al. (2006) revealed a dominance of ECM ascomycetes on the roots of bare-root grown *Pinus* and *Picea*. Although differences in species composition may exist

Use of inoculum: techniques and success

Inoculum for use in tree nurseries

Various types of natural and laboratory-produced inocula and several application techniques have been developed over the years, which now allow the successful inoculation of nursery seedlings. Unsterilized forest soil, humus, excised ectomycorrhizas, crushed carpophores, pure cultures, and spores have all been used as sources of inoculum (Mosse et al. 1981); however, these different techniques have had mixed success. The most convenient, practical, and controllable way of inoculating seedlings under commercial nursery conditions is to incorporate the inoculum, either as a solid mycelium or a liquid mycelial slurry, into the growth substrate just prior to seed sowing. This method is currently being used to inoculate *Pinus banksiana* Lamb. (jack pine) and *P. glauca* seedlings in Alberta, and solid and liquid inocula were both found to be equally effective under commercial nursery conditions (Quoreshi et al. 2005). There are also several types of commercial inocula available on the market, which consist of free spores, gel-incorporated spores, mycelium-colonized vermiculite–peat mixes, liquid mycelial slurries, and root fragments. Traditionally, modified Melin–Norkran's (MMN) medium (Marx 1969) has been used to produce ectomycorrhizal fungal inoculum in pure culture. However, recently, it has been shown that using glucose – yeast – malt extract (GYME) medium is more effective than MMN for the large-scale production of ectomycorrhizal inoculum in bioreactors (Hambling et al. 2000).

Inoculation in commercial nurseries: factors determining success

The extent of mycorrhizal colonization in nurseries depends on several factors including the nursery's operating system, the type of growth substrate used and its pH, the fertilizer application rates, the use of pesticides and fungicides, and the degree of substrate waterlogging (Väre 1990; Cordell and Marx 1994; O'Neill and Mitchell 2000; Kernaghan et al. 2003; Sundari and Adholeya 2003). Although several studies have reported that high fertilizer application rates in nurseries limit mycorrhizal formation (Ruehle and Marx 1977; Shaw et al. 1982; Gagnon et al. 1988; Hunt 1988; Chakravarty and Chatarpaul 1990; Le Tacon et al. 1997), this is not always the case (Molina and Chamard 1983; Danielson et al. 1984; Castellano et al. 1985; Khasa et al. 2001). However, to obtain higher levels of root colonization by ECM fungi on nursery-grown seedlings, modification of the fertilizer application schedules used by commercial nurseries may be needed. Hunt (1992) recommended limiting the application of soluble fertilizers to between 80 and 100 ppm N and between 30 and 35 ppm P. Recently, a new technique to integrate mycorrhizal biotechnology into commercial tree production was developed by Quoreshi and Timmer (1998, 2000). Using this method, the authors obtained adequate formation of ectomycorrhizas on *Picea mariana* (Mill.) BSP (black spruce) subjected to a high-dose exponential fertilization regime. However, different host species may respond differently to inoculation depending on the species of fungus and the prevailing nursery conditions (Molina and Chamard 1983). Khasa et al. (2001) reported that colonization of conifer roots by ECM fungi was not affected by different fertilizer application levels (i.e., low, medium, and high) under commercial nursery conditions. However, a current inoculation program for *P. banksiana* and *P. glauca* has shown successful formation of ectomycorrhizas under commercial nursery conditions but following a slight modification to the regular nursery fertilizer application schedule (Quoreshi et al. 2005). In this study, which tested both mycelial slurries and mycelium grown in solid substrates, commercial nursery production systems and fertilizer application schedules did not inhibit mycorrhizal formation. Following inoculation with either *Hebeloma crustuliniforme* Bull (Quel.), *Laccaria bicolor* (R. Mre.) Orton, *A. byssoides*, or *Suillus tomentosus* (Kauffman) Singler, Snell & Dick, between 51% and 90% of short roots were colonized (Quoreshi et al. 2005).

In another study, *Populus tremuloides* Michx. (aspen) and *Populus balsamifera* L. (balsam poplar) seedlings were inoculated with six species of ectomycorrhizal fungi (Quoreshi and Khasa 2008). Seedlings were inoculated with a mycelial slurry at planting and grown under three fertilizer application levels (i.e., 33%, 67%, and 100% of the recommended nursery fertilizer application regime). Four of the species, *Hebeloma longicaudum* Pers. Fr., *L. bicolor*, *Paxillus involutus* Batsch: Fr., and *Pisolithus tinctorius* (Pers.) Coker & Couch, formed adequate mycorrhizal associations with *P. tremuloides* and *P. balsamifera*. The fertilizer application rates used in this study did not significantly affect the mycorrhizal colonization levels.

Nursery inoculation success

There are numerous examples demonstrating the ability of selected, artificially inoculated, mycorrhizal fungi to enhance the growth and nutritional status of tree seedlings under nursery conditions and in the field following outplanting (Danielson and Visser 1989; Kropp and Langlois 1990; Villeneuve et al. 1991; Browning and Whitney 1992, 1993; Le Tacon et al. 1992, Gagné et al. 2006). In many parts of the world, unsuccessful attempts at afforestation and reclamation have been attributed to the absence of appropriate mycorrhizal fungi (Mikola 1970; Ruehle 1982; Smith and Read 1997). Therefore, inoculation may be particularly beneficial on reforestation sites where soil is poor and weather conditions are extreme. Furthermore, in cases where soilborne mycorrhizal inoculum is insufficient, for example, following severe disturbance and where fungal dispersal (via either spore rain or animal vectors) is inadequate (see also Chapter 11), outplanting of seedlings preinoculated with appropriate ECM fungal strains should be encouraged. The soil found on degraded sites is frequently low in available nutrients, mycorrhizal fungi, and other beneficial microorganisms (Cooke and Lefor 1990), and inoculation with selected ECM fungi can significantly improve survival, growth, and establishment of young trees (Perry et al. 1987; Le Tacon et al. 1992; Pera et al. 1999; Ortega et al. 2004). Furthermore, preinoculation may enhance establishment under unfavourable environmental and soil conditions (Jones and Hutchinson 1988). Although the inoculation of nursery seedlings with selected ECM fungi is a technique generally considered for use during the production of seedlings to be outplanted in soils with low inoculum potential, inoculation has also proven beneficial for routine reforestation (Marx 1980; Stenström and Ek 1990; Browning and Whitney 1992).

Field performance of inoculated seedlings

The fundamental means of assessing the effectiveness of inoculating nursery seedlings is their survival and growth in the field (see reviews by Trappe 1977; Le Tacon et al. 1992; Castellano 1996). In Europe, forest tree seedlings inoculated with *L. bicolor* S238N still showed a significant increase in stem volume 8 years after outplanting. Furthermore, in the United States, the success of *P. tinctorius* in improving seedling growth and survival after outplanting is well recognized. Successful nursery inoculation of seedlings with different mycorrhizal fungi has been widely reported in Canada (Gagnon and Langlois 1987; Danielson and Visser 1988; Gagnon et al. 1988, 1995; Browning and Whitney 1993; Quoreshi and Timmer 1998; Khasa et al. 2001). However, little is known concerning the survival and growth of outplanted mycorrhizal seedlings (Kropp and Langlois 1990), and only a few studies (Table 9.1) have investigated their long-term survival and success (Berch and Roth 1993; Campbell et al. 2003).

A study aimed at evaluating the possible need for nursery inoculation of seedlings produced for afforestation work in the prairies was done by Whitney et al. (1972). The ectomycorrhizal inoculum potential of soils from the Canadian Prairie Provinces was examined using *P. banksiana* seedlings planted at various locations. Although the presence of inoculum was confirmed, the study suggested that seedlings planted in Saskatchewan and Manitoba would require artificial inoculation for satisfactory root colonization. Particularly, the prairie soils of Saskatchewan and Manitoba are often characterized by a high pH and frequently suffer from long periods of drought; both factors would potentially limit ectomycorrhizal development. In the above study, the authors observed improved growth of *P. banksiana* seedlings following inoculation with a mixture of ectomycorrhizal root tips and forest humus.

A study done in Québec investigated the competitive performance of selected ectomycorrhizal isolates following outplanting (McAfee and Fortin 1986). *P. banksiana* seedlings were inoculated with *P. tinctorius*, *L. bicolor*, and *Rhizopogon rubescens* Tul. and were outplanted in the boreal forest near La Grande (James Bay, Québec). *Laccaria bicolor* seemed to be the most efficient of the three isolates tested, being an excellent competitor and colonizer on all but one site. In a second study, McAfee and Fortin (1989) conducted a field trial to investigate the occurrence and abundance of indigenous ectomycorrhizal fungi associated with *P. banksiana* and *P. mariana* seedlings in a number of reforested sites across Québec. Six-week-old nonmycorrhizal seedlings were outplanted on sites affected by different forms of forestry-associated disturbance and under different ecological conditions. Colonization by indigenous ectomycorrhizal fungi was lower on

Table 9.1. Evaluation of mycorrhizal inoculation and outplanting performance of forest tree seedlings in Canada.

Location	Plant species	Mycorrhizal fungi	Field performance	Reference
Boreal forest, La Grand, Québec	Jack pine	*Laccaria bicolor*, *Pisolithus tinctorius*, *Rhizopogon rubescens*	Competitive interactions between inoculated fungi tested under field conditions; *Laccaria bicolor* was the most promising mycobiont on most sites tested	McAfee and Fortin 1986
Syncrude Canada Ltd. site, Fort McMurray, Alberta	Jack pine	*Hebeloma* sp., *Laccaria proxima*, *Telephora terrestris*, E-strain, *P. tinctorius*, *Cenococcum geophilum*, *Lactarius paradoxus*, *Astraeus hygrometricus*, *Sphaerosporella brunnea*, *Amphinema byssoides*, *Hydnum imbricatum*, *Tricholoma flavovirens*	The majority of introduced fungi did not colonize new roots; inoculation with E-strain and *T. terrestris* resulted in a two- to three-fold increase in shoot biomass after two years; inoculation was unsuccessful with *A. byssoides*, *H. imbricatum*, and *T. flavovirens*	Danielson and Visser 1989
Sault Sainte Marie, Ontario	Jack pine, black spruce	*L. proxima*, *L. bicolor*, *T. terrestris*, *Hebeloma cylindrosporum*, *P. tinctorius*, *R. rubescens*	Inoculation with selected ectomycorrhizal fungi improved field performance; growth stimulation of jack pine was noted after the first growing season but delayed in black spruce	Browning and Whitney 1992
Southern Interior of British Columbia	Douglas-fir, white spruce, lodgepole pine	*Laccaria* sp., *Rhizopogon* sp., E-strain, *A. byssoides*, *T. terrestris*	Overall, inoculation with selected mycorrhizal fungi did not significantly improve field performance	Hunt 1992
Vancouver Island, British Columbia	Douglas-fir	*Rhizopogon vinicolor*	The inoculated fungus improved growth in the nursery and survived at least one year on roots after outplanting; no benefit observed in the field due to browsing	Berch and Roth 1993
Three sites near Sault Sainte Marie, Ontario	Jack pine, black spruce	*L. proxima*, *L. bicolor*, *T. terrestris*	Seedling growth was not significantly affected by the inoculation treatment; *Laccaria proxima* enhanced the drought tolerance of jack pine seedlings to a greater extent than the others	Browning and Whitney 1993
Iroquois Falls, Ontario	Black spruce	*L. bicolor*	Mycorrhizal seedlings given nutrient fertilizer showed enhanced growth and nutrient status compared with controls	Quoreshi and Timmer 2000
Thompson Plateau near Falkland, and PRT Red Rock, Prince George, British Columbia	Lodgepole pine	*Hebeloma longicaudum*, *R. rubescens*	The growth of inoculated seedling was not significantly different from noninoculated controls; planting environment had the greatest influence on seedling performance	Campbell et al. 2003

Table 9.1 (*concluded*).

Location	Plant species	Mycorrhizal fungi	Field performance	Reference
Cynthia site, Drayton Valley, and Wandering River site, Alberta	White spruce, black spruce, lodgepole pine	*H. longicaudum, L. bicolor, Paxillus involutus, P. tinctorius, R. vinicolor, Suillus tomentosus*	Significant differences in plot volume index were found with certain plant–fungus–site combinations	Gagné et al. 2006
Wandering River, Alberta; Drayton Valley, Alberta; Prince Albert, Saskatchewan	White spruce, black spruce, lodgepole pine, Siberian larch	*H. longicaudum, L. bicolor, P. involutus, P. tinctorius, R. vinicolor, S. tomentosus*	Significantly increased stem volume and plot volume index with certain plant–fungus combinations	Quoreshi et al. 2008
Boyle, Alberta	Various poplars	*H. longicaudum, L. bicolor, P. involutus, P. tinctorius, R. vinicolor, S. tomentosus, Glomus intraradices*	The effect of nursery inoculation was limited and did not affect field performance	Quoreshi et al. 2008

P. mariana than on *P. banksiana*, suggesting that inoculation with selected mycorrhizal fungi could be of greater benefit for outplanted *P. mariana* seedlings (McAfee and Fortin 1989).

During the early 1990s, Browning and Whitney (1992, 1993) investigated the benefits of ectomycorrhizal inoculation on the survival and growth of *P. mariana* and *P. banksiana* seedlings outplanted on reforestation sites in Ontario. In one study, both species were inoculated with *L. bicolor*, *Laccaria proxima* (Bond.) Pat., *Hebeloma cylindrosporum* Romagnesi, *P. tinctorius*, or *R. rubescens*. These fungi had previously exhibited the potential to rapidly colonize host roots and to improve survival and growth on reforested sites. Each tree species was planted on two three-year-old clear-cut sites contrasting in their available soil nutrient levels (Browning and Whitney 1992). The growth of inoculated *P. banksiana* was enhanced after the first growing season; however, positive differences between inoculated and control *P. mariana* seedlings were not observed until after the second growing season. Overall, inoculation improved growth and nutrition, but the results varied depending on the plant species, fungal species, and site. The study concluded that inoculation of container-grown seedlings of these two species with selected ECM fungi could improve outplanting performance on routine reforestation sites. Although *P. tinctorius* failed to form mycorrhizas, the other fungi persisted on the root systems of outplanted seedlings for two growing seasons but declined after the second year. *Laccaria bicolor* was found to be the most persistent mycobiont on root system of both tree species and improved the growth of *P. banksiana*. However, *L. proxima* stimulated the growth of both tree species on both sites. The study concluded that inoculation of containerized *P. mariana* and *P. banksiana* with selected mycorrhizal fungi can improve outplanting performance on some sites.

In another study, field growth and survival of inoculated *P. banksiana* and *P. mariana* seedlings were compared with seedlings that had been naturally colonized by *T. americana* in a nursery (Browning and Whitney 1993). Both tree species were inoculated with *L. bicolor* or *L. proxima*. The presence of both these species prevented *T. americana* from colonizing significant amounts of the roots of the seedlings. Unlike previous field trials (Browning and Whitney 1992), there was no significant difference between inoculated plants and noninoculated controls (Browning and Whitney 1993). This was perhaps due to the heavy colonization of the noninoculated control seedlings by *T. americana*. However, the seedlings were outplanted just prior to a month-long drought, and the results of the experiment indicated that *L. proxima* inferred a greater drought resistance to *P. banksiana* seedlings than did *T. americana* or *L. bicolor*.

Published research from several countries demonstrates that appropriate mycorrhizal fungi can enhance growth and survival of plants on disturbed sites, leading to successful reclamation (Cordell et al. 1991; Marx 1991; Malajczuk et al. 1994). Danielson and Visser (1989), working on oil sand tailings in Alberta, inoculated *P. banksiana* seedlings with nine different mycorrhizal fungi; however, two years after outplanting, only plants colonized by E-strain and *T. americana* showed a greater shoot mass than the noninoculated controls. In this study, most of the inoculated fungi failed to colonize new roots growing in the oil sands – peat mixture, highlighting the fact that only certain fungi can adapt to certain site-specific soil environments. Furthermore, this suggests that it may be advantageous to identify and use both host-specific and site-specific plant–fungal combinations. Such an approach may lead to a greater success following outplanting.

Hunt (1992) examined the potential benefits of ectomycorrhizal fungi in nurseries, as well as after outplanting on standard sites in the southern interior of British Colombia. *Pseudotsuga menziesii* (Mirb.) Franco. var. *glauca* (Beissn.) (Douglas-fir), *Picea engelmannii* Parry ex Engelm. (Engelmann spruce), *P. glauca*, and *Pinus contorta* Doug. ex Loud. var. *latifolia* Engelm. (lodgepole pine) seedlings were colonized by a given ectomycorrhizal fungal species, either by artificial inoculation or natural colonization. *Amphinema byssoides*, E-strain, and *T. americana* were the main fungal species found on naturally colonized nursery-grown seedlings. Artificial inoculation was done using a culture of a *Laccaria* sp. or a spore suspension of a *Rhizopogon* sp. Two years after outplanting, seedlings colonized by E-strain or *A. byssoides* had relatively taller shoots with larger root collar diameters than their respective controls. However, shoots of artificially inoculated seedlings were shorter than the controls, probably because of smaller size at the time of outplanting. Inoculation with selected mycorrhizal fungi did not affect tree survival on the less rigorous sites used in this study. Overall, inoculation with selected mycorrhizal fungi failed to significantly increase plantation performance. However, the plantations had relatively good soil conditions and low drought stress, and the control seedlings were spontaneously inoculated by indigenous fungi, which confounded the results.

A recent study by Campbell et al. (2003) investigated the field performance of *P. contorta* seedlings inoculated with either *H. longicaudum* or *R. rubescens* and planted into different rooting environments at two locations in British Columbia. In the first experiment, seedlings were outplanted onto fully rehabilitated log landings, ripped log landings, and unprepared cut blocks. After two growing seasons, there were no significant differences in seedling size between inoculated and noninoculated seedlings. However, seedlings on fully rehabilitated log landings had better height and diameter growth compared with other sites, suggesting that the planting environment had the greatest influence on growth. In the second experiment, seedlings were either outplanted in boot screefed cutblocks or directly into the undisturbed forest floor. The results showed a minor growth response to inoculation.

Berch and Roth (1993) investigated whether *Rhizopogon vinicolor* Smith could improve the growth of *P. menziesii* seedlings in the nursery and in the field one year after ouplanting. The results indicate that the fungus improved seedling growth in the nursery and survived on the roots for at least one growing season following outplanting. Nursery adapted *T. americana* colonized one-half of the control seedlings and at least 10% of inoculated seedlings prior to outplanting. After one season in the field, both inoculated and control seedlings were colonized by several other fungal species and the relative abundance of *T. americana* decreased. However, because the plants were heavily browsed in the field, it was not possible to determine whether the advantage achieved in the nursery was maintained in the field.

In a more recent study, Gagné et al. (2006) investigated the long-term survival after outplanting on clear-cut sites of six ectomycorrhizal fungal species inoculated onto the roots of three conifer species under nursery conditions. Morphological and molecular characterization indicated that the most abundant ECM fungi present on the roots were ascomycetes and that the most common basidiomycete was *A. byssoides*. The authors demonstrated that, of the six fungi used to inoculate the seedlings, only the strain of *L. bicolor* was still present after years 5 and 6. However, significant differences in the plot volume index (PVI) of trees were observed between inoculated and control treatments, suggesting a positive responses to nursery inoculation. The PVI is directly related to plant survival and seedling volume and is considered to be an excellent indication of the field performance of seedlings (Marx et al. 1977, 1991).

Mikro-Tek, a Canadian biotechnology company, conducted trials in the Canadian boreal forest to determine the potential benefits of mycorrhizal technology to forestry. The preliminary results showed that inoculation enhanced tree seedling survival by up to 50% on harsh industrial reclamation sites and that early growth was increased by between 20% and 30% on a variety of other reforestation sites (Smith and Mohammed 1997). Another Canadian biotechnology company, SymbioTech Research Inc., is currently concentrating on the development of environmentally sustainable techniques for the production of forest tree seedlings using effective indigenous fungi.

The company has developed Treeboost®, an ECM fungal inoculum that has the potential to improve afforestation and reforestation activities in the Prairies Provinces. The product is proving particularly useful on difficult sites and is currently being tested on the nutrient-poor, highly alkaline, saline spoil left after the surface mining of oil sands by the tar sands industry in Alberta (Quoreshi et al. 2005; see also Chapter 10).

Mycorrhizal fungi are not the only microorganisms to colonize plant roots (see Chapter 4); in a recent study, the field performance of different conifer and deciduous softwood species inoculated with a range of different soil organisms was evaluated at either three or five years after outplanting on forest sites in the Prairie Provinces (Quoreshi et al. 2008). Four field trials, three with conifer species (*P. glauca*, *P. mariana*, *P. contorta*, and *Larix sibirica* Ledeb. (Siberian larch)) and one with deciduous softwoods (*Populus alba* L. (white aspen) and four poplar clones), were established at different locations in Alberta and Saskatchewan. The conifer species were inoculated with one of six ECM fungal species, *H. longicaudum*, *L. bicolor*, *P. involutus*, *P. tinctorius*, *R. vinicolor*, and *S. tomentosus*. The deciduous softwood trial was established using the same six fungal species, one AM fungus, two species of bacteria, a growth hormone, coinoculation with *P. tinctorius*, and a control (Quoreshi and Khasa 2008). The field growth and survival data of these field trials were converted into estimated seedling volume and PVI (Table 9.2).

The PVI of *P. glauca* on one of the sites was significantly higher (33%–84%) in all inoculation treatments when compared with the controls. The results from a second site show an increased PVI for *P. mariana* in all inoculation treatments, except with *R. vinicolor*. Moreover, the seedlings inoculated with *S. tomentosus* exhibited approximately an 80% increase in PVI, and those inoculated with *L. bicolor* showed approximately a 95% increase compared with their respective controls. However, the PVI of *P. glauca* on the same site was not improved by the inoculation treatments but showed a higher PVI than the controls in certain treatments. In the case of *P. contorta*, the PVI generally increased with all treatments compared with the control. However, only the *P. involutus* treatment showed a significantly greater PVI (approximately 235%) compared with the controls. In the field trial site in Saskatchewan, *L. sibirica* seedlings showed a significantly greater PVI in almost all inoculation treatments (103%–534%) compared with the controls. These results demonstrate that certain inoculated plant–fungus combinations play a positive role during the initial establishment of seedlings in the field. In the deciduous softwood trial, none of the inoculation treatments had a significant influence on seedling performance, even though significant differences were observed between the different clones.

The differences in the outplanting performance of the control and inoculated seedlings in the above experiment would probably have been greater if the seedlings had been planted in disturbed sites where the ECM inoculum potential was low. The lack of improvement observed in some cases may also be due to a greater competitiveness of certain site-indigenous mycorrhizal fungi (McAfee and Fortin 1986). Nevertheless, nursery-grown tree seedlings with adequate mycorrhizal root systems are more biologically preconditioned than nonmycorrhizal seedlings and have a generally greater ability to exploit the outplanting site for the resources required for growth (Quoreshi and Timmer 2000; Quoreshi 2003). It must be borne in mind that mycorrhizal inoculation may not always provide positive results and that the fungal strain, tree species or genotype, and site characteristics are all important factors determining the level of success of an inoculation program. Small-scale experiments should always be done before the large-scale use of a given fungus–host combination on a given site.

Further experiments are required to test the effectiveness of seedlings inoculated with more site-adapted, stress-tolerant mycorrhizal fungi for outplanting

Table 9.2. Effects of different ectomycorrhizal inoculation treatments on seedling mean plot volume index (PVI).

Plant species	Inoculation treatment	PVI (cm^3)
Wandering River, Alberta		
Black spruce	Control	796b
	Hebeloma longicaudum	1237ab
	Laccaria bicolor	1552a
	Paxillus involutus	1236ab
	Pisolithus tinctorius	1184ab
	Rhizopogon vinicolor	719b
	Suillus tomentosus	1435a
White spruce	Control	1249a
	H. longicaudum	742abc
	L. bicolor	465c
	P. involutus	683bc
	P. tinctorius	630bc
	R. vinicolor	719abc
	S. tomentosus	1111ab
Lodgepole pine	Control	1679b
	H. longicaudum	2557b
	L. bicolor	2418b
	P. involutus	5622a
	P. tinctorius	2831ab
	R. vinicolor	2759b
	S. tomentosus	2543b
Drayton Valley, Alberta		
White spruce	Control	504c
	H. longicaudum	783a
	L. bicolor	924a
	P. involutus	672ab
	P. tinctorius	746ab
	R. vinicolor	906a
	S. tomentosus	702b
Prince Albert, Saskatchewan		
Siberian larch	Control	80d
	H. longicaudum	343ab
	L. bicolor	163c
	P. involutus	209bc
	P. tinctorius	402a
	R. vinicolor	507ab
	S. tomentosus	120d

Note: The PVIs were calculated five years after initial outplanting, except for the Prince Albert site, where they were calculated after three years. For each site and plant species, means within a column followed by the same letter are not significantly different (LSD test, $p < 0.05$).

in heavily disturbed sites. Advances in inoculum production are needed to provide readily available and economically viable commercial mycorrhizal inocula. Furthermore, when possible, indigenous site-adapted fungal strains should be used, because these may enhance reclamation and restoration. The further development and application of molecular techniques to monitor introduced microbial consortia and characterize their activities and persistence are also required.

References

Allen, M.F. 1991. The ecology of mycorrhizae. Cambridge University Press, Cambridge, UK.

Barrow, J.R., and Osuna, P. 2002. Phosphorus solubilization and uptake by dark septate fungi in fourwing saltbush, *Atriplex canescens* (Pursh) Nutt. J. Arid Environ. **51**: 449–459.

Berch, S.M., and Roth, A.L. 1993. Ectomycorrhizae and growth of Douglas-fir seedlings preinoculated with *Rhizopogon vinicolor* and outplanted on eastern Vancouver Island. Can. J. For. Res. **23**: 1711–1715.

Browning, M.H.R., and Whitney, R.D. 1992. Field performance of black spruce and jack pine inoculated with selected species of ectomycorrhizal fungi. Can. J. For. Res. **22**: 1974–1982.

Browning, M.H.R., and Whitney, R.D. 1993. Infection of containerized jack pine and black spruce by *Laccaria* species and *Thelephora terrestris* and seedling survival and growth after outplanting. Can. J. For. Res. **23**: 330–333.

Campbell, D.B., Jones, M.D., Kiiskila, S., and Bulmer, C. 2003. Two-year field performance of lodgepole pine seedlings: effects of container type, mycorrhizal fungal inoculants, and site preparation [online]. B.C. J. Ecosyst. Manage. **3**(2): article 2. Available from www.forrex.org/JEM/ISS23/vol3_no2_art2.pdf

Castellano, M.A. 1996. Outplanting performance of mycorrhizal inoculated seedlings. *In* Concepts in mycorrhizal research. *Edited by* K.G. Mukerji. Kluwer Academic Publisher, Dordrecht, the Netherlands. pp. 223–301.

Castellano, M.A., and Trappe, J.M. 1985. Ectomycorrhizal formation and plantation performance of Douglas-fir nursery stock inoculated with *Rhizopogon* spores. Can. J. For. Res. **15**: 613–617.

Castellano, M.A., Trappe, J.M., and Molina, R. 1985. Inoculation of container-grown Douglas-fir seedlings with basidiospores of *Rhizopogon vinicolor* and *Rhizopogon colossus*: effects of fertility and spore application rate. Can. J. For. Res. **15**: 10–13.

Cázares, E., and Smith, J.E. 1995. Occurrence of vesicular–arbuscular mycorrhizae in *Pseudotsuga menziesii* and *Tsuga heterophylla* seedlings grown in Oregon Coast Range soils. Mycorrhiza, **6**: 65–67.

Chakravarty, P., and Chatarpaul, L. 1990. Effect of fertilization on seedling growth, ectomycorrhizal symbiosis, and nutrient uptake in *Larix laricina*. Can. J. For. Res. **20**: 245–248.

Colpaert, J.V., and van Assche, J.A. 1987. Heavy metal tolerance in some ectomycorrhizal fungi. Funct. Ecol. **1**: 415–421.

Cooke, J.C., and Lefor, M.W. 1990. Comparison of vesicular–arbuscular mycorrhizae in plants from disturbed and adjacent undisturbed regions of a coastal salt marsh in Clinton, Connecticut, USA. Environ. Manage. **14**: 131–137.

Cordell, C.E., and Marx, D.H. 1994. Effects of nursery cultural practices on management of specific ectomycorrhizae on bareroot tree seedlings. *In* Mycorrhizae and plant health. *Edited by* F.L. Pfleger and R.G. Linderman. American Phytopathological Society Press, St. Paul, Minn. pp. 133–152.

Cordell, C.E., Marx, D.H., and Caldwell, C. 1991. Operational application of specific ectomycorrhizal fungi in mineland reclamation. *In* Procedings of the 1991 National Meetings of the American Society of Surface Mining and Reclamation, 14–17 May 1991, Durango, Colo. Edited by W. Oaks and J. Bowden. American Society of Surface Mining and Reclamation, Lexington, Ky. pp. 641–648.

Danielson, R.M. 1991. Temporal changes and effects of amendments on the occurrence of sheathing (ecto-) mycorrhizas of conifers growing in oil sands tailings and coal spoil. Agric. Ecosyst. Environ. **35**: 261–281.

Danielson, R.M., and Visser, S. 1988. Ectomycorrhizae of jack pine and green alder: assessment of the need for inoculation, development of inoculation techniques and outplanting trials on oil sand tailings. Alberta Land Conservation and Reclamation Council, Edmonton, Alta Rep. RRTAC 88-5.

Danielson, R.M., and Visser, S. 1989. Host response to inoculation and behavior of introduced and indigenous ectomycorrhizal fungi of jack pine grown on oil-sands tailings. Can. J. For. Res. **19**: 1412–1421.

Danielson, R.M., and Visser, S. 1990. The mycorrhizal and nodulation status of container-grown trees and shrubs reared in commercial nurseries. Can. J. For. Res. **20**: 609–614.

Danielson, R.M., Griffith, C.L., and Parkinson, D. 1984. Effects of fertilization on the growth and mycorrhizal development of container-grown jack pine seedlings. For. Sci. **30**: 828–835.

Duchesne, L.C., Peterson, R.L., and Ellis, B.E. 1988. Interaction between the ectomycorrhizal fungus *Paxillus involutus* and *Pinus resinosa* induces resistance to *Fusarium oxyporum*. Can. J. Bot. **66**: 558–562.

Duddridge, J.A., Malibari, A., and Read, D.J. 1980. Structure and function of mycorrhizal rhizomorphs with special reference to water transport. Nature (Lond.), **287**: 834–836.

Dunabeitia, M., Rodriguez, N., Salcedo, I., and Sarrionandia, E. 2004. Field mycorrhization and its influence on the establishment and development of the seedlings in a broadleaf plantation in the Basque country. For. Ecol. Manage. **195**: 129–139.

Egger, K.N. 1996. Molecular systematics of E-strain mycorrhizal fungi: *Wilcoxina* and its relationship to *Tricharina* (Pezizales). Can. J. Bot. **74**: 773–779.

El Karkouri, K., Martin, F., Douzery, J.P., and Mousain, D. 2005. Diversity of ectomycorrhizal fungi naturally established on containerised *Pinus* seedlings in nursery conditions. Microbiol. Res. **160**: 47–52.

Gagné, A., Jany, J.-L., Bousquet, J., and Khasa, D.P. 2006. Ectomycorrhizal fungal communities of nursery-inoculated seedlings outplanted on clear-cut sites in northern Alberta. Can. J. For. Res. **36**: 1684–1694.

Gagnon, J., and Langlois, C.-G. 1987. Growth of containerized jack pine seedlings inoculated with different ectomycorrhizal fungi under a controlled fertilization schedule. Can. J. For. Res. **17**: 840–845.

Gagnon, J., Langlois, C.-G., and Fortin, J.A. 1988. Growth and ectomycorrhizal formation of containerized black spruce seedlings as affected by nitrogen fertilization, inoculum type, and symbiont. Can. J. For. Res. **18**: 922–929.

Gagnon, J., Langlois, C.-G., Bouchard, D., and Le Tacon, F. 1995. Growth and ectomycorrhizal formation of container-grown Douglas-fir seedlings inoculated with *Laccaria bicolor* under four levels of nitrogen fertilization. Can. J. For. Res. **25**: 1953–1961.

Hambling, B., Andrewes, C., and Khasa, P.D. 2000. Development of ectomycorrhizal fungal inoculum for pre-commercial production. Final report for SymbioTech Research Inc. submitted to National Research Council (IRAP Program), Alberta and Northwest Territories office. SymbioTech Research Inc., Edmonton, Alta.

Hunt, G.A. 1988. Effect of controlled-release fertilizers on formation of ectomycorrhizae and growth of container-grown Engelmann spruce. *In* Proceedings of the Combined Western Forest Nursery Council and Intermountain Nursery association meeting, 8–11 August 1988, Vernon, British Columbia. Edited by T.D. Landis and D. Thomas. USDA For. Serv. Rocky Mountain For. Range Exp. Stn. RM-167.

Hunt, G.A. 1992. Effects of mycorrhizal fungi on quality of nursery stock and plantation performance in the southern interior of British Columbia. Forestry Canada, Ottwa, Ont., and Province of British Columbia, Victoria, B,C. FRDA Rep. 185.

Iwanski, M., Rudawska, M., and Leski, T. 2006. Mycorrhizal associations of nursery grown Scots pine (*Pinus sylvestris* L.) seedlings in Poland. Ann. For. Sci. **63**: 715–723.

Jones, M.D., and Hutchinson, T.C. 1988. Nickel toxicity in mycorrhizal birch seedlings infected with *Lactarius rufus* or *Scleroderma flavidum*. I. Effects on growth, photosynthesis, respiration and transpiration. New Phytol. **108**: 451–459.

Jones, M.D., Hagerman, S.M., and Gillespie, M. 2002. Ectomycorrhizal colonization and richness of previously colonized, containerized *Picea engelmannii* does not vary across clearcuts when planted in mechanically site-prepared mounds. Can. J. For. Res. **32**: 1425–1433.

Jumpponen, A., and Trappe, J.M. 1998. Dark septate endophytes: a review of facultative biotrophic root-colonizing fungi. New Phytol. **140**: 295–319.

Kernaghan, G., Sigler, L., and Khasa, D. 2003. Mycorrhizal and root endophytic fungi of containerized *Picea glauca* seedlings assessed by rDNA sequence analysis. Microb. Ecol. **45**: 128–136.

Khasa, D.P., Sigler, L., Chakravarty, P., Dancik, B.P., Erickson, L., and McCurdy, D. 2001. Effect of fertilization on growth and ectomycorrhizal development of container-grown and bare-root nursery conifer seedlings. New For. **22**: 179–197.

Kropp, B.R., and Fortin, J.A. 1988. The incompatibility system and relative ectomycorrhizal performance of monokaryons and reconstituted dikaryons of *Laccaria bicolor*. Can. J. Bot. **66**: 289–294.

Kropp, B.R., and Langlois, C.G. 1990. Ectomycorrhizae in reforestation. Can. J. For. Res. **20**: 438–451.

Laiho, O. 1965. Further studies on the ectendotrophic mycorrhiza. Acta For. Fenn. **79**: 1–35.

Le Tacon, F., Álvarez, I.F., Bouchard, D., Henrion, B., Jackson, M.R., Luff, S., Parladé, I.J., Pera, J., Stenström, E., Villeneuve, N., and Walker, C. 1992. Variations in field response of forest trees to nursery ectomycorrhizal inoculation in Europe. *In* Mycorrhizas in ecosystems. Edited by D.J. Read, H.D. Lewis, A.H. Fitter, and I.J. Alexander. CAB International, Wallingford, UK. pp. 119–134.

Le Tacon, F., Mousain, D., Garbaye, J., Bouchard, D., Churin, J.L., Argillier, C., Amirault, J.M., and Genere, B. 1997.

Mycorhizes, pépinières et plantations forestières en France. Rev. For. Fr. **49**(No. Spec.): 131–134.

Lodge, D.J. 2000. Ecto- or arbuscular mycorrhizas—which are best? New Phytol. **146**: 353–354.

Malajczuk, N., Redell, P., and Brundrett, M. 1994. Role of ectomycorrhizal fungi in minesite reclamation. *In* Mycorrhizae and plant health. *Edited by* F.L. Pfleger and R.G. Linderman. American Phytopathological Society Press, St. Paul, Minn. pp. 83–100.

Marx, D.H. 1969. The influence of ectotrophic mycorrhizal fungi on the resistance of pine roots to pathogenic infections. I. Antagonism of mycorrhizal fungi to root pathogenic fungi and soil bacteria. Phytopathology, **59**: 153–163.

Marx, D.H. 1980. Ectomycorrhiza fungus inoculations: a tool for improving forestation practices. *In* Tropical mycorrhiza research. *Edited by* P. Mikola. Oxford University Press, Oxford, UK. pp. 13–71.

Marx, D.H. 1991. The practical significance of ectomycorrhizae in forest establishment. *In* Ecophysiology of ectomycorrhizae of forest trees. Wallenberg Foundation, Stockholm. pp. 54–90.

Marx, D.H., Bryan, W.C., and Cordell, C.E. 1977. Survival and growth of pine seedlings with *Pisolithus* ectomycorrhizae after two years on reforestation sites in North Carolina and Florida. For. Sci. **23**: 363–373.

Marx, D.H., Hedin, A., and Toe, S.F.P., IV. 1985. Field performance of *Pinus caribaea* var. *hondurensis* seedlings with specific ectomycorrhizae and fertilizer after three years on a savannah site in Liberia. For. Ecol. Manage. **13**: 1–25.

Marx, D.H., Ruehle, J.L., and Cordell, C.E. 1991. Methods for studying nursery and field response of trees to specific ectomycorrhizae. *In* Techniques for mycorrhizal research. *Edited by* J.R. Norris, D. Read, and A.K. Varma. Academic Press, San Diego, Calif. pp. 384–411.

Massicotte, H.B., Melville, L.H., Peterson, R.L., and Molina, R. 1999. Biology of the ectomycorrhizal fungal genus, *Rhizopogon*. IV. Comparative morphology and anatomy of ectomycorrhizas synthesized between several *Rhizopogon* species on ponderosa pine (*Pinus ponderosa*). New Phytol. **142**: 355–370.

McAfee, B.J., and Fortin, J.A. 1986. Competitive interactions of ectomycorrhizal mycobionts under field conditions. Can. J. Bot. **64**: 848–852.

McAfee, B.J., and Fortin, J.A. 1989. Ectomycorrhizal colonization on black spruce and jack pine seedlings outplanted in reforestation sites. Plant Soil, **116**: 9–17.

Menkis, A., Vasiliauskas, R., Taylor, A.F.S., Stenlid, J., and Finlay, R. 2005. Fungal communities in mycorrhizal roots of conifer seedlings in forest nurseries under different cultivation systems, assessed by morphotyping, direct sequencing and mycelial isolation. Mycorrhiza, **16**: 33–41.

Menkis, A., Vasiliauskas, R., Taylor, A.F.S., Stenström, E., Stenlid, J., and Finlay, R. 2006. Fungi in decayed roots of conifer seedlings in forest nurseries, afforested clear-cuts and abandoned farmland. Plant Pathol. **55**: 117–129.

Mikola, P. 1965. Studies on the ectendotrophic mycorrhiza of pine. Acta For. Fenn. **79**: 1–56.

Mikola, P. 1970. Mycorrhizal inoculation in afforestation. Int. Rev. For. Res. **3**: 123–196.

Mikola, P. 1988. Ectendomycorrhiza of conifers. Silva Fenn. **22**: 19–27.

Miller, R.M., and Jastrow, J.D. 1992. Extraradical hyphal development of vesicular–arbuscular mycorrhizal fungi in a chronosequence of prairie restorations. *In* Mycorrhizas in ecosystems. *Edited by* D.J. Read, H.D. Lewis, A.H. Fitter, and I.J. Alexander. CAB International, Wallingford, UK. pp. 171–176.

Molina, R., and Chamard, J. 1983. Use of the ectomycorrhizal fungus *Laccaria laccata* in forestry. II. Effects of fertilizer forms and levels on ectomycorrhizal development and growth of container-grown Douglas-fir and ponderosa pine seedlings. Can. J. For. Res. **13**: 89–95.

Molina, R., and Trappe, J.M. 1982. Applied aspects of ectomycorrhizae. *In* Advances in agricultural microbiology. *Edited by* R. Subba and N.S. Oxford. IBH Publishing Co., New Delhi. pp. 305–324.

Molina, R., and Trappe, J.M. 1984. Mycorrhiza management in bareroot nurseries. *In* Forest nursery manual: production of bareroot seedlings. *Edited by* M.L. Duryea and T.D. Landis. Martinus Nijhoff/Dr. W. Junk Publishers, The Hague. pp. 211–233.

Molina, R., Massicotte, H., and Trappe, J.M. 1992. Specificity phenomena in mycorrhizal symbioses: community-ecological consequences and practical implications. *In* Mycorrhizal functioning: an integrative plant–fungal process. *Edited by* M.J. Allen. Chapman & Hall, New York. pp. 357–423.

Mosse, B., Stribley, D.P., and Le Tacon, F. 1981. Ecology of mycorrhizae and mycorrhizal fungi. Adv. Microb. Ecol. **5**: 137–210.

O'Neill, J.J.M., and Mitchell, D.T. 2000. Effects of benomyl and captan on growth and mycorrhizal colonization of Sitka spruce (*Picea sitchensis*) and ash (*Fraxinus excelsior*) in Irish nursery soil. For. Pathol. **30**: 165–174.

Ortega, U., Dunabeitia, M., Menendez, S., Gonzalez-Muria, C., and Majada, J. 2004. Effectiveness of mycorrhizal inoculation in the nursery on growth and water relations of *Pinus radiata* in different water regimes. Tree Physiol. **24**: 64–73.

Pera, J., Álvarez, I.F., Rincon, A.M., and Parladé, J. 1999. Field performance in northern Spain of Douglas-fir seedlings inoculated with ectomycorrhizal fungi. Mycorrhiza, **9**: 77–84.

Perry, D.A., Molina, R., and Amaranthus, M.P. 1987. Mycorrhizae, mycorrhizospheres, and reforestation: current knowledge and research needs. Can. J. For. Res. **17**: 929–940.

Peterson, R.L., Wagg, C., and Pautler, M. 2008. Associations between microfungal endophytes and roots: do structural features indicate function? Botany, **86**: 445–456.

Quoreshi, A.M. 2003. Nutritional preconditioning and ectomycorrhizal formation of *Picea mariana* (Mill.) B.S.P. seedlings. Euras. J. For. Res. **6**: 1–63.

Quoreshi, A.M., and Khasa, D.P. 2008. Effectiveness of mycorrhizal inoculation in the nursery on root colonization, growth, and nutrient uptake of aspen and balsam poplar. Biomass Bioenergy, **32**(5): 381–391.

Quoreshi, A.M., and Timmer, V.R. 1998. Exponential fertilization increases nutrient uptake and ectomycorrhizal development of black spruce seedlings. Can. J. For. Res. **28**: 674–682.

Chapter 9. Mycorrhizal fungi in Canadian forest nurseries and field performance of inoculated seedlings

Quoreshi, A.M., and Timmer, V.R. 2000. Early outplanting performance of nutrient-loaded containerized black spruce seedlings inoculated with *Laccaria bicolor*: a bioassay study. Can. J. For. Res. **30**: 744–752.

Quoreshi, A.M., Khasa, D.P., Bois, G., Jany, J.-L., Begrand, E., McCurdy, D., and Fung, M. 2005. Mycorrhizal biotechnology for reclamation of oil sand composite tailings and tailings sand in Alberta. *In* The Thin Green Line: A Symposium on the State-of-the-Art in Reforestation Proceedings, 26–28 July 2005, Thunder Bay, Ont. *Edited by* S.J. Colombo. Ontario Ministry of Natural Resources, Ontario Forest Research Institute, Sault Ste. Marie, Ont. For. Res. Inf. Pap. 160. pp. 122–127.

Quoreshi, A.M., Piché, Y., and Khasa, D.P. 2008. Field performance of conifer and hardwood species 5 years after nursery inoculation in the Canadian Prairie Provinces. New For. **35**: 235–253.

Rivera-Becerril, F., Calantzis, C., Turnau, K., Caussanel, J.P., Belimov, A.A., Gianinazzi, S., Straaser, R.J., and Gianinazzi-Pearson, V. 2002. Cadmium accumulation and buffering of cadmium-induced stress by arbuscular mycorrhiza in three *Pisum sativum* L. genotypes. J. Exp. Bot. **53**: 1177–1185.

Rudawska, M., Leski, T., Trocha, L.K., and Gornowicz, R. 2006. Ectomycorrhizal status of Norway spruce seedlings from bare-root forest nurseries. For. Ecol. Manage. **236**: 375–384.

Ruehle, J.L. 1982. Field performance of container-grown loblolly pine seedlings with specific ectomycorrhizae on a reforestation site in South Carolina. South. J. Appl. For. **6**: 30–33.

Ruehle, J.L., and Marx, D.H. 1977. Developing ectomycorrhizae on containerized pine seedlings. USDA Forest Service Southeast For. Range Exp. Stn. Res. Note SE-242.

Schelkle, M., and Peterson, R.L. 1997. Suppression of common root pathogens by helper bacteria and ectomycorrhizal fungi *in vitro*. Mycorrhiza, **6**: 481–485.

Shaw, G.C., III, Molina, R., and Walden, J. 1982. Development of ectomycorrhizae following inoculation of containerized Sitka and white spruce seedlings. Can. J. For. Res. **12**: 191–195.

Smith, S.M., and Read, D.J. 1997. Mycorrhizal symbiosis 2nd ed. Academic Press, London.

Smith, W., and Mohammed, G.H. 1997. Inoculation with mycorrhizal fungi (*Heboloma* spp.) can increase drought stress resistance and improve field performance of jack pine, black spruce, and white spruce. Ontario Ministry of Natural Resources, Ontario Forest Research Institute, Sault Ste. Marie, Ont. For. Res. Rep. 145. pp. 1–10.

Staddon, P.L. 2005. Mycorrhizal fungi and environmental change: the need for a mycocentric approach. New Phytol. **167**: 635–637.

Stenström, E., and Ek, M. 1990. Field growth of *Pinus sylvestris* following nursery inoculation with mycorrhizal fungi. Can. J. For. Res. **20**: 914–918.

Sundari, S.K., and Adholeya, A. 2003. Growth profile of ectomycorrhizal fungal mycelium: emphasis on substrate pH influence. Antonie van Leeuwenhoek, **83**: 209–214.

Trappe, J.M. 1977. Selection of fungi for ectomycorrhizal inoculation in nurseries. Annu. Rev. Phytopathol. **15**: 203–222.

Trappe, J.M., and Strand, R.F. 1969. Mycorrhizal deficiency in a Douglas-fir region nursery. For. Sci. **15**: 381–389.

Treseder, K.K., and Allen, M.F. 2000. Mycorrhizal fungi have a potential role in soil carbon storage under elevated CO_2 and nitrogen deposition. New Phytol. **147**: 189–200.

Treseder, K.K., Turner, K.M., and Mack, M.C. 2007. Mycorrhizal responses to nitrogen fertilization in boreal ecosystems: potential consequences for soil carbon storage. Glob. Change Biol. **13**: 78–88.

Trocha, L.K., Rudawska, M., Leski, T., and Dabert, M. 2006. Genetic diversity of naturally established ectomycorrhizal fungi on Norway spruce seedlings under nursery conditions. Microb. Ecol. **52**: 418–425.

Ursic, M., and Peterson, R.L. 1997. Morphological and anatomical characterization of ectomycorrhizas and ectendomycorrhizas on *Pinus strobus* seedlings in a southern Ontario nursery. Can. J. Bot. **75**: 2057–2072.

Valdes, M. 1986. Survival and growth of pines with specific ectomycorrhizae after 3 years on a highly eroded site. Can. J. Bot. **64**: 885–888.

Väre, H. 1990. Effects of soil fertility on root colonization and plant growth of *Pinus sylvestris* nursery seedlings inoculated with different ectomycorrhizal fungi. Scand. J. For. Res. **5**: 493–499.

Villeneuve, N., Le Tacon, F., and Bouchard, D. 1991. Survival of inoculated *Laccaria bicolor* in competition with native ectomycorrhizal fungi and effects on the growth of outplanted Douglas-fir seedlings. Plant Soil, **135**: 95–107.

Wagg, C., Pautler, M., Massicotte, H.B., and Peterson, R.L. 2008. The co-occurrence of ectomycorrhizal, arbuscular mycorrhizal, and dark septate fungi in seedlings of four members of the Pinaceae. Mycorrhiza, **18**: 103–110.

Whitney, R.D., Bohaychuk, W.P., and Briant, M.A. 1972. Mycorrhizae of jack pine seedlings in Saskatchewan and Manitoba. Can. J. For. Res. **2**: 228–235.

Yu, T.E.J.-C., Egger, K.N., and Peterson, R.L. 2001. Ectendomycorrhizal associations: characteristics and functions. Mycorrhiza, **11**: 167–177.

Chapter 10
Ectomycorrhizal inoculation for boreal forest ecosystem restoration following oil sand extraction: the need for an initial three-step screening process

Grégory Bois and Andrew P. Coughlan

Introduction

In northeastern Alberta (Canada), opencast mining for oil sands involves the deforestation of vast areas of boreal forest (Fig. 10.1) and generates large volumes of spoil. In addition, the subsequent oil extraction process using NaOH produces large quantities of sand tailings (e.g., between 1978 and 1998, Syncrude Canada Ltd., one of the major companies mining oil sands in Alberta, disturbed over 40 000 ha of land (Fung and Macyk 2000) and produced more than 75×10^6 m^3 of tailings (Li and Fung 1998)).

Once exhausted, mining pits (30–80 m deep) are filled with a mix of tailings and saline release water generated during the extraction process. This mix is capped with a reconstructed "soil" (a substrate capable of supporting restoration work), which consists of a 20–40 cm layer of overburden spoil (naturally highly saline) overlain with 20–30 cm of peat and (or) mineral soil (Fig. 10.2). Various phytostabilization and ecosystem-rehabilitation techniques are then employed to accelerate the reestablishment of a boreal ecosystem on the site. Typically, during the first year, the arbuscular mycorrhizal host *Hordeum vulgare* L. (barley) is used to reduce wind and water erosion and stabilize the site. Subsequent plantings comprise nursery-grown seedlings of indigenous boreal forest tree species (e.g., *Pinus banksiana* Lamb., jack pine, and *Picea glauca* (Moench) Voss, white spruce; Fung and Macyk 2000), which typically have associated ectomycorrhizas.

Despite the peat and (or) mineral soil amendment, the fertility and physical stability of the reconstructed soil remain low, which constitutes a major problem for revegetation work. Plant establishment is further hindered by the presence of hydrocarbon residues, heavy metals, and high concentrations of various salts. The latter include NaCl, a natural component of Albertan oil sands, together with the NaOH used to increase the oil extraction yield (Fung and Macyk 2000; Alberta Environment 2001) and, in certain cases, $CaSO_4 \cdot 2H_2O$, which is added to fine tailings to flocculate out clay particles to reduce the final volume of tailing produced (Li and Fung 1998; see below and Franklin et al. (2002) for the chemical composition of release water from sand tailings treated with $CaSO_4 \cdot 2H_2O$).

In addition to the aforementioned physicochemical constraints, the extraction process effectively sterilizes the tailings (Bois et al. 2005). Furthermore, stockpiling and physical disturbance of the amendment materials and their subsequent dilution with tailings and overburden reduce the number of active mycorrhizal fungal propagules per unit volume (Danielson et al. 1983; Malajczuk et al. 1994; Pfleger et al. 1994). As a result, the sites for restoration exhibit a low mycorrhizal inoculum potential (Bois et al. 2005). This situation must be palliated to achieve an acceptable rate of plant survival following site revegetation (Bois et al. 2005). Different species and isolates of mycorrhizal fungi have been shown to reduce the negative effect of osmotic or drought stress (Lamhamedi et al. 1992; Bois et al. 2006b) and the toxicity of certain elements (e.g., heavy metals) on the health of their host plants (Kropp and Langlois 1990; Kottke 1992; Malajczuk et al. 1994;

G. Bois.[1] PHYTOREM® S.A, 30, avenue Charles de Gaulle, 13140 Miramas, France.
A.P. Coughlan. Centre d'étude de la forêt, Pavillon C.-E.-Marchand, Université Laval, Québec, QC G1V 0A6, Canada.
[1]Corresponding author (e-mail: boisgregory@gmail.com).

Fig. 10.1. Deforestation prior to oil sand mining (Syncrude Canada Ltd. mining site, Fort McMurray, Alberta, Canada).

Marschner and Dell 1994; Pfleger et al. 1994; Smith and Read 1997; Jentschke and Goldbold 2000). Therefore, the mycorrhizal fungal component cannot be neglected in reclamation programs because most plants rely on them for survival, growth, and completion of their life cycle.

Because of the saline nature of the reconstructed postmining substrate, the phyto- (host plant) and myco-bionts (mycorrhizal fungi) used in revegetation work must be able to cope with drought stress and excess salt toxicity. The organisms employed must be able either to tolerate the aforementioned stresses or increase the osmolyte content of their tissues to counteract osmotic stress. Some organisms may achieve this by using compatible solutes (e.g., polyols, amino acids, and K), whereas others may use the salts present in excess (e.g., NaCl). Because boreal forest plants and mycorrhizal fungi are not necessarily resistant to the stresses found in reconstructed postmining substrates, recent research efforts have been focused on understanding and improving plant resistance to this multistress situation; particular attention has been paid to plant (Khasa et al. 2002) and mycorrhizal fungal selection (Kernaghan et al. 2002; Bois et al. 2006a) and to improving soil reconstruction techniques (e.g., Li and Fung 1998).

This chapter provides details concerning the selection of mycorrhizal fungi that are capable of forming associations with coniferous species typical of the boreal forest region and that have the potential to increase survival of host plants growing on reconstructed saline soils. The revegetation approach outlined is known as "phytobial-remediation" (Lynch and Moffat 2005), which is "bioremediation assisted by plants" (Salt et al. 1995). This is a more natural and powerful approach than the use of bioremediation and phytoremediation techniques alone.

The selection of suitable mycobionts should ideally consist of three steps: preliminary in vitro experiments, greenhouse trials, and small-scale field trials prior to large-scale deployment. The in vitro screening for potentially salt stress tolerant mycorrhizal fungi is necessary to compare the physiological response of the each candidate strain under axenic conditions. This step enables large numbers of species and strains to be tested, narrows down the number of candidate strains, and provides valuable information that may help interpret the response of mycorrhizal seedlings during subsequent greenhouse trials. The greenhouse experiments assess the salt-stress tolerance of different plant–fungus combinations under controlled conditions and allows the selection of appropriate symbiotic systems for testing in the field. The last step, small-scale field trial under realistic site conditions, is used to confirm the results of greenhouse trials prior to large-scale deployment of the selected mycorrhizal system. The chapter concludes with some suggested directions for future work in this challenging field.

In vitro selection of salt stress tolerant ectomycorrhizal fungi

In this section, we outline the process typically used to select salt stress tolerant mycorrhizal fungal isolates. Although the response of mycorrhizal fungi grown in axenic culture may differ from that observed when in symbiosis, in vitro selection is an important step in the understanding of the potential

Chapter 10. Ectomycorrhizal inoculation for boreal forest ecosystem restoration following oil sand extraction

Fig. 10.2. Amended tailing sands at the edge of an undisturbed forest stand. The top amendment layer is muskeg peat (Syncrude Canada Ltd. mining site, Fort McMurray, Alberta, Canada).

stress response of the symbiotic plant–fungal system, and it may indicate the potential in situ persistence of a given fungal isolate on a host plant's root system.

Kernaghan et al. (2002) were among the first researchers to use in vitro studies to assess the tolerance of ectomycorrhizal fungal isolates to different saline–alkaline conditions. The 17 isolates tested were chosen from among nine species occurring naturally within the Canadian boreal forest that Hutchison (1990) had shown to be potentially resistant to NaCl. Briefly, the isolates were grown on modified Melin–Norkran medium (MNM) (Marx 1969) amended either with NaCl, Na_2SO_4, $CaCl_2$, or $CaSO_4 \cdot 2H_2O$ to give final concentrations of 100 or 200 mmol/L or with undiluted saline–alkaline tailing release water. To distinguish between the pH effect and the effects of the different salts, the isolates were also grown on MNM adjusted to pH 4.4, 5.0, 6.0, and 8.0. The authors concluded that saline–alkaline resistance varies at the species level and that isolates selected for revegetation work should be tolerant of both excess levels of salt and high pH. This study highlighted two potential isolates for use on saline sites: *Laccaria bicolor* Maire (Orton) UAMH 8232 and *Hebeloma crustuliniforme* Bull (Quel.) UAMH 5247. The former exhibited rapid growth and an overall salt tolerance, and the latter showed the highest biomass accumulation in the release water treatment. In addition, both species were relatively alkaliphilic with growth increasing up to pH 6.8.

The above technique allowed selection of salt-tolerant isolates from a culture collection. However, it is reasonable to expect that mycobionts growing in association with host plants that have either succeeded in naturally recolonizing exploited sites or that have colonized outplanted tree seedlings will show an even greater level of tolerance. To compare the salt tolerance of site-indigenous fungal isolates with that of *L. bicolor* UAMH 8232 and *H. crustuliniforme* UAMH 5247, Bois et al. (2006a) compared the growth and salt-stress tolerance strategies of the two former species with those of three isolates collected from the roots of ectomycorrhizal host species growing on a reconstructed saline soil at Syncrude Canada Ltd. These were *Suillus tomentosus* (Kauff) Sing., Snell & Dick, a species of *Hymenoscyphus*, and a species of *Phialocephala*. The latter two are from ascomycete genera containing pioneer species that exhibit a high degree of plasticity, but for which the symbiotic relationship with a given host plant is still poorly understood (but see Chapter 9). The authors hypothesized that the site-indigenous fungi would exhibit a similar or greater tolerance to salt stress than species selected from a culture collection (Kernaghan et al. 2002).

Tests showed that the five isolates were able to grow at all the NaCl concentrations used (Bois et al. 2006a). Nevertheless, the comparison of fungal resistance based on radial growth alone may provide misleading results because stress resistance strategies are often complex and can rarely be determined by a single parameter. Ideally, the selection process should be based on the results of a wide range of tests, including biochemical tissue analyses. The latter provide useful information concerning the salt-stress response of a given fungus. Biochemical analyses done by Bois et al. (2006a), showed that the species isolated from the field exhibited the highest stress resistance. This was manifest by a much lower investment in osmoprotection (i.e., the production and accumulation of compatible solutes such as trehalose, mannitol, glycerol, glucose, and proline) while maintaining a similar or higher growth rate than *H. crustuliniforme* or *L. bicolor*. In contrast to the suite of fungi tested by Kernaghan et al. (2002), Bois et al. (2006a) found *L. bicolor* to be the most salt-stress sensitive of the five species they tested. Finally, among the three species isolated from the field, the ascomycetes were the most resistant; exhibiting the highest radial growth, dry mass (DM), and water content in the presence of NaCl.

Because of the uncertain nature of the symbiotic relationship of the two ascomycete species with a given host plant, the result obtained in the above study suggest that, if putatively ectomycorrhizal fungi are desired, *S. tomentosus* and *H. crustuliniforme* are likely the most interesting candidate species for improving the salt-stress resistance of coniferous seedlings growing under saline conditions. However, to confirm this, the physiological response of plants colonized by the above fungi and subjected to excess NaCl must be analyzed in vivo.

The physiological response of inoculated seedlings to single stressors under controlled conditions

Before carrying out small-scale field experiments, it is important to assess the physiological response of mycorrhizal seedling to single stressors under controlled conditions. This short-term approach provides important information that allows the prediction of the likely long-term response of the plant–fungus system in the field. In the following section, the physiological response of greenhouse-grown mycorrhizal conifer seedlings to excess NaCl is outlined.

Muhsin and Zwiazek (2002) evaluated the NaCl sensitivity of *P. glauca* seedlings inoculated with *H. crustuliniforme* UAMH 5247. The authors hypothesized that this mycobiont would increase water flow to the roots of the host under salt stress and, thus, facilitate the maintenance of the plant's water balance. Inoculated and noninoculated control seedlings were grown in a peat:sand (3:1, v/v) mix for ten months. Ten weeks before harvest, the seedlings were subjected to treatments of either 0 or 25 mmol/L NaCl. Overall, inoculated seedlings exhibited a higher biomass than noninoculated ones, and *H. crustuliniforme* reduced the Na content of shoots and roots of inoculated plants. The roots of mycorrhizal seedlings exhibited a hydraulic conductance four times that of noninoculated plants. This was possibly because, once taken up, the water flux through the hyphae is not subjected to the resistance offered by the external saline solution.

Bois et al. (2006b) conducted a similar experiment but used a wider NaCl concentration range (0, 50, 100, or 200 mmol/L) than Muhsin and Zwiazek (2002) to more closely reflect conditions inherent on their study site. According to the plant-stress concept proposed by Lichtenthaler (1996), the damage from long-term low-stress exposure is similar to that from short-term high-stress exposure once the threshold of stress resistance has been exceeded. Therefore, the concentrations used by Bois et al. (2006b) allowed the authors to contrast the adaptive response induced by a low level of stress, with the physiological disturbance induced by a high level of stress. The authors used differences in chlorophyll *a* fluorescence (ChFL) levels to measure the stress response of the seedlings. This nondestructive approach indicates the degree of photochemical perturbations (i.e., F_v/F_m ratio), which is a direct reflection of plant health. Coupling these measurements to biochemical data provides information on overall stress resistance or sensibility.

In their experiment, Bois et al. (2006b) used *P. banksiana* and *P. glauca* inoculated with *L. bicolor*, *H. crustuliniforme*, or *S. tomentosus* (80%–100% root colonization) and nonmycorrhizal control seedlings. After six months of growth and a three-month period of dormancy, the seedlings were transplanted into pots containing a 1:1 (v/v) sand:Turface® mix (Profile™, Buffalo Grove, Illinois). Seedling dormancy was broken, and the NaCl treatments were applied by immersing the pots in solutions containing 0, 50, 100, or 200 mmol/L NaCl. Because of the high levels of NaCl used (seawater reaches 500 mmol/L), the authors reduced the time course of their experiments from the ten weeks used by Muhsin and Zwiazek (2002) to four weeks to prevent seedling death prior to harvesting. The 50 mmol/L NaCl treatment substrate had an $EC_{1:5}$ (soil solution electrical conductivity) and a $SAR_{1:5}$ (sodium absorption ratio) value above the threshold used for characterizing saline–sodic substrates (Slavich and Petterson 1993;

Sumner 1993; Sumner et al. 1998), and these values increased with each increase in NaCl concentration.

Inoculation improved growth or reduced NaCl stress in both host plant species. The F_v/F_m ratio showed that *P. banksiana* was the more sensitive of the two plants to NaCl. The F_v/F_m ratio decreased (indicating photochemical perturbations) with increasing NaCl concentrations for all *P. banksiana* seedlings, except those inoculated with *H. crustuliniforme*. In part, this was explained by a higher control of Na distribution in the plants colonized by this species. *Hebeloma crustuliniforme* induced a higher accumulation of Na in roots than in shoots compared with the two other fungi. The analysis of the F_v/F_m ratio showed that a physiological shift exists in the *P. banksiana* response at a threshold concentration situated between 100 and 200 mmol/L NaCl, which corresponds to a shoot tissue Na concentration close to 2.5% (see also Bois 2005). Compared with the two other fungi, *H. crustuliniforme* was the only one to confer resistance to *P. banksiana* seedlings in treatments above 100 mmol/L NaCl. Below 100 mmol/L NaCl, *P. banksiana* seedlings inoculated with *S. tomentosus* exhibited the highest biomass and a low level of photochemical disturbance. Thus, the latter two fungi are potential candidates for *P. banksiana* inoculation but under different site conditions. This threshold was not observed in *P. glauca* seedlings; instead of exhibiting physiological perturbations, these seedlings showed an adaptation to excess NaCl. This species showed a higher Na content in roots than in shoots in all NaCl treatments, and in part, this distribution of excess Na may explain its apparently greater resistance. The results of this experiment showed that *S. tomentosus* was the most interesting symbiont for use with *P. glauca* because this plant–fungus combination resulted in the highest biomass accumulation over the range of NaCl concentrations tested.

For both conifer species, the overall strategy in the presence of excess NaCl in the soil solution was to accumulate a certain amount of Na in their tissues. Coupled with this was an increase in the concentration of osmoprotectant solutes (e.g., proline and polyol), which probably served to protect cell membranes, organelles, and general metabolism from excess Na accumulation. The *S. tomentosus* isolate induced the greatest changes in the quantity and diversity of accumulated organic osmotica in *P. banksiana* seedlings, and *H. crustuliniforme* played a similar role in *P. glauca* seedlings. This observation indicated a specific plant–fungus combination response and demonstrated the importance of the choice of a specific fungus (or suite of fungi) for a given plant species.

Under salt stress, plants invest carbon resources in some form of salt-management strategy, and this necessarily reduces their growth potential (Levitt 1980; Cheeseman 1988; Munns 1993, 2005; Kozlowski 1997; Hasegawa et al. 2000; Orcutt and Nilsen 2000; Zhu 2001, 2002). Thus, enhanced growth and stress effect reduction by fungal inoculation are important for successful reclamation. The above greenhouse experiments show that *L. bicolor*, *H. crustuliniforme*, and *S. tomentosus* are, for different reasons, all potential candidate species for use in reclamation programs. However, conditions at the site of interest will determine which fungus is likely to be the most appropriate for use with a given plant species. *P. glauca* seedlings are more resistant to excess NaCl than *P. banksiana* seedlings. Therefore, the results of the above experiment show that *P. glauca* should be inoculated with the fungus capable of inducing the greatest growth response, which in this case was either *L. bicolor* or *S. tomentosus*. Preferably, *P. banksiana* seedlings should be inoculated with *H. crustuliniforme* in extreme saline site conditions; however, in conditions below the NaCl concentration threshold detected (between 100 and 200 mmol/L), *L. bicolor* or *S. tomentosus* should be used, because they are capable of inferring good growth responses under mildly saline conditions.

Physiology of inoculated seedlings exposed to saline oil sand tailings in the greenhouse and the field

In this section, we summarize the results of preliminary experiments by Bois (2005) to assess the response of mycorrhizal seedlings exposed to conditions similar to those in the field but under greenhouse conditions and the response of mycorrhizal seedlings in a small-scale planting on the Syncrude Canada Ltd. site (Alberta).

To understand the physiological response of inoculated seedlings under conditions resembling those in the field, Bois (2005) exposed seedlings grown under the conditions outlined in Bois et al. (2006b) to release water from sand tailings. Briefly, after dormancy (see above), individual plants were transplanted into polyvinyl chloride tubes (closed at one end by nylon net) containing a Sand–Turface® mix (see above). Each tube was partly buried in a plas-

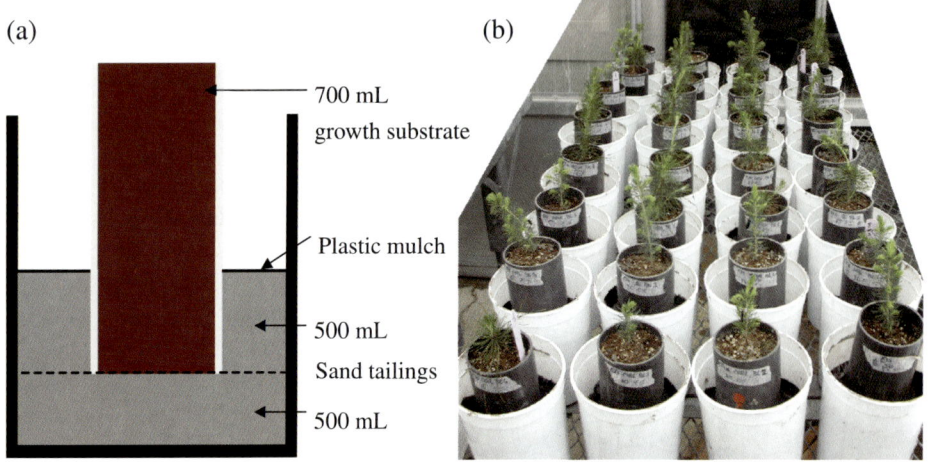

Fig. 10.3. (*a*) Cross section of an experimental unit and (*b*) photograph of the experimental setup.

tic container containing 1 L tailing sand (Fig. 10.3). Plants were watered weekly with 300 mL of saline–alkaline release water (pH_{H2O} 8, EC = 3.25 dS/m; 51.5 mmol/L Na^+; 30.3 mmol/L Cl^-; 0.05 mmol/L K^+; 74 mmol/L Ca^{2+}) that was applied to the tailings. The tailings were covered with a plastic mulch to prevent evaporation and to preferentially drive the saline water through the substrate in the PVC tube. This system avoided direct exposure of plants to sand tailings and simulated the potential rise of saline waters into the rooting zone of plants growing on amended reclamation substrates.

After a four-week exposure period, the growth substrate was highly saline ($EC_{1:5}$ = 0.8 dS/m) and sodic ($SAR_{1:5}$ = 16). Root and shoot biomasses of inoculated seedlings were greater than those of noninoculated seedlings. *Pinus banksiana* seedlings showed a similar biomass for all inoculation treatments. The biomass of *P. glauca* seedlings inoculated with *S. tomentosus* was greater than that of seedlings inoculated with *L. bicolor*, which was greater than that of seedlings inoculated with *H. crustuliniforme*. In this experiment, the results obtained for the stress parameters investigated (i.e., water relations, ChFL measurement, and proline and polyol content) suggested that this treatment only induced a low level of stress. This may have been due to a slower than anticipated increase in salinity. Nevertheless, all plant–fungus combinations exhibited a higher polyol content than that observed for noninoculated seedlings. Therefore, it is probable that mycorrhizal seedlings would exhibit a better short-term adaptation potential to salt stress under field conditions and a higher transplant shock resistance. However, field trials are necessary to confirm this.

Prior to any field test, it is necessary to assess the possibility of large-scale inoculation of the selected host plants with the selected mycobiont under commercial nursery conditions. Although the results of recent studies using *L. bicolor* UAMH 8232 show this strain to be less efficient than strains of other ectomycorrhizal fungal species at elevated NaCl levels, this fungus is fast growing, is easy to culture and inoculate, and is considered to be a model species for ectomycorrhizal fungal studies. Furthermore, it performed well in the experiment conducted by Kernaghan et al. (2002). Therefore, Bois (2005) inoculated *P. banksiana* and *P. glauca* seedlings with this strain to evaluate the possibility of inoculating eight-month-old seedlings grown under commercial tree nursery conditions. Inoculated seedlings were well colonized by *L. bicolor* (60%–80% root colonization) and noninoculated seedlings were colonized by *Thelephora americana* Lloyd, an ectomycorrhizal fungus common in tree nurseries. In contrast to the findings of Bois et al. (2006*b*), *L. bicolor* reduced the size and biomass of inoculated seedlings under tree nursery conditions. This may have been due to slightly different experimental conditions and a genetically different plant stock. Egger and Hibbett (2004) showed that certain isolates of *L. bicolor* have the potential to exploit the mycorrhizal system: draining carbon from host plants for their own growth, while providing little in terms of enhanced plant nutrition. As a result, under nonlimiting or overly stressful conditions, the fungus is possibly able to dominate the resource exchange with the host plant and use its host's resources for the benefit of its own growth and survival. Bearing in mind the potential exploitative behavior of *L. bicolor*, the use of *S. tomentosus* is perhaps a

Fig. 10.4. Illustration of the experimental site: (*a*) top of the slope at the boundary between raw sand tailings and the reconstructed soil and (*b*) surface waterlogging of plot 2.

better choice (see previous section). Furthermore, the *S. tomentosus* isolate was indigenous to the site under study and is probably better adapted to other site-specific stresses. Nevertheless, species of *Suillus* generally exhibit poor growth in stirred liquid culture (A.P. Coughlan, unpublished data).

Following the large-scale inoculation experiment, Bois (2005) used the resulting *L. bicolor* inoculated *P. banksiana* and *P. glauca* seedlings to establish a small-scale planting on an experimental reclamation site in Alberta. The objective was to monitor long-term survival and growth of inoculated seedlings under typical field conditions (i.e., reconstructed saline substrate capped with a thin peat layer). Two experimental plots were established on the edge of the slope of a large sand tailing deposit (Fig. 10.4). This site was not highly saline ($EC_{1:5}$ = 1.1 dS/m and $SAR_{1:5}$ = 8), and plants were exposed to a number of common stresses encountered on reclaimed sites: wind and water erosion and direct exposure to wind, intense sunlight, frost heave, waterlogging, and periods of drought.

One year after outplanting, mortality of inoculated and noninoculated seedlings was high and reached 100% at the top of the slope. *Picea glauca* was more resistant than *P. banksiana* to saline conditions (Fig. 10.5), and inoculation with *L. bicolor* increased survival on both experimental plots. By contrast, inoculated *P. banksiana* seedlings showed higher survival on plot 1 but lower survival on plot 2, which experienced a higher degree of waterlogging (Fig. 10.4). The *L. bicolor* isolate was found to be ill suited for revegetation programs on heavily stressed sites, particularly with the host species *P. banksiana*, thus confirming the results of the in vitro and greenhouse experiments and highlighting the need for preliminary small-scale plantings.

Conclusion and future research

This chapter outlines a three-step process aimed at selecting suitable symbiont combinations for the revegetation of tailing sands, based on their physiological strategy of resistance to salt stress. The first two ex situ steps provide valuable scientific information that clearly confirm the possibility of using a given mycorrhizal fungi to alleviate the impact of a given stress on a host plant. In ex situ experiments, two main responses are of interest: enhanced stress resistance and growth of the host plant. Nonetheless, as shown in the third section, the selection of ectomycorrhizal fungal isolates for use in the field cannot be based solely on in vitro work and controlled greenhouse conditions bioassays. The ability of a given isolate to grow in the presence of a given stressor under axenic conditions does not infer that the isolate will confer tolerance to that stressor to a host plant. Future studies should consider using the above strains of *S. tomentosus* and *H. crustuliniforme* in field trials on the Syncrude Canada Ltd. site.

It should be noted that, for industrial or commercial objectives, it is only the results of the third step in the selection process (i.e., direct exposure to site stressors) that are required. However, large-scale nursery inoculation followed by outplanting experiments to test several host species and mycobionts is expensive. Therefore, prescreening of a large number of strains with as many isolates as possible from the intended outplanting site is advisable. Furthermore, given the range of stressors present at a single site, multispecies inocula of ectomycorrhizal fungi or the use of tree seedlings inoculated with one of several different species could be a useful way forward, allowing the best adapted plant–fungal association to survive.

Future research should also include fungal species that may not be of primary interest, such as the ascomycetes used in the first section of this chap-

Fig. 10.5. (*a* and *b*) *Picea glauca* seedlings and (*c*) alive and (*d*) dead *Pinus banksiana* seedlings one year after outplanting.

ter. Their symbiotic relationship is still unknown, and we are currently unable to predict their value in the field. Furthermore, endomycorrhizal and ericoid mycorrhizal associations are also key players in the boreal region and may favor the reestablishment of cover grasses and shrubs, which may help survival of coniferous seedlings. Ericoid mycorrhizal fungi, like many ectomycorrhizal fungi, can be grown under axenic conditions; therefore, the techniques used for ectomycorrhizal fungal selection should be applicable to them. In the context of the revegetation of oil sand tailings to recreate a functional boreal forest ecosystem, the pioneer work done by Zak and Parkinson (1982, 1983), Zak et al. (1982), and Visser et al. (1984) should to be revisited and expanded on.

Acknowledgements

These studies were funded by Syncrude Canada Ltd. and the Natural Sciences and Engineering Research Council (grant CRDPJ 250448-01 to D.P. Khasa). The authors thank Dr. Khasa, Dr. Piché, Dr. Bigras, Dr. Bertrand, Dr. Zwiazek, Dr. Jany, Dr. Guérin-Laguette, and Martin Fung for their help and advice. The authors also thank Yves Dubuc for all his help with plant physiological measurements, Lucette Chouinard and Pierre Lechasseur for biochemical analyses, Alain Brousseau for elemental analyses, and Michèle Bernier-Cardou for advice with statistical analyses. The authors are grateful to Erin Bergrand and to Claude Fortin for technical help and advice.

References

Alberta Environment. 2001. Salt contamination assessment and remediation guidelines. Alberta Environmnet, Edmonton, Alta. Guideline ENV-190-OP. Available from www.gov.ab.ca/env/ [accessed 1 December 2004].

Bois, G. 2005. Ecophysiology of ectomycorrhizal coniferous seedlings in saline and sodic conditions. Ph.D. thesis, Département des sciences du bois et de la forêt, Université Laval, Québec, Que.

Bois, G., Piché, Y., Fung, M.Y.P., and Khasa, D.P. 2005. Mycorrhizal inoculum potentials of pure reclamation materials and revegetated tailing sands from the Canadian oil sand industry. Mycorrhiza, **15**: 149–158.

Bois, G., Bertrand, A., Piché, Y., Fung, M.Y.P., and Khasa, D.P. 2006*a*. Growth, compatible solute and salt accumulation of five mycorrhizal fungal species grown over a range of NaCl concentrations. Mycorrhiza, **16**: 99–109.

Bois, G., Bigras, F., Bertrand, A., Piché, Y., Fung, M.Y.P., and Khasa, D.P. 2006*b*. Ectomycorrhizal fungi affect the physiological responses of *Picea glauca* and *Pinus banksiana* seedlings exposed to an NaCl gradient. Tree Physiol. **26**: 1185–1196.

Cheeseman, J.M. 1988. Mechanisms of salinity tolerance in plants. Plant Physiol. **87**: 547–550.

Danielson, R.M., Visser, S., and Parkinson, D. 1983. Microbial activity and mycorrhizal potential of four overburden types used in the reclamation of extracted oil sands. Can. J. Soil Sci. **63**: 363–375.

Egger, K.N., and Hibbett, D.S. 2004. The evolutionary implications of exploitation in mycorrhizas. Can. J. Bot. **82**: 1110–1121.

Franklin, J.A., Renault, S., Croser, C., Zwiazek, J.J., and MacKinnon, M. 2002. Jack pine growth and elemental composition are affected by saline tailings water. J. Environ. Qual. **31**: 648–653.

Fung, M.Y.P., and Macyk, T.M. 2000. Reclamation of oil sand mining areas. *In* Reclamation of drastically disturbed lands. 2nd ed. *Edited by* R.I. Barnhisel, R.G. Darmody, and

W.L. Daniels. American Society of Agronomy, Madison, Wisc. pp. 755–774.

Hasegawa, P.M., Bressan, R.A., Zhu, J.-K., and Bohnert, H.J. 2000. Plant cellular and molecular responses to high salinity. Annu. Rev. Plant Physiol. Plant Mol. Biol. **51**: 463–499.

Hutchison, L.J. 1990. Studies on the systematics of ectomycorrhizal fungi in axenic culture. IV. The effect of some selected fungitoxic compounds upon linear growth. Can. J. Bot. **68**: 2172–2178.

Jentschke, G., and Goldbold, D.L. 2000. Metal toxicity and ectomycorrhizas. Physiol. Plant. **109**: 107–116.

Kernaghan, G., Hambling, B., Fung, M.Y.P., and Khasa, D.P. 2002. In vitro selection of boreal ectomycorrhizal fungi for use in reclamation of saline–alkaline habitats. Restor. Ecol. **10**: 43–51.

Khasa, D.P., Hambling, B., Kernaghan, G., Fung, M., and Ngimbi, E. 2002. Genetic variability in salt tolerance of selected boreal woody seedlings. For. Ecol. Manage. **165**: 257–269.

Kottke, I. 1992. Ectomycorrhizas—organs for uptake and filtering of cations. *In* Mycorrhizas in ecosystems. *Edited by* D.J. Read, D.H. Lewis, A.H. Fitter, and I.J. Alexander. CAB International, Wallingford, UK. pp. 316–322.

Kozlowski, T.T. 1997. Response of woody plants to flooding and salinity. Tree Physiol. **1**: 1–29.

Kropp, B.R., and Langlois, C.-G. 1990. Ectomycorrhizae in reforestation. Can. J. For. Res. **20**: 438–451.

Lamhamedi, M.S., Bernier, P.Y., and Fortin, J.A. 1992. Growth, nutrition and response to water stress of *Pinus pinaster* inoculated with ten dikaryotic strains of *Pisolithus* sp. Tree Physiol. **10**: 153–167.

Levitt, J. 1980. Response of plants to environmental stresses. Vol. 2. Water, radiation, salt and other stresses. 2nd ed. Academic Press, London.

Li, X., and Fung, M.Y.P. 1998. Creating soil-like materials for plant growth using tailings sand and fine tails. J. Can. Petrol. Technol. **37**: 44–47.

Lichtenthaler, H.K. 1996. Vegetation stress: an introduction to the stress concept in plants. J. Plant Physiol. **148**: 4–14.

Lynch, J.M., and Moffat, A.J. 2005. Bioremediation—prospects for the future application of innovative applied biological research. Ann. Appl. Biol. **146**: 217–221.

Malajczuk, N., Reddell, P., and Brundrett, M. 1994. Role of ectomycorrhizal fungi in minesite reclamation. *In* Mycorrhizae and plant health. *Edited by* F.L. Pfleger and R.G. Linderman. American Phytopathological Society Press, St. Paul, Minn. pp. 83–100.

Marschner, H., and Dell, B. 1994. Nutrient uptake in mycorrhizal symbiosis. Plant Soil, **159**: 89–102.

Marx, D.H. 1969. The influence of ectotrophic mycorrhizal fungi on the resistance of fine roots to pathogenic infections. I. Antagonism of mycorrhizal fungi to root pathogenic fungi and soil bacteria. Phytopathology, **59**: 153–163.

Muhsin, T.M., and Zwiazek, J.J. 2002. Colonization with *Hebeloma crustuliniforme* increases water conductance and limits shoot sodium uptake in white spruce (*Picea glauca*) seedlings. Plant Soil, **238**: 217–225.

Munns, R. 1993. Physiological processes limiting plant growth in saline soils: some dogmas and hypotheses. Plant Cell Environ. **16**: 15–24.

Munns, R. 2005. Genes and salt tolerance: bringing them together. New Phytol. **167**: 645–663.

Orcutt, D.M., and Nilsen, E.T. 2000. Physiology of plants under stress. John Wiley & Sons Inc., New York.

Pfleger, F.L., Stewart, E.L., and Noyd, R.K. 1994. Role of VAM fungi in mine land revegetation. *In* Mycorrhizae and plant health. *Edited by* F.L. Pfleger and R.G. Linderman. American Phytopathological Society Press, St Paul, Minn. pp. 47–81.

Salt, D.E., Blaylock, M., Kumar, N.P.B.A., Dushenkov, V., Ensley, B.D., Chet, I., and Raskin, L. 1995. Phytoremediation: a novel strategy for the removal of toxic metals from the environment using plants. Bio/Technology, **13**: 468–474.

Slavich, P.G., and Petterson, G.H. 1993. Estimating the electrical conductivity of saturated paste extracts from 1:5 soil:water suspensions and texture. Aust. J. Soil Res. **31**: 73–81.

Smith, S.E., and Read, D.J. 1997. Mycorrhizal symbiosis. 2nd ed. Academic Press, San Diego, Calif.

Sumner, M.E. 1993. Sodic soils: new perspectives. Aust. J. Soil Res. **31**: 683–750.

Sumner, M.E., Rengasamy, P., and Naidu, R. 1998. Sodic soils: a reappraisal. *In* Sodic soils: distribution, properties, management and environmental consequences. *Edited by* M.E. Sumner and R. Naidu. Oxford University Press, New York. pp. 3–17.

Visser, S., Griffiths, C.L., and Parkinson, D. 1984. Topsoil storage effects on primary production and rates of vesicular–arbuscular mycorrhizal development in *Agropyron trachycaulum*. Plant Soil, **82**: 51–60.

Zak, J.C., and Parkinson, D. 1982. Initial vesicular–arbuscular mycorrhizal development of slender wheatgrass on two amended mine spoils. Can. J. Bot. **60**: 2241–2248.

Zak, J.C., and Parkinson, D. 1983. Effects of surface amendation of two mine spoils in Alberta, Canada, on vesicular–arbuscular mycorrhizal development of slender wheatgrass: a 4-year study. Can. J. Bot. **61**: 798–803.

Zak, J.C., Danielson, R.M., and Parkinson, D. 1982. Mycorrhizal fungal spore numbers and species occurrence in two amended mine spoils in Alberta, Canada. Mycologia, **74**: 785–792.

Zhu, J.-K. 2001. Plant salt tolerance. Trends Plant Sci. **6**: 66–71.

Zhu, J.-K. 2002. Salt and drought stress signal transduction in plants. Annu. Rev. Plant Biol. **53**: 247–273.

Chapter 11
Technological transfer: the use of ectomycorrhizal fungi in conventional and modern forest tree nurseries in northern Africa

Mohammed S. Lamhamedi, Mohammed Abourouh, and J. André Fortin

Introduction

Forests are of great social and economic importance in the majority of developing countries, providing food, fuel, medicinal plants, and revenue for much of the rural population. In recent years, the growing pressure exerted on natural forests and plantations has become a major preoccupation for many of these countries. Furthermore, in regions where drought cycles have become more frequent, tree growth and forest health are declining. Many developing countries are now investing considerable resources in reforestation programs in an attempt to reduce forest ecosystem degradation, soil erosion, and desertification. Unfortunately, survival and growth of outplanted tree seedlings has fallen short of expectations.

In northern Africa, the failures observed have largely been due to the poor morphophysiological quality of the seedlings used (Lamhamedi et al. 2000). In many cases, this can be linked to the use of outdated infrastructure and production techniques (e.g., poor-quality seed, perforated polyethylene seedling containers, nonstandardized substrates, poor water quality and irrigation equipment, inadequate shade structures, and the lack of species-specific production criteria and seedling quality norms). To ameliorate the situation and help achieve reforestation goals, many countries are now introducing new tree seedling production techniques into conventional, new, and modernized nurseries. These techniques include the use of more standardized substrates (i.e., composted hardwood branches, leaves, and bark), rigid seedling containers on raised structures, automated irrigation and fertilization systems, and inoculation with suitable mycorrhizal fungi to enhance nutrient and water uptake.

Artificial inoculation with mycorrhizal fungi is being advocated because many native species of ectomycorrhizal (ECM) fungi do not fruit abundantly in arid and semiarid zones. As a consequence, the chance of root colonization by airborne ECM fungal spores is extremely low. The technique of artificial inoculation is becoming widely adopted in modern and conventional nurseries (Abourouh et al. 1995; Abourouh 2000; Lamhamedi et al. 2006). Establishment of this symbiosis prior to outplanting is particularly important for seedlings produced for use on sites subject to severe environmental stresses, particularly those characterized by nutrient-poor soils and an impoverished soil fauna and flora (Meyer 1973; Mikola 1973; Marx 1980; Delwaulle et al. 1987; Mousain et al. 1994).

Much research into the use of ecologically adapted site-specific ECM fungi has been done in Europe, North America, and Australia (e.g., Lalonde and Piché 1988; Le Tacon et al. 1992; Castellano 1996). By contrast, relatively few studies have been done in developing countries (Marx 1980; Castellano 1996). In the majority of cases, the work that has been done in these countries has been conducted within the framework of short-term technical cooperations or development projects. Nevertheless, controlled trials have been done in forest nurseries in the Congo, Ghana, Korea, Malawi, Morocco,

---

M.S. Lamhamedi.[1] Direction de la recherche forestière, ministère des Ressources naturelles et de la Faune, 2700, rue Einstein, Québec, QC G1P 3W8, Canada.

M. Abourouh. Centre de Recherche Forestière, Boîte postale 763, Agdal, Rabat, Maroc.

J.A. Fortin. Centre d'étude de la forêt, Pavillion C.-E.-Marchard, Université Laval, Québec, QC G1V 0A6, Canada.

[1]Corresponding author (e-mail: mohammed.lamhamedi@mrnf.gouv.qc.ca).

Nigeria, Thailand, and Tunisia (Mikola 1980; Ruehle et al. 1981; Delwaulle et al. 1982; Marx et al. 1985; Marx and Cordell 1988; Abourouh 1992, 2000; Abourouh et al. 1995; Lamhamedi et al. 1997), and the techniques used have been documented in several publications (Mikola 1980; Marx et al. 1982, 1984; Le Tacon et al. 1983; Molina and Trappe 1984; Gagnon et al. 1987, 1988; Castellano and Molina 1989; Brundrett et al. 1996).

Despite these studies, little information exists regarding the use of ECM fungi to improve seedling quality and correct micronutrient deficiencies in large-scale forest nurseries in developing countries. In many North African countries, iron (Fe) deficiency is a common problem encountered during the production of coniferous species. This is because the compost-based substrates and the irrigation water used often have pH values approaching or exceeding neutrality. The irrigation water may also contain high levels of bicarbonate and calcium carbonate, especially in zones where limestone is the dominant underlying parent material (Lamhamedi et al. 2006).

In this chapter, we describe (*i*) the operational use of ECM fungi in conventional forest nurseries in Morocco and modern forest nurseries in Tunisia, (*ii*) the use of ECM fungi to correct Fe deficiency under strict operational constraints, and (*iii*) the positive effects of ECM fungal inoculation on the survival of seedlings on challenging reforestation sites.

Seedling production and mycorrhizal formation in conventional forest nurseries

Standard seedling production techniques in conventional nurseries

The techniques used and the associated production problems are similar in most forest nurseries in developing countries. In northern Africa, conventional government-run forest nurseries each produce, on average, between 1.5×10^6 and 3×10^6 seedlings, depending on the regional and national reforestation strategies of the individual country. Relatively few seedlings are produced in regional, private, or community nurseries.

In tree nurseries in countries within the northern temperate or boreal zone, the substrate used for container-grown seedlings is usually a peat–vermiculite mix. However, in many developing countries, peat is scarce; although growing media composed of standardized mixtures of composted bark and branches are beginning to be used (Miller and Jones 1995; Lamhamedi et al. 2000; Ammari et al. 2003), many nurseries in arid and semiarid zones still use agricultural or forest soil (Abourouh 1994a, 2000; Lamhamedi et al. 2006).

Generally, nurseries in close proximity to forest stands use substrates made from the remains of charcoal fires (charcoal soil) or well-decomposed humus collected from forest stands mixed with sand (forest soil). Occasionally, manure is used. The type, proportion, and quality of each constituent used vary among nurseries and interannually because of changes in availability (Lamhamedi et al. 2000). Use of dense (often >1.8 g/cm^3), non-standardized, nutrient-poor substrates (Fig. 11.1*a*) negatively affects root architecture and development, and overall seedling quality (Lamhamedi and Fortin 1994; Lamhamedi et al. 1997). Compacted substrates limit root penetration (Fig. 11.1*b*) (Lamhamedi et al. 2000); when ECM fungi are present on the roots of such plants, hyphal growth within the mycorrhizosphere is also reduced. Under such conditions, the mycorrhizal roots and the extraradical phase of the fungus tend to develop at the substrate–container interface rather than inside the poorly oxygenated soil plug.

The cultural practices used in conventional forest nurseries in most developing countries have remained relatively unchanged. For example, irrigation is generally done manually, using watering cans (Fig. 11.1*c*). Few nurseries are equipped with sprinkler or boom-type automatic irrigation systems, and misting is only used for rooted cuttings and cloning of fast-growing plants, such as species of *Eucalyptus* (Zine El Abidine and Lamhamedi 1994). In developing countries, most tree seedlings are grown in perforated polyethylene bags (7 cm in diameter and 30 cm tall). The bags are normally placed directly on the ground or on cement slabs equipped with a drain. The use of polyethylene bags for seedling production encourages root deformation (Fig. 11.1*d*), which may increase vulnerability to environmental stresses and may lead to mortality even several years after outplanting. Furthermore, little information is available regarding the nutrition of nursery-grown forest tree seedlings and substrate fertility, and the use of fertilization schedules that link nutritional demands to a given seedling growth stage is practically nonexistent. The most common cultural operations are manual weeding and the application of pesticides (Abourouh 1994a).

Fig. 11.1. (*a*) Use of a compacted soil for seedling production in a private nursery (India). (*b*) Root development at the substrate–container interface rather than within the soil plug (diameter, 7 cm) resulting from the high density of the volcanic sand-based substrate (Nicaragua). (*c*) Use of a watering can to irrigate seedlings grown in perforated polyethylene containers, which causes a marked heterogeneity in substrate water content and seedling growth (India). (*d*) Root deformation and spiralling caused by the lack of solid walls and grooves in perforated polyethylene container (Tunisia).

Seedling inoculation in conventional forest nurseries: the Moroccan example

Since the late 1970s, long-term studies in Morocco have aimed to evaluate the extent of natural mycorrhization and the efficiency of different seedling inoculation techniques under both semicontrolled and standard forest nursery conditions (Abourouh 1983, 2000). Other studies have focused on genetic selection of different mycobionts using controlled crosses between compatible monokaryotic cultures of species of *Pisolithus*. This has been done with the view of selecting dikaryons capable of conferring superior drought tolerance and disease resistance to the seedlings (Lamhamedi et al. 1991).

Natural mycorrhization or spontaneous inoculation in conventional forest nurseries

The quality of a forest tree seedling depends on the presence of mycorrhizal fungi on its root system. Natural mycorrhization occurs as a result of the deposition of ECM fungal spores by wind or small mammals or by the presence of spores, sclerotia, or mycorrhiza in the growing media (Marx 1980; Castellano and Molina 1989; Abourouh 2000). This natural mycorrhization can have beneficial effects on seedlings in the nursery (Fig. 11.2*a*). In certain conventional forest tree nurseries, ECM fungi naturally colonize roots, whereas these mycobionts are absent in others (Abourouh 1994*b*, 2000). In Mediterranean tree species, the presence of ectomycor-

Fig. 11.2. (*a*) *Pinus halepensis* seedlings in a nursery in Khouribga, Morocco: naturally mycorrhizal (*Suillus* type) (left) and nonmycorrhizal (right) seedlings. (*b*) *Pinus pinaster* var. *atlantica* seedlings inoculated with *Pisolithus* sp. spores and (*c*) *P. pinaster* var. *atlantica* seedlings inoculated with ground *Rhizopogon* sp. carpophores. The seedlings were grown in a disinfected charcoal-based substrate in plastic bags (7 cm × 30 cm). (*d*) Carpophore produced by *Hebeloma mesophaeum* (cap diameter: 30 mm) in association with a *Quercus suber* seedling.

rhizas varies enormously, depending on the species in question. For example, despite the fact that, in Morocco and Algeria, *Cedrus atlantica* (Manetti ex Endl.) forms symbiotic associations with more than 15 ECM fungi (Lepoutre 1963*a*, 1963*b*; Hocine et al. 1990; Abourouh 2000), Abourouh (2000) found only 4% of nursery-grown *C. atlantica* seedlings to be naturally ectomycorrhizal. In another study, Marx (1979) observed a natural colonization rate of 10%–20% on roots of *C. atlantica* under nursery conditions. The poor natural mycorrhization of *C. atlantica* seems to be due to intrinsic features of the species rather than the environment in which the seedlings are produced. The root system of this species is dominated by long roots, which are rich in phenolics (Abourouh 1983, 1994*c*), factors that impede colonization by mycorrhizal fungi (Theodorou 1980). Poor natural mycorrhization may also result from unfavourable conditions for ectomycorrhizal formation, such as high bicarbonate and carbonate concentrations, salinity, and pH values of the growing media and irrigation water. Recently, in an inoculation experiment involving *C. atlantica* seedlings and the ECM fungus *Tricholoma tridentinum* (Sing.) grown in a peat–vermiculite substrate (1:4, v/v) under an optimal fertilization regime, Boukcim and Mousain (2001) found at least one ectomycorrhiza of the fungus in question on each root system.

Abourouh (2000) observed natural ECM fungal colonization levels ranging from 0% to 100% on pine (*Pinus pinaster* var. *atlantica* Ait., *Pinus pinaster* var. *moghrebiana* H. del Villar, *Pinus halepensis* Mill., and *Pinus canariensis* Sweet ex K. Spreng.) seedlings. One of the most probable explanations for these divergent results is the use of nonstandardized substrate mixtures and the subsequent lack of consistency in terms of substrate fertility and physicochemical properties.

However, it must be borne in mind that the ECM fungi that spontaneously colonize seedling roots are not necessarily those best suited to a given tree species (Fortin and Pineau 1971) and that artificial inoculation may help improve seedling growth.

Artificial mycorrhization

Many different materials have been used to inoculate seedlings with mycorrhizal fungi under nursery conditions (Mikola 1973; Trappe 1977). These include forest soil or humus, excised mycorrhiza, ground dried fruiting bodies, spore suspensions, or pure mycelial cultures, which can all be added to the growth substrate either at seeding or at the start of germination. In this section, we concentrate on inoculation techniques that are easy for nursery managers to use and that have been evaluated under pilot projects and operational conditions.

Use of forest soil
Because of its simplicity, the use of forest soil as a source of mycorrhizal inoculum is one of the most common techniques used in forest nurseries in developing countries. Métro (1957) recommended that nurseries in Morocco incorporate a proportion of forest soil into bare-root seedling beds and potting substrates. The author specified that the soil should be taken from zones where the species being propagated grows well and regenerates naturally. Soil inoculum, which normally represents 10%–20% of the substrate mixture, has been successfully used to increase the percentage of ECM *C. atlantica* seedlings (Abourouh 1983). However, in a subsequent study with *Quercus suber* (L.) seedlings, this technique did not significantly increase the percentage of mycorrhizal seedlings (Abourouh 1998). The principal disadvantages of this technique are that it does not allow the use of a specific mycorrhizal fungal species, and it may introduce potential pathogens into the nursery.

Inoculation with spores of **Pisolithus** *sp.*
At least five ecologically distinct forms of *Pisolithus* occur in the Mediterranean region. These appear to show a degree of host and site specificity (Diez et al. 2001). Demoulin and Dring (1975) identified the presence of two forms of *Pisolithus* in Morocco. The first form was associated with *Eucalyptus* and could be identified in the field by the white peridium of young specimens. The second form, which is slightly darker, was found in association with *Q. suber* and *Pinus* spp. The authors suggested that the former is probably native to Australia and that it was introduced into the region on the seeds of imported *Eucalyptus*. The principal advantages of using *Pisolithus* for artificial inoculation of nursery-grown tree seedlings is that carpophores can be easily obtained during a large part of the year and that spores can be readily extracted from the carpophores using a fine sieve. It is possible to collect 10 carpophores an hour in *P. pinaster* var. *atlantica* stands, and each carpophore yields approximately 20 g of spores (Abourouh 2000). After extraction, the spores may be stored in airtight containers at 4 °C (Abourouh 2000).

Artificial inoculation of disinfected charcoal-based substrates and the direct coating of seeds with spores of *Pisolithus* have resulted in the successful colonization of seedlings of *P. pinaster* var. *atlantica* (Fig. 11.2*b*), *P. pinaster* var. *moghrebiana*, *P. halepensis*, and *P. canariensis*. However, during the same study, ectomycorrhizas failed to develop on the roots of seedlings of *C. atlantica*, *Eucalyptus camaldulensis* Dehnh., *Q. suber*, or species of *Casuarina* (Abourouh 1992, 1998, 2000). The percentage of ECM seedlings and ectomycorrhizas per seedling varied among species. Our results suggest that, under standard nursery conditions, spores from *Pisolithus* carpophores collected from stands dominated by *Pinus* spp. are best suited to the inoculation of pines, rather than other Mediterranean forest species. The use of spores as an inoculation technique is highly recommended because it guarantees a high genetic diversity while avoiding both the selection processes and problems associated with mycelial culture, such as the aging of isolates.

Inoculation with ground **Rhizopogon vulgaris** *carpophores*
Fungi within the genus *Rhizopogon* form ectomycorrhizas with pine seedlings; in many cases, their presence is essential for plantation success (Theodorou and Bowen 1970) (Fig. 11.2*c*). *Rhizopogon vulgaris* (Vitt.) M. Lange produces semihypogeous carpophores in early autumn in natural pine stands and plantations (Abourouh 2000). This production is positively correlated with rainfall (Abourouh

2000), and the carpophores can be easily collected, dried at ambient temperature for several weeks, ground, and stored at 4 °C prior to use.

During inoculation, ground *R. vulgaris* carpophores are mixed with the substrate and moistened seeds. The most spectacular results have been obtained with *P. pinaster* var. *atlantica* (Fig. 11.2c) and *P. pinaster* var. *moghrebiana* seedlings grown in polyethylene containers. The inoculated seedlings began to produce *R. vulgaris* carpophores the following autumn.

Use of pure fungal cultures

The use of pure fungal cultures is considered one of the best inoculation techniques. Its principal advantage is that the inoculum can be produced from an isolated ectomycorrhiza or the carpophore of a given fungus that has proven to be efficient, competitive, and well adapted to either nursery conditions, reforestation sites, or the tree species being grown (Mousain et al. 1994).

Pure culture inoculum is produced either in a peat and vermiculite substrate moistened with nutrient solution (solid inoculum) or in a liquid media. Solid inoculum is generally incubated at room temperature for two and a half to three months, and stirred liquid inoculum can be produced in four to five weeks. The solid inoculum is normally rinsed before being incorporated into the growth substrate in a 1:5 ratio (v/v). Another successful technique is to place a layer of rinsed solid inoculum in the middle of each pot in a 1:8 ratio (v/v). In this form, the inoculum can be dried and stored at 4 °C.

During the production of stirred liquid inoculum, porcelain beads are frequently used to continually break up the mycelial masses. However, if an unstirred liquid media is used, the mycelial mass must be ground in a blender for several seconds before being mixed into the substrate or injected directly into the rhizosphere (Abourouh 2000). If the injection method is used, 10 mL of liquid inoculum, with an approximate concentration of 8.8×10^5 propagules·mL^{-1}, should be injected into each pot or bag.

Most attempts to inoculate pine and cedar seedlings under traditional nursery conditions using either liquid or solid inocula of *Pisolithus tinctorius* (Pers.) Coker & Couch have failed. However, placing 100 mL of solid inoculum in the centre of 800 mL polyethylene containers led to successful formation of ECM on *Q. suber* seedlings (Abourouh 1998, 2000). It appears that, unlike other species (e.g., species of *Laccaria*), this fungus does not tolerate fragmentation. Under a different set of growth conditions, Ruehle et al. (1981) successfully used a pure mycelial culture of *Pisolithus* to colonize *P. pinaster* var. *atlantica*, *P. pinaster* var. *moghrebiana*, and *P. halepensis*, but attempts to colonize *C. atlantica* failed.

Other studies have shown the ectomycorrhizal fungus *Hebeloma mesophaeum* (Persoon ex Fr.) Quélet., which is naturally associated with *P. pinaster* var. *atlantica*, produces a high percentage of mycorrhizal seedlings (70%) and high levels of ectomycorrhizal short roots (>50%) when solid inoculum is placed in a layer in the centre of growth containers (1:8, v/v), or when liquid inoculum is injected into the rhizosphere (Abourouh 2000). This species also produced carpophores in the nursery (Fig. 11.2d).

Use of dikaryotic mycelial cultures obtained from selected crosses

Inoculation of tree seedlings with ectomycorrhizal fungi under operational nursery conditions generally involves a naturally occurring indigenous strain with an unknown genotype. However, the in vitro germination of spores of fungi, such as *Pisolithus*, means that it is now possible to select for certain fungal characteristics (Kope and Fortin 1990a). Briefly, during selection of a desired physiological criterion, a monokaryotic culture (issued from a single spore) is crossed with another compatible monokaryon, resulting in a dikaryotic culture. Because of this method and the large genetic variability between dikaryons, it has been possible to select cultures that rapidly colonize roots and improve plant growth (Lamhamedi et al. 1990; Rosado et al. 1994a, 1994b), that show a greater extension of the extraradical phase and development of mycelial strands (Lamhamedi and Fortin 1991; Rosado et al. 1994b), that produce antifungal substances (Kope and Fortin 1990b), or that improve drought tolerance of *P. pinaster* seedlings (Lamhamedi et al. 1991, 1992a, 1992b). This variability in the performance of *Pisolithus* strains also exists among the naturally occurring populations and can vary between hosts (Cairney and Chambers 1997).

Inoculation of *P. pinaster* seedlings grown in sandy or presterilized peat substrates under greenhouse conditions with different *Pisolithus* dikaryons permitted selection of cultures that produced good ECM colonization and increased drought tolerance. However, when a similar experiment was done in a Moroccan nursery where *Pisolithus* dikaryons were incorporated into a nonsterilized charcoal-based substrate, the approach was unsuccessful. This was probably due to the use of nonstandardized substrates with undefined texture, fertility, or soil flora.

In terms of the variability in plant quality and the level of mycorrhization, the results observed in forest nurseries suggest that a significant improvement in plant quality and a good level of root colonization are unlikely to be achieved without the modernization of forest nursery techniques. Notable improvements needed include the use of organic substrates with well-defined physicochemical properties that remain relatively stable throughout seedling production.

Seedling production and mycorrhizal establishment in modern forest nurseries in Tunisia

Standard seedling production techniques in modern forest nurseries

Between 1995 and 2001, 16 modern forest tree nurseries were established in different bioclimatic zones in Tunisia. This modernization involved the installation of new infrastructure and forest tree seedling production techniques (e.g., the use of shade structures; fertigation; solid-walled seedling containers; stable organic substrates; automatic potting and irrigation equipment; mycorrhizal fungi; and computerized monitoring of cultural techniques, environmental conditions, and plant growth) (Lamhamedi et al. 2000, 2006) (Fig. 11.3a). One of the main breakthroughs has been the replacement of traditional growing media with local composted plant tissue (i.e., chipped deciduous branches, bark, and cork) (Lamhamedi et al. 2000, 2006; Ammari et al. 2003, 2007). This move has allowed nurseries to decrease their dependence on forest soils and imported peat-based substrates. Peat is subject to increasingly stricter harvest guidelines, and consequently, peat-based substrates are becoming more expensive on the international market.

The problem of iron deficiency in nurseries in semiarid and arid zones

One of the main problems encountered during tree seedling production in Tunisia is chlorosis resulting from Fe deficiency. The availability of this trace element is pH dependent and is severely reduced at neutral and alkaline pHs. Therefore, chlorosis is generally more frequent in areas predominated by calcarous soils (Brown 1956; Smith and Mitchell 1977; Mengel 1994). Although several studies have highlighted the role that ECM fungi play in Fe absorption (e.g., Lapeyrie 1990; Hauer and Dawson 1996), to the best of our knowledge, no technology transfer project has advocated the use of these fungi in forest tree nurseries to correct chlorosis.

For optimal seedling growth, the rhizosphere pH should be maintained at 5.5 for conifers and at 6.5 for broadleaf species (Landis et al. 1989). In semiarid and arid zones, the irrigation water used in most forest nurseries contains significant quantities of bicarbonate (HCO_3^-) and carbonate (CO_3^{2-}) ions and is strongly alkaline (pH > 7). Irrigation water with a bicarbonate concentration greater than 60 ppm can cause a significant increase in the pH of the growing medium (Mathers 1998), especially towards the end of plant production (Lamhamedi et al. 2006). In Tunisia, the pH of irrigation water often ranges between 7.6 and 8.1, and the bicarbonate concentration between 146 and 207 mg/L. Chlorosis induced by iron deficiency tends to first appear in the overlap zone between two successive sprinklers, where there is a higher accumulation of bicarbonate ions and a subsequently higher rhizosphere pH level (Fig. 11.3b). Chlorosis typically occurs in the young apical needles of conifers (Fig. 11.3c) (Landis et al. 1989) and in young leaves of deciduous species (Gogorcena et al. 2001; Alcántara et al. 2003). Because the substrate volume in each growth container is relatively small, the increase in pH occurs more rapidly than it would in an agricultural field.

The rate of increase of the substrate pH depends on its composition, initial pH, and the fertigation regime. In areas where the concentration of bicarbonates in irrigation water is high, Fe deficiency rapidly appears in substrates that had an initial pH close to neutrality (Figs. 11.3d–3f). It should also be noted that the Fe requirement and sensitivity to Fe deficiency vary among species. For example, when identical substrates and fertigation schedules were used, *Pinus pinea* L. seedlings showed signs of Fe deficiency earlier than *Pinus halepensis* Mill. seedlings (Fig. 11.3f), whereas *Cupressus sempervirens* L. seedlings exhibited no visible symptoms (Lamhamedi et al. 1997).

If Fe deficiency is not corrected, needles on the terminal and lateral shoots turn white, which affects photosynthesis and plant growth (see Chaney (1984) and Korcak (1987) for detailed reviews). In severe cases, tissue desiccation and seedling death occurs (Figs. 11.4a and 11.4b) (Lamhamedi et al. 2006).

Fig. 11.3. (*a*) General view of a modern forest tree nursery in Tunisia equipped with a shade cover. (*b*) Appearance of chlorosis in the area of irrigation overlap between two sprinklers. (*c*) Initial appearance of chlorosis on young apical needles of *Pinus pinea* seedlings grown in a compost-based substrate. Note the absence of chlorosis in the seedlings in the background, which were grown in a peat–vermiculite substrate. (*d*) Premature appearance of chlorosis in plants grown in a compost-based substrate with an initial pH close to neutral (indicated by arrow). Each row of containers represents a different substrate. (*e*) Chlorosis may become generalized and affect seedling quality. Nonchlorotic seedlings were grown in a peat–vermiculite substrate. The white line marks the boundary between seedlings produced in a peat–vermiculite substrate and those produced in compost (at the back). (*f*) Premature appearance of needle chlorosis in *P. pinea* plants produced in compost-based substrates (centre). Note the growth and absence of chlorosis in seedlings of the same species produced in a peat–vermiculite substrate (indicated by arrow). Note also the absence of chlorosis in *Pinus halepensis* (right) and *Cupressus sempervirens* (left) seedlings at this stage of development. Each row of containers represents a different substrate composition.

Inoculation of seedlings with ectomycorrhizal fungi to correct iron deficiency

Mycorrhizal fungi have been shown to facilitate Fe absorption and availability (Rodriguez et al. 1984; Cress et al. 1986; Clark and Zeto 1996). Initially, Fe uptake can be enhanced by a simple increase of the absorptive surface area of the seedling's root system. However, ECM fungal hyphae also release organic acids, particularly in the presence of nitrates, calcium, and bicarbonates (Arvieu et al. 2003), which acidify the mycorrhizosphere and increase Fe availability (Lapeyrie 1990; Plassard 1996). Ectomycorrhizal fungi may also release Fe-chelating siderophores (e.g., hydroxamate) (Szaniszlo et al. 1981), which are 10^{30} times more stable

Fig. 11.4. (*a*) If iron deficiency is not corrected, chlorosis may reach the apices of the lateral branches (right, arrow). If the chlorosis becomes severe, the apices turn white and desiccate (centre, double arrow). On the left is a mycorrhizal *Pinus pinea* seedling with no symptoms of iron deficiency (Tunisia). (*b*) Desiccation of the chlorotic apices of *Pinus oocarpa* seedlings (Nicaragua). (*c*) Formation of mycelial strands and good root plug cohesion of *P. pinea* seedlings (Tunisia). (*d*) Mycorrhizal *P. pinea* seedlings with good root development and showing no symptoms of iron deficiency (Tunisia).

than fertilizers (Powell et al. 1982) and further increase Fe availability.

In an attempt to reduce Fe deficiency and improve tree seedling quality, we conducted an inoculation experiment using compost-based substrates that were relatively sterile and contained no ECM fungal propagules. Prior to inoculation with a spore-based inoculum (either *Rhizopogon* sp. or *Pisolithus* sp.), the fertilization schedules were optimized, taking into account the pH of the irrigation water and the different compost-based substrates. Phosphoric acid (85%) was used to decrease the pH of the irrigation water, and the fertilizer Sequestrene (NaFeEDDHA, containing 6% Fe that remains available across the full pH range) was used as a source of chelated Fe.

After taking into account the possible negative effect of elevated concentrations of mineral nutrients, most notably phosphorus, on the degree of ECM colonization, seedlings were inoculated at the beginning of autumn (mid-August). This coincided with the hardening period when shade structures are removed and fertilizer application is significant reduced. It also coincided with the appearance of symptoms of Fe deficiency and the period of increased root growth. It should be noted that species that do not have high nutrient requirements can be inoculated as soon as the seeds have germinated.

Use of a spore-based inoculum is convenient because it requires little equipment and is easily integrated into a standard production schedule. At the time of inoculation, the spores were sieved to eliminate larger particles that might obstruct the sprinkler system and the existing fertigation system was used for the inoculation process. The spores are so small that they easily pass through the filters and the sprinkler nozzles. Between 40 and 60 mL of spores is sufficient to inoculate 100 000 seedlings. The necessary quantity of spores was mixed with 2 L of water and refrigerated to assure a good dispersion of the spores in the initial solution. Before the spore solution was injected into the fertigation system, the seedlings were watered for about 2 min to prevent the spores from sticking to the foliage. The seedlings were rinsed for another 2 min after the inoculation procedure to wash the spores into the substrate. Using this technique, the first ECM began to appear two to three weeks after inoculation.

Aleppo (*P. halepensis*) and Italian stone (*P. pinea*) pines inoculated with spores of *Rhizopogon* sp. developed heavily colonized root systems with abundant extraradical hyphal growth, and Fe deficiency induced chlorosis was corrected (Figs. 11.4*a*–4*d*). Furthermore, use of a more porous organic substrate and inoculation with *Rhizopogon* sp. favoured root growth and plug cohesion (Figs. 11.4*c* and 11.4*d*).

The effects of nursery inoculation of seedlings in relation to their performance on reforestation sites in northern Africa

Trials in Morocco showed that *P. pinaster* var. *atlantica* seedlings inoculated with *P. tinctorius* had a higher survival rate (80%) than controls (20%) (Abourouh 2000). These differences were most noticeable when the percentage of colonized short roots was >60%. In addition to good root colonization, the ECM fungus must remain on the root system after outplanting. In certain situations, for example in subhumid environments dominated by *Arbutus unedo* L., *P. tinctorius* was not competitive and was rapidly replaced by indigenous ECM species (Abourouh 2000). No beneficial effect was observed either when seedlings were inoculated with *R. vulgaris*, despite good root colonization in the nursery. However, one year after outplanting, carpophores of this species appeared around certain seedlings (Abourouh 2000). This suggests that the effect of ECM fungi on seedling performance may only be apparent in the presence of very severe environmental stresses.

In Tunisia, experiments have shown that mycorrhizal forest tree seedlings produced in modern forest nurseries have a higher growth rate than noninoculated seedlings grown in conventional nurseries (Figs. 11.5*a* and 11.5*b*) (Lamhamedi et al. 2006).

Conclusion

Production of high-quality tree seedlings in northern Africa cannot be guaranteed unless conventional nurseries modernize their operations. This requires infrastructure reinforcement and the implementation of modern forest nursery production techniques.

In the studies presented, controlled mycorrhization with a number of different ectomycorrhizal fungal species and using different techniques was mastered equally well in modern and conventional nurseries. However, the use of ECM fungal spores is the easiest method to use in both types of nursery. Artificial inoculation significantly increased seedling quality, alleviated chlorosis related to Fe deficiency, and improved seedling survival and growth following outplanting. Nevertheless, the success of mycorrhization is intimately linked to optimizing and mastering of the principal cultural techniques affecting fungal colonization and growth, notably the relationship between the physicochemical properties of growing media and fertigation techniques.

Acknowledgements

The results presented in this chapter are the synthesis of several research projects, funded by a

Fig. 11.5. (*a*) Enhanced performance of ectomycorrhizal seedlings produced in modern forest tree nurseries in Tunisia (left), in comparison with seedlings produced under conventional nursery conditions (right), on sites that received no mechanical site preparation. (*b*) Enhanced growth was also observed on mechanically prepared sites.

variety of development organizations, most notably the International Development Research Centre (phase I: 3-P-85-1007 and phase II: 3-P-90-0063), the Canadian International Development Agency (project: 660/13328), the International Foundation for Science, le Fonds International de Coopération Universitaire, l'Agence Universitaire de la Francophonie (project 2000/PAS/15), the World Bank (project: BIRD 3601, loan granted to the government of Tunisia), and the Nordic Development Fund (project NIB/NDF). We are grateful to Mario Renaud (ministère des Ressources naturelles et de la Faune du Québec), Debra C. Stowe, Andrew Coughlan, Dr. Yves Piché (Université Laval), and two anonymous reviewers for editorial comments.

References

Abourouh, M. 1983. Essais de mycorhization en pépinière de *Cedrus atlantica* Manetti (Cèdre de l'Atlas). Thèse de 3e cycle, Université Claude Bernard-Lyon 1, Villeurbanne, France.

Abourouh, M. 1992. Essai de mycorhization en pépinière par les spores de *Pisolithus tinctorius*. Ann. Rech. For. Maroc, **26**: 127–138.

Abourouh, M. 1994*a*. Les plants forestiers : normes de qualité et techniques culturales. *In* Actes de la première journée nationale sur les plants forestiers, 3 June 1992, Salé, Morocco. *Edited by* M. Abourouh. Division Recherches et Expérimentations Forestières, Agdal, Rabat, Morocco. pp. 17–34.

Abourouh, M. 1994*b*. Inoculation artificielle des plants en pépinière : cas de deux champignons mycorhiziens. *In* Actes de la première journée nationale sur les plants forestiers, 3 June 1992, Salé, Morocco. *Edited by* M. Abourouh. Division

Recherches et Expérimentations Forestières, Agdal, Rabat, Morocco. pp. 135–144.

Abourouh, M. 1994c. Les ectomycorhizes du cèdre de l'Atlas : état des connaissances et perspectives. *In* Le cèdre de l'Atlas, *Cedrus atlantica* (Manetti) : actes du séminaire international sur le cèdre de l'Atlas, 7–11 June 1993, Ifrane, Morocco. *Edited by* O. Mhirit, A. Samih, and M. Malagnoux. Ann. Rech. For. Maroc, **27**(No. Spec.): 337–348.

Abourouh, M. 1998. Les ectomycorhizes du chêne-liège : caractérisation et rôle possible dans la régénération. *In* Actes du séminaire méditérranéen sur la régénération des forêts du chêne-liège, 22–24 Oct. 1996, Tabarka, Tunisia. Ann. ENGREF (No. Spec.): 164–175.

Abourouh, M. 2000. Ectomycorhizes et mycorhization des principales essences forestières du Maroc. Thèse d'Etat, Faculté des Sciences de Rabat, Rabat, Morocco.

Abourouh, M., Lamhamedi, M.S., and Fortin, J.A. 1995. Techniques de mycorhization en pépinière des plants forestiers. Centre National de la Recherche Forestière, Rabat, Morocco.

Alcántara, E., Cordeiro, A.M., and Barranco, D. 2003. Selection of olive varieties for tolerance to iron chlorosis. J. Plant Physiol. **160**: 1467–1472.

Ammari, Y., Lamhamedi, M.S., Akrimi, N., and Zine El Abidine, A. 2003. Compostage de la biomasse forestière et son utilisation comme substrat de croissance pour la production de plants en pépinières forestières modernes. Ann. Inst. Natl. Agron. Tunis. **18**: 99–119.

Ammari, Y., Lamhamedi, M.S., Zine El Abidine, A., and Aprimi, N. 2007. Production et croissance des plants résineux dans différents substrats à base de compost dans une pépinière forestière moderne en Tunisie. Rev. For. Fr. **4**: 339–358.

Arvieu, J.C., Leprince, F., and Plassard, C. 2003. Release of oxalate and protons by ectomycorrhizal fungi in response to P-deficiency and calcium carbonate in nutrient solution. Ann. For. Sci. **60**: 815–821.

Boukcim, H., and Mousain, D. 2001. Effets de la fertilisation phosphatée sur la mycorhization, la croissance et la nutrition en phosphore et en azote de semis de Cèdre (*Cedrus atlantica* Manetti) inoculés en pépinière par *Tricholoma tridentinum* Sing. var. *cedretorum* Bon. Ann. For. Sci. **58**: 289–300.

Brown, J.C. 1956. Iron chlorosis. Annu. Rev. Plant Physiol. **7**: 171–190.

Brundrett, M., Bougher, N., Dell, B., Grove, T., and Malajczuk, N. 1996. Working with mycorrhizas in forestry and agriculture. Australian Centre for International Agricultural Research, Canberra, Australia. ACIAR Monogr. 32.

Cairney, J.W.G., and Chambers, S.M. 1997. Interactions between *Pisolithus tinctorius* and its hosts: a review of current knowledge. Mycorrhiza, **7**: 117–131.

Castellano, M.A. 1996. Outplanting performance of mycorrhizal inoculated seedlings. *In* Concepts in mycorrhizal research. *Edited by* K.G. Mukerji. Kluwer Academic Publishers, Dordrecht, the Netherlands. pp. 223–301.

Castellano, M.A., and Molina, R. 1989. Mycorrhizae. *In* The container tree nursery manual. Vol. 5. *Edited by* T.D. Landis, R.W. Tinus, S.E. McDonald, and J.P. Barnett. US Dep. Agric. Agric. Handb. 674. pp. 101–167.

Chaney, R.L. 1984. Diagnostic practices to identify iron deficiency in higher plants. J. Plant Nutr. **7**: 47–67.

Clark, R.B., and Zeto, S.K. 1996. Iron acquisition by mycorrhizal maize grown on alkaline soil. J. Plant Nutr. **19**: 247–264.

Cress, W.A., Johnson, G.V., and Barton, L.L. 1986. The role of endomycorrhizal fungi in iron uptake by *Hilaria jamesii*. J. Plant Nutr. **9**: 547–556.

Delwaulle, J.C., Garbaye, J., and Okombi, G. 1982. Stimulation de la croissance initiale de *Pinus caribea* Morelet dans une plantation du Congo par contrôle de la mycorhization. Revue Bois et Forêts des Tropiques, **196**: 25–32.

Delwaulle, J.C., Diangana, D., and Garbaye, J. 1987. Augmentation de la production du pin des Caraïbes dans la région côtière du Congo par introduction du champignon ectomycorhizien *Pisolithus tinctorius*. Rev. For. Fr. **39**: 409–418.

Demoulin, V., and Dring, D.M. 1975. Gasteromycetes of Kivu (Zaire), Rwanda and Burundi. Bull. Jardin Bot. Natl. Belg. **45**: 339–372.

Diez, J., Anta, B., Manj'on, J.L., and Honrubia, M. 2001. Genetic variability of *Pisolithus* associated with native hosts and exotic eucalyptus in the western Mediterranean region. New Phytol. **149**: 577–587.

Fortin, J.A., and Pineau, M. 1971. L'inoculation mycorhizienne dans la production des plants forestiers. For. Chron. **47**: 1–4.

Gagnon, J., Langlois, C.-G., and Fortin, J.A. 1987. Growth of containerized jack pine seedlings inoculated with different ectomycorrhizal fungi under a controlled fertilization schedule. Can. J. For. Res. **17**: 840–845.

Gagnon, J., Langlois, C.-G., and Fortin, J.A. 1988. Growth and ectomycorrhiza formation of containerized black spruce seedlings in function of nitrogen fertilization, inoculation type and symbiont. Can. J. For. Res. **18**: 922–929.

Gogorcena, Y., Molias, N., Larbi, A., Abadia, J., and Abadia, A. 2001. Characterization of the response of cork oak (*Quercus suber*) to iron deficiency. Tree Physiol. **21**: 1335–1340.

Hauer, R.J., and Dawson, J.O. 1996. Growth and iron sequestring of pine oak (*Quercus palustris*) seedlings inoculated with soil containing ectomycorrhizal fungi. J. Arboric. **22**: 122–130.

Hocine, H., Perrin, R.L., and Belarbi, R. 1990. Variation of mycorrhizal association of *Cedrus atlantica* Manetti, the example of Tala Guilef Forest (Djurdjura). *In* Abstracts of the 8th Congress of the Mediterranean Phythopathological Union, Agadir, Morocco. IAV Hassan II. p. 441. [Abstr.]

Kope, H., and Fortin, J.A. 1990a. Germination and comparative morphology of basidiospores of *Pisolithus arhizus*. Mycologia, **82**: 350–357.

Kope, H., and Fortin, J.A. 1990b. Antifungal activity in culture filtrates of the ectomycorrhizal fungus *Pisolithus tinctorius*. Can. J. Bot. **68**: 1254–1259.

Korcak, R.F. 1987. Iron chlorosis. Hortic. Rev. **9**: 133–186.

Lalonde, M., and Piché, Y. (*Editors*). 1988. Canadian Workshop on Mycorrhizae in Forestry. Centre de recherche en biologie forestière, Faculté de Foresterie et de Géodésie, Université Laval, Québec.

Lamhamedi, M.S., and Fortin, J.A. 1991. Genetic variations of ectomycorrhizal fungi: extramatrical phase of *Pisolithus* sp. Can. J. Bot. **69**: 1927–1934.

Lamhamedi, M.S., and Fortin, J.A. 1994. La qualité des plants forestiers: critères d'évaluation et performances dans les sites de reboisement. *In* Actes de la première journée nationale sur les plants forestiers, 3 June 1992, Salé, Morocco. *Edited by* M. Abourouh. Division Recherches et Expérimentations Forestières, Agdal, Rabat, Morocco. pp. 35–50.

Lamhamedi, M.S., Fortin, J.A., Kope, H.H., and Kropp, B.R. 1990. Genetic variation in ectomycorrhiza formation by *Pisolithus arhizus* on *Pinus pinaster* and *Pinus banksiana*. New Phytol. **115**: 689–697.

Lamhamedi, M.S., Fortin, J.A., and Bernier, P.Y. 1991. La génétique de *Pisolithus* sp.: une nouvelle approche de biotechnologie forestière pour assurer une meilleure survie des plants en conditions de sécheresse. Sécheresse, **2**: 251–258.

Lamhamedi, M.S., Bernier, P.Y., and Fortin, J.A. 1992*a*. Growth, nutrition and water stress tolerance of *Pinus pinaster* inoculated with ten dikaryotic strains of *Pisolithus* sp. Tree Physiol. **10**: 153–167.

Lamhamedi, M.S., Bernier, P.Y., and Fortin, J.A. 1992*b*. Hydraulic conductance and soil water potential at the soil–root interface of *Pinus pinaster* seedlings inoculated with different dikaryons of *Pisolithus* sp. Tree Physiol. **10**: 231–244.

Lamhamedi, M.S., Fortin, J.A., Ammari, Y., Benjelloun, S., Poirier, M., Fecteau, B., Bougacha, A., and Godin, L. 1997. Évaluation des composts, des substrats et de la qualité des plants (*Pinus pinea*, *Pinus halepensis*, *Cupressus sempervirens* and *Quercus suber*) élevés en conteneurs. Pampev Internationale, Montréal, Québec. Projet Banque mondiale 3601.

Lamhamedi, M.S., Ammari, Y., Fecteau, B., Fortin, J.A., and Margolis, H. 2000. Problématique des pépinières forestières en Afrique du Nord et stratégies d'orientation. Cah. Agric. **9**: 369–380.

Lamhamedi, M.S., Fecteau, B., Godin, L., and Gingras, C. 2006. Guide pratique de production en hors sol de plants forestiers, pastoraux et ornementaux en Tunisie. Pampev Internationale, Montréal, Québec.

Landis, T.D., Tinus, R.W., and Barnett, J.P. 1989. The container tree nursery manual. Vol. 4. US Dep. Agric. Agric. Handb. 674.

Lapeyrie, F. 1990. The role of ectomycorrhizal fungi in calcareous soil tolerance by "symbiocalcicole" woody plants. Ann. Sci. For. **47**: 579–589.

Le Tacon, F., Jung, G., Michelot, P., and Mugnier, M. 1983. Efficacité en pépinière forestière d'un inoculum de champignon ectomycorhizien produit en fermenteur et inclus dans une matrice de polymères. Ann. For. Sci. **40**: 165–176.

Le Tacon, F., Alvarez, I.F., Bouchard, D., Henrion, B., Jackson, R.M., Luff, S., Parlade, J.I., Pera, J., Stenström, E., Villeneuve, N., and Walker, C. 1992. Variations in field response of forest trees to nursery ectomycorrhizal inoculation in Europe. *In* Mycorrhizas in ecosystems. *Edited by* D.J. Read, D.H. Lewis, A.H. Fitter, and I.J. Alexander. CAB International, Wallingford, UK. pp. 119–134.

Lepoutre, B. 1963*a*. Premier essai de synthèse sur le mécanisme de régénération du cèdre dans le Moyen Atlas marocain. Ann. Rech. For. Maroc, **7**: 55–163.

Lepoutre, B. 1963*b*. Suite d'observations sur la régénération du cèdre par tâches. Ann. Rech. For. Maroc, **7**: 1–20.

Marx, D.H. 1979. Ectomycorrhizal status of tree seedlings in nurseries and field sites in northern Morocco. Report for the Centre de Recherche Forestière, Rabat, Morocco.

Marx, D.H. 1980. Ectomycorrhizal fungus inoculation: a tool for improving forestation practices. *In* Tropical mycorrhizal research. *Edited by* P. Mikola. Clarendon Press, Oxford, UK. pp. 13–71.

Marx, D.H., and Cordell, C.E. 1988. Specific ectomycorrhizae improve reforestation and reclamation in the eastern United States. *In* Canadian Workshop on Mycorrhizae in Forestry. *Edited by* M. Lalonde and Y. Piché. Centre de recherche en biologie forestière, Faculté de Foresterie et de Géodésie, Université Laval, Québec. pp. 75–86.

Marx, D.H., Ruehle, J.L., Kenny, D.S., Cordell, C.E., Riffle, J.W., Molina, R.J., Pawuck, W.H., Navratil, S., Tinus, R.W., and Goodwin, O.C. 1982. Commercial vegetative inoculum of *Pisolithus tinctorius* and inoculation techniques for development of ectomycorrhizae on container grown tree seedlings. For. Sci. **28**: 373–400.

Marx, D.H., Cordell, C.E., Kenny, D.S., Mexal, J.G., Arthan, J.D., Riffle, J.W., and Molina, R.J. 1984. Commercial vegetative inoculum of *Pisolithus tinctorius* and inoculation techniques for development of ectomycorrhizae on bareroot tree seedlings. For. Sci. Monogr. **25**: 1–101.

Marx, D.H., Hedin, A., and Toe, S.F.P., IV. 1985. Field performance of *Pinus caribea* var. *honduras* seedlings with specific ectomycorrhizae and fertilizer after three years on a savanna site in Liberia. For. Ecol. Manage. **13**: 1–25.

Mathers, H. 1998. Water quality: thinking about pH and alkalinity. Digger, **6**: 34–36.

Mengel, K. 1994. Iron availability in plant tissues—iron chlorosis on calcareous soils. Plant Soil, **165**: 275–283.

Métro, A.E. 1957. Les pépinières forestières au Maroc. *Edited by* F. Moncho. Station de Recherche des Eaux et Forêts, Rabat, Morocco. Brochure Tech. 1.

Meyer, F.H. 1973. Distribution of ectomycorrhizae in native and manmade forests. *In* Ectomycorrhizae: their ecology and physiology. *Edited by* G.C. Marks and T.T. Kozlowski. Academic Press, Inc., New York. pp. 79–105.

Mikola, P. 1973. Application of mycorrhizal symbioses in forestry practice. *In* Ectomycorrhizae: their ecology and physiology. *Edited by* G.C. Marks and T.T. Kozlowski. Academic Press, Inc., New York. pp. 383–411.

Mikola, P. (*Editor*). 1980. Tropical mycorrhiza research. Clarendon Press, Oxford, UK.

Miller, J.H., and Jones, N. 1995. Organic and compost-based growing media for tree seedlings nurseries. World Bank Tech. Pap. 264.

Molina, R., and Trappe, J.M. 1984. Mycorrhiza management in bareroot nurseries. *In* Forest nursery manual: production of bareroot seedlings. *Edited by* M. Duryea and T.D. Landis. Nijhoff/Junk, The Hague. pp. 211–223.

Mousain, D., Plassard, C., Argillier, C., Sardin, T., Leprince, F., El Kerkouri, K., Arvieu, J.C., and Cleyet-Marel, J.C. 1994. Stratégie d'amélioration de la qualité des plants forestiers et des reboisements méditerranéens par utilisation de la mycorhization contrôlée en pépinière. Acta Bot. Gallica, **141**: 571–580.

Plassard, C. 1996. La mycorhization des plantes forestières en milieu aride et semi-aride: nutrition minérale en terrains calcaires. Cahiers Options Méditerranéennes, **20**: 27–32.

Powell, P.E., Szaniszlo, P.J., Cline, G.R., and Reid, C.P.P. 1982. Hydroxamate siderophores in the iron nutrition of plants. J. Plant Nutr. **5**: 653–673.

Rodriguez, R.K., Dwight, J.K., and Barton, L.L. 1984. Iron metabolism by an ectomycorrhizal fungus, *Cenococcum geophilum*. J. Plant Nutr. **7**: 459–468.

Rosado, S.C.S., Kropp, B.R., and Piché, Y. 1994a. Genetics of ectomycorrhizal symbiosis. I. Host plant variability and heritability of ectomycorrhizal fungi. New Phytol. **126**: 105–110.

Rosado, S.C.S., Kropp, B.R., and Piché, Y. 1994b. Genetics of ectomycorrhizal symbiosis. II. Fungal variability and heritability of ectomycorrhizal traits. New Phytol. **126**: 111–117.

Ruehle, J.L., Marx, D.H., and Abourouh, M. 1981. Development of *Pisolithus tinctorius* and *Thelephora terrestris* ectomycorrhizae on seedlings of coniferous trees important to Morocco. Ann. Rech. For. Maroc, **21**: 281–296.

Smith, E.M., and Mitchell, C.D. 1977. Eastern white pine iron deficiency. J Arboric. **3**: 129–130.

Szaniszlo, P.J., Powell, P.E., Reid, C.P.P., and Cline, G.R. 1981. Production of hydroxamate siderophore iron chelators by ectomycorrhizal fungi. Mycologia, **73**: 1158–1174.

Theodorou, C. 1980. The sequence of mycorrhizal infection of *Pinus radiata* D. Don. following inoculation with *Rhizopogon luteolus* Fr., and Nordh. Aust. For. Res. **10**: 381–387.

Theodorou, C., and Bowen, G.D. 1970. Mycorrhizal responses of radiata pine in experiments with different fungi. Aust. For. **34**: 183–191.

Trappe, J.M. 1977. Selection of fungi for ectomycorrhizal inoculation in nursery. Annu. Rev. Phytopathol. **15**: 203–222.

Zine El Abidine, A., and Lamhamedi, M.S. 1994. Production des plants et techniques de plantation des Eucalyptus au Maroc. *In* Les Eucalyptus au Maroc. *Edited by* M. Fechtal and E.K. Achhal. Caisse National de Crédit Agricole, Morocco. pp. 84–104.

Chapter 12
Ectomycorrhizas in the neotropics with emphasis on lowland forests

Bradley R. Kropp

Introduction

Much of what is known about ectomycorrhizas in the tropical regions of the Americas (neotropics) comes from work that has been done in the oak (*Quercus*, Fagaceae) and pine (*Pinus*, Pinaceae) forests of this area. Pine forests extend from the boreal and temperate regions of North America as far south as Nicaragua via the cordilleras of Mexico. Palynological evidence indicates that they had moved southward into Central America by the middle Miocene (Millar 1998), roughly 15 million years ago. As pines migrated south, their ectomycorrhizal symbionts would have moved with them. Therefore, pine-associated fungi would have existed in Central America for an equally long period of time. Reports of the ectomycorrhizal fungi associated with pines in Belize indicate that there is a significant biogeographical relationship with fungi found along the Gulf Coast and in the southeastern part of the United States, and a large number of species occur in both areas (Singer et al. 1983; Kropp 2001).

On the other hand, oaks, appear to have arrived much more recently. Pollen deposits indicate that they were not present in Colombia until around 330 000 years ago (Hoogiemstra and Cleef 1995). Like those of pines, the ectomycorrhizal associates of neotropical oaks, are biogeographically linked to temperate North America. Halling and Mueller (2002) report that, at the genus level, the boletes in these oak forests have a 94% similarity to those in north temperate forests. However, at the species level, much less overlap occurs, and a relatively high number of endemic species are found. At least one of the species that occurs both in Central America and the northern temperate zone, *Laccaria trichodermophora* Mueller, is sexually compatible across this geographical range (Mueller and Strack 1992), indicating that either it arrived in Central America relatively recently or that continued genetic exchange occurs between northern temperate and Central American populations of this species.

Ectomycorrhizas in moist lowland or submontane forests

In contrast with neotropical oak and pine forests, we know relatively little about the ectomycorrhizal fungi in lowland or submontane neotropical forests from which these hosts are lacking. The work done on these fungi has been somewhat sporadic and consists largely of taxonomic reports or long-term studies of community structure that were set up after the first reports of ectomycorrhizal associations in these forests were published (Kreisel 1970; Singer and Araujo 1979). Prior to this, it was generally believed that, with the exception of certain montane environments, disturbed areas, or plantations of introduced ectomycorrhizal tree species, these neotropical forests lacked ectomycorrhiza (Singer and Araujo 1979).

In many cases, lowland or submontane forests in the neotropics are indeed dominated by host plants forming arbuscular mycorrhiza. For example, root surveys done in Brazil and Venezuela showed that mycorrhizal colonization was largely dominated by arbuscular mycorrhizal (AM) fungi (St. John 1980; St. John and Uhl 1983). A more recent survey done in French Guyana confirmed that trees within these forest types almost exclusively formed AM associations (Bereau et al. 1997). However, all the above studies reported that ectomycorrhizas were also present but at low frequencies. Other studies indi-

B.R. Kropp. Biology Department, Utah State University, Logan, UT 84322, USA (e-mail: brkropp@biology.usu.edu).

cate that ectomycorrhizas are common in some neotropical forests. The groundbreaking work of Singer and Araujo (1979) showed that ectomycorrhizas are abundant on the roots of dominant tree species in certain types of forest in the Amazon Basin. They reported that the ectomycorrhizal symbiosis was a major component of "campinarana" forests on nutrient poor white-sand podzols (oxisols) and that they were also common in both "campina" and periodically flooded "igapó" forests. In contrast, the "terra firma" forest that they studied was largely nonectomycorrhizal with a minor ectomycorrhizal component that was found only after an intensive search of the area. Moyersoen (1993) reported a similar situation in a "caatinga" forest in Venezuela, where ectomycorrhizal trees were codominant in "bana alta" forest stands, but scarce in adjacent "bana abierta" and "caatinga alta" stands encountered at different places along a gentle topographic gradient.

Singer and Araujo (1979) were probably the first to describe these predominantly nonectomycorrhizal rainforests as being "intermittent with interspersed islands of ectotroph" forest. These authors were of the opinion that ectomycorrhizal forests occur wherever edaphic or climatic conditions favor ectomycorrhizal tree species over those AM-forming species present in the surrounding stands. In certain situations, ectomycorrhizal hosts occur in sharply defined stands within a largely nonectomycorrhizal forest matrix (Henkel et al. 2002). In such cases, the ectomycorrhizal trees form largely single-species dominated (i.e., monodominant) stands and their associated ectomycorrhizal fungi are restricted to these stands. In other situations, ectomycorrhizal tree species are either codominant in the stands (Moyersoen 1993), or sparsely scattered throughout the forest. In a study done by B.R. Kropp (unpublished data) in a moist submontane broad-leaved forest in Belize, ectomycorrhizal trees occurred either individually or in small, scattered clusters rather than in monodominant stands. This is somewhat similar to the situation described by Singer and Araujo (1979) for the "terra firma" forest in the Amazon Basin, where ectomycorrhizal hosts were present but scarce.

Development of monodominant stands

In cases where ectomycorrhizal trees form single-species stands, it has been hypothesized that it is the ectomycorrhizas themselves that help drive their establishment. Janos (1985) suggested that sites with pulsed water and mineral availability, homogeneous nutrient-poor soils and slow decomposition rates, potentially favor ectomycorrhizal fungi. Situations that favour the formation of ectomycorrhizas could lead to host monodominance, especially if the ectomycorrhizal relationships are well adapted to the site and are exclusive of other hosts. However, little research has been done to test this hypothesis. Nevertheless, the little information that is available, suggests that, at least in the tropics of Africa and Asia (paleotropics), it may not be the ectomycorrhizas that drive the formation of these stands. For example, Torti and Coley (1999), investigating monodominant trees in Africa, concluded that there was no direct relationship between ectomycorrhizal fungi and and the development of single-species stands. In this study, two of the dominant species *Gilbertiodendron* sp. and *Julbernardia* sp. (Fabaceae), formed both ectomycorrhizal and AM associations, whereas another Fabaceae, a *Cynometra* sp., formed only arbuscular mycorrhizas. Further work by Torti et al. (2001) showed that the monodominance of *Gilbertiodendron* sp. results from a combination of site characteristics (e.g., deep shade and deep, slowly decomposing litter) and the large seed size of this species, which favored seedling establishment.

The influence of edaphic factors on development of ectomycorrhizal stands is also unclear. The scarce information that is available on this subject implies that those edaphic factors that have been suggested to play a role in their development do not correlate with the presence of ectomycorrhizal stands. For example, Moyersoen (1993), studying two sites characterized by pulsed water availability in Venezuela, showed that ectomycorrhizal trees dominated one of the sites, whereas AM trees dominated the other. However, conditions were somewhat drier in the AM host-dominated site and this factor might have favored the formation of arbuscular mycorrhizas over that of ectomycorrhizas (Lodge 1989; Lodge and Wentworth 1990). Furthermore, the monodominant stands of the ectomycorrhizal tree species *Dicymbe corymbosa* Spruce ex Benth studied by Henkel et al. (2002) in Guyana were found on widely varying soils; once again, this suggests that edaphic factors, alone do not explain the formation of monodominant stands.

Without further studies, it will be difficult to understand the formation of monodominant patches of ectomycorrhizal trees in the lowland tropics. Nonetheless, these patches probably form in response to a complex of environmental and host biological factors. Ectomycorrhizal fungi appear to be only one

Chapter 12. Ectomycorrhizas in the neotropics with emphasis on lowland forests

Table 12.1. Genera of ectomycorrhizal trees reported from the neotropics.

Host genus and family	Mycorrhizal status*		References
	ECM	AM	
Fabaceae			
Aldina	C	C	Moyersoen 1993; Henkel et al. 2002
Andira	C	C	Lodge 1996; Bereau et al. 1997
Copaifera	F	C	Norris 1969; Carneiro et al. 1998
Dicymbe	C	—	Henkel et al. 2002
Eperua	F	C	St. John and Uhl 1983; Bereau et al. 1997
Hymenaea	F	C	Bereau et al. 1997; Miller et al. 2000
Mora	F	—	Norris 1969
Swartzia	F	C	Bereau et al. 1997; Buyck and Ovrebo 2002
Gnetaceae			
Gnetum	C	—	Singer and Araujo 1979; St. John 1980
Nyctaginaceae			
Guapira	C	C	Moyersoen 1993; Haug et al. 2005
Neea	C	C	Bereau et al. 1997; Haug et al. 2005
Pisonia	C	—	Cairney et al. 1994; Lodge 1996
Polygonaceae			
Coccoloba	C	C	Kreisel 1970; Bereau et al. 1997
Sapotaceae			
Glycoxylon	C	—	Singer and Araujo 1979
Unknown			
"Macure"	F	—	St. John and Uhl 1983
"Cabari"	F	—	St. John and Uhl 1983

Note: The neotropics exclude the pine and oak forests of Central America and Colombia.
*ECM, ectomycorrhizal; AM, arbuscular mycorrhizal; C, mycorrhizal status confirmed by microscopic analysis; F, mycorrhizal status based on field observations or method unclear; missing letters indicate that the mycorrhizal status was not specified. All mycorrhizal types forming mantles are classified here as ectomycorrhizal.

of the factors involved, and the dynamics of patch formation probably varies for different species and locations.

Potential hosts and mycorrhizal status

A variety of mantle-forming mycorrhizal types has been reported from neotropical plants including ect-endomycorrhizas observed on roots of species of *Andira* (Fabaceae) and *Neea* (Nyctaginaceae) Moyersoen 1993; Lodge 1996; Miller et al. 2000), peritrophic mycorrhizas that develop a mantle but no clear Hartig net (Moyersoen 1993), and a mycorrhiza on *Neea obovata* Spruce ex Heimerl & Ruprechtia roots (Moyersoen 1993) that exhibited a mantle and apparent transfer cells of the type reported from *Pisonia grandis* R. Br. in Oceania (Ashford and Allaway 1982). For the purposes of this article, all mantle-forming mycorrhizas are treated as ectomycorrhizas.

Thus far, two unidentified plant species and species from 14 identified plant genera have been reported to form ectomycorrhizal associations in lowland neotropical forests. This is similar to the numbers obtained for African forests, where 18 ectomycorrhizal plant genera have been identified (Thoen 1993). Taxonomically, the reported neotropical hosts are rather diverse and distributed among five different families (Table 12.1); however, most of the genera are either legumes (Fabaceae) or belong to the Nyctaginaceae. In a number of the existing reports of ectomycorrhizas from neotropical forests, the information is poorly documented and ectomycorrhizal status was not confirmed by microscopic studies of the roots. For example, in an early study, Norris (1969) noted that roots of two *Eperua* (Fabaceae) species in South America were "brown, covered stubby outgrowths, probably mycorrhiza" or that the roots of *Copaifera* (Fabaceae) species were brown with "much mycorrhiza," but no microscopic study was done to confirm the presumed ectomycorrhizal status of the

Table 12.2. Species of *Coccoloba* reported to form ectomycorrhizas.

Species	Location	References
C. latifolia	French Guyana	Bereau et al. 1997
C. mollis	French Guyana	Bereau et al. 1997
C. pyrifolia	Puerto Rico	Lodge 1996
C. diversifolia	Puerto Rico	Lodge 1996
C. mansanillensis	Panama	Buyck and Ovrebo 2002
C. excelsa	Venezuela	Moyersoen 1993
C. uvifera	Cuba	Kreisel 1970
C. belizensis	Belize	B.R. Kropp, unpublished data

roots. The report by the same author of ectomycorrhizas in roots of species of *Mora* (Fabaceae) is equally ambiguous. In a number of other studies, the authors did not clearly explain how the mycorrhizal status of the trees was determined or relied on the cooccurrence of ectomycorrhizal fungal genera with certain trees in the field. More work is needed to either confirm or disprove the ectomycorrhizal status of some of the genera in Table 12.1. In addition, given the enormous diversity of tree species in lowland neotropical forests, the large geographical area they cover, and the relatively little work that has been done on their fungal symbionts, additional ectomycorrhizal hosts undoubtedly remain to be discovered in these forests.

Ectomycorrhizas versus arbuscular mycorrhizas

A number of the published reports on the mycorrhizal status of neotropical tree genera are contradictory. For example, Norris (1969) found *Eperua* species to be ectomycorrhizal, St. John and Uhl (1983) reported that they formed both ectomycorrhizas and arbuscular mycorrhizas, and Moyersoen (1993) and Bereau et al. (1997) found them to be solely colonized by AM fungi. Although the study by Norris (1969) was not based on an anatomical investigation of the roots and, thus, is open to question, St. John and Uhl (1983) used a microscope to assess AM formation and, presumably, also did so with the ectomycorrhizal roots. Similarly, Bereau et al. (1997) found only arbuscular mycorrhizas on roots of *Andira* sp. (Fabaceae), while Lodge (1996) reported that roots of *Andira inermis* (Wright) Kunth ex DC. are colonized by non-necrotic intracellular infections. These were later identified as ectendomycorrhizas (Miller et al. 2000). In addition, Carneiro et al. (1998) found that members of the genus *Swartzia* (Fabaceae) formed arbuscular mycorrhizas, whereas Buyck and Ovrebo (2002) recorded the presence of sporocarps of ectomycorrhizal fungi in a dense stand of a species of *Swartzia* in Panama, suggesting that *Swartzia* species can form ectomycorrhizas.

Given the different methods used to determine mycorrhizal status, it is not surprising that discrepancies exist between published reports of their occurrence. Nevertheless, some of the conflicting reports probably accurately reflect the mycorrhizal status of the plants. Some of these reports were made on different species within a genus in different geographical regions. For example, St. John and Uhl (1983) reported ectomycorrhizas from the roots of *Eperua leucantha* Benth. and *Eperua purpurea* Benth. in Venezuela, whereas Bereau et al. (1997) found that *Eperua falcata* Aublet and *Eperua grandiflora* (Aublet) Benth. in French Guyana form only arbuscular mycorrhizas. Thus, it is possible that different species within a genus form different kinds of mycorrhizal associations. Yet, what we know about ectomycorrhizas indicates that the ability to form this symbiosis often tends to be consistent within a genus: for example, all species of *Pinus* and *Quercus* have the potential to form ectomycorrhizas. The mycorrhizal relationships of neotropical trees are not as well known, but a few have been studied well enough to indicate that the ability to form ectomycorrhizas might also be consistent within a given genus. The genus *Coccoloba* (Polygonaceae) is probably the best example of this. The mycorrhizal status of eight different *Coccoloba* species has been studied over a wide geographical area, and all have been reported to form ectomycorrhizas (Table 12.2).

Another explanation for the discrepancies among the reports of mycorrhizal status is that the trees might be forming ectomycorrhizas or arbuscular mycorrhizas at different times. If this is the case, the type of mycorrhiza reported to be present on the roots would depend on when the sample was taken. That certain plants form both of these mycorrhizal associations is relatively well known, and Smith and Read (1997) list 30 genera that are known do so. Ectomycorrhizas and arbuscular mycorrhizas can be formed simultaneously or at different times under varying conditions. For example, Lapeyrie and Chilvers (1985) and Chilvers et al. (1987), found that seedlings of *Eucalyptus dumosa* Cunn. ex Oxley first developed arbuscular mycorrhizas but that these were later replaced by ectomycorrhizas. In addition, the type of mycorrhiza that formed on temperate *Salix* and *Populus* species (Salicaceae) varied depending on soil moisture levels. The ectomycorrhizal fungi associated with these trees appeared to compete better at sites of intermediate moisture levels, whereas arbuscular mycorrhizas were more prevalent at either wetter or drier sites (Lodge 1989; Lodge and Wentworth 1990).

In the neotropics, species of *Neea*, *Coccoloba*, *Guapira* (Nyctaginaceae), and *Aldina* (Fabaceae) have the capacity to form double symbioses with both ectomycorrhizal and AM fungi (mycorrhiza types confirmed anatomically) (Moyersoen 1993; Bereau et al. 1997) (Table 12.1). It is probable that, in at least some of the tree species of the above genera, ectomycorrhizas and arbuscular mycorrhizas interact with one another and with their environment in a dynamic fashion. Most of the field studies on neotropical mycorrhizas have been done at a fixed point in time. Therefore, it is currently impossible to determine whether the mycorrhizal status of the trees in question is static or whether it varies under certain environmental conditions. For instance, we do not know whether the frequency of arbuscular mycorrhizas versus ectomycorrhizas in the ectomycorrhiza-dominated Venezuelan "bana alta" forest studied by Moyersoen (1993) changes with pulses of moisture in the soil.

The fungi

Three major works have provided much of what is known about the taxonomy of ectomycorrhizal fungi in neotropical regions. The first of these was a mycofloristic treatment of Venezuela and adjacent areas by Dennis (1970), which included a number of putatively ectomycorrhizal fungi. Later, Pegler (1983) published a treatise on the mycota of the Lesser Antilles that included many ectomycorrhizal species, and Singer et al. (1983) produced a major review of ectomycorrhizal fungi in the neotropics. Although the treatment by Singer et al. (1983) also covers fungi from the pine and oak forests of Central America, most of the species treated are found in lowland forests of South America. Although other taxonomic reports also cover this region and many

Table 12.3. Fungal genera containing putatively ectomycorrhizal species in the neotropics.

Family*	Genus
Amanitaceae	*Amanita*
Boletaceae	*Boletus*
	Pulveroboletus
Cantharellaceae	*Cantharellus*
Craterellaceae	*Craterellus*
Cortinariaceae	*Cortinarius*
	Inocybe
Gyrodontaceae	*Gyrodon*
	Gyroporus
	Phlebopus
Gomphaceae	*Gomphus*
Hymenochaetaceae	*Coltricia*†
Paxillaceae	*Neopaxillus*
Ramariaceae	*Ramaria*
Russulaceae	*Lactarius*
	Russula
Sclerodermataceae	*Scleroderma*
Scutigeraceae	*Scutiger*
Strobilomycetaceae	*Austroboletus*
	Chalciporus
	Fistulinella
	Strobilomyces
	Tylopilus
Thelephoraceae	*Thelephora*
	Sarcodon
Tricholomataceae	*Laccaria*
	Tricholoma
Xerocomaceae	*Boletellus*
	Phylloporus
	Xerocomus

Note: The neotropics exclude the pine and oak forests of Central America and Colombia (Bas 1978; Pegler 1983; Singer et al. 1983; Miller et al. 2000; Henkel et al. 2002; B.R. Kropp, unpublished data).

*Families are based on Hawksworth et al. (1995).

†*Coltricia* is often overlooked in surveys of ectomycorrhizal fungi, but certain members of the genus have been shown experimentally to form ectomycorrhiza (see Danielson 1984).

presently undescribed ectomycorrhizal fungal species will undoubtedly be added to the neotropical list, the works mentioned above allow a few general conclusions to be drawn.

For the most part, the ectomycorrhizal genera found in northern temperate and neotropical pine or oak forests, also occur in the lowland neotropics (Table 12.3); however, relatively little overlap occurs at the species level. Many of the species found in the lowland forests are unique to these forests. In addition, certain groups of ectomycorrhizal fungi that are very prominent in much of the temperate zone, for example the Cortinariaceae, are poorly represented in the lowland neotropics. *Cortinarius* is one of the most common and species-rich ectomycorrhizal fungal genera in many temperate regions; it is also abundant throughout the *Nothofagus* (Nothofagaceae) forest region of southern South America and occurs in the oak forests found in parts of Central and northern South America (Moser and Horak 1975; Halling and Mueller 2005). However, no species have been reported from the Lesser Antilles, and only a handful of species have been reported from ectomycorrhizal forests in the lowland neotropics (Singer et al. 1983). Likewise, fungi of the genus *Inocybe* are abundant in northern temperate areas and relatively common in the *Nothofagus* forest region of South America (Horak 1979) but are quite scarce in the lowland neotropics (Pegler 1983; Singer et al. 1983). Ectomycorrhizal members of the Tricholomataceae, such as *Laccaria* and *Tricholoma*, are also common in temperate forests and occur in the oak forests of Central America but are scarce in the Lesser Antilles and the Amazon Basin (Pegler 1983; Singer et al. 1983). By contrast, members of the Russulaceae are important members of the northern temperate and Central American ectomycorrhizal floras, and they are also particularly abundant throughout the lowland neotropics. Nearly one-half of the floristic treatment of Singer et al. (1983) is devoted to the ectomycorrhizal fungal genera *Russula* and *Lactarius*, and additional species have recently been reported from *Dicymbe* (Fabaceae) forests in Guyana (Miller et al. 2002). The boletes are also especially species-rich in the neotropics, with 12 genera represented. Finally, *Amanita*, the sole genus of the Amanitaceae present in the lowland neotropics, is another example that is well represented by species in both temperate and neotropical areas (Bas 1978; Simmons et al. 2002).

Dispersal

That ectomycorrhizal fungi in the neotropics occur either with scattered individual trees or with trees forming islands of monodominant or codominant species within forests dominated by AM-forming species, implies that long distance dispersal is essential to their survival. Within dominant or codominant stands, mycelial growth alone might suffice to allow the fungi in question to spread from tree to tree. However, to move between patches of ectomycorrhizal trees or to come into contact with isolated hosts within the forest, the fungi need suitable dispersal mechanisms to cover much greater distances. Spore dispersal is the most likely mechanism by which this is accomplished. In northern temperate forests, mycophagy by animals provides one such means of spore dispersal (Castellano et al. 1989). Although ectomycorrhizal fungal spore dispersal by animals has not been documented in the neotropics, a recent report of mycophagy by primates in South America (Hanson et al. 2003) demonstrates that at least some neotropical animals consume fungi. Thus, the potential for dispersal of ectomycorrhizal fungal spores by animals in the tropics exists and merits further study.

Host specificity

Much work has been done on host specificity of ectomycorrhizal fungi in temperate systems and a wide range of specificity has been found (Molina et al. 1992). Given the species richness of most neotropical lowland forests, the distance between ectomycorrhizal hosts is likely to be relatively great, compared with that in less diverse temperate forests. Therefore, suitable hosts are relatively scarce, and the chance of an ectomycorrhizal fungal spore encountering a host root is much reduced. As a consequence, host specificity might be selected against in these forests (Lodge and Cantrell 1995). On the other hand, airborne spores are a very effective way for fungi to disperse (Gregory 1973); in reality, given the large numbers of spores generally produced by ectomycorrhizal fungi, finding the right host might not be a problem for host-specific fungi with widely spaced hosts. Furthermore, even if selection does favour broad host ranges in neotropical ectomycorrhizal fungi, host-specific fungi might still thrive in monodominant stands where they would have relatively little problem contacting suitable hosts either by spore dispersal or mycelial growth.

In tropical Africa, a small number of host-specificity studies have been done. Thoen and Ba (1989) reported that host specificity appears to be strong in seasonally dry forests containing ectomycorrhizal *Afzelia* (Fabaceae) and *Uapaca* (Phyllanthaceae) species. Although these hosts occurred in close proximity to one another, the authors observed that only 6 of the 43 potentially ectomycorrhizal fungi appeared to be shared by both tree genera. When cultures of fungi isolated from *Uapaca* roots were inoculated onto the roots of *Afzelia* species in an in vitro study, none formed mycorrhizas (Ba and Thoen 1990).

Virtually no experimental work of this type has been done on the host specificity of neotropical ectomycorrhizal fungi. However, some insight into host specificity can be gleaned from the data available in taxonomic treatments of neotropical fungi. An examination of the habitat and geographical distributions for the ectomycorrhizal fungi treated by Singer et al. (1983), indicates that, at least, some of them have wide host ranges. *Inocybe neotropicalis* Singer et al. is an example of a broad host range fungus that occurs from Florida to Costa Rica, where it forms ectomycorrhizas with hosts as diverse as *Quercus* (Fagaceae) and *Coccoloba* (Polygonaceae). Another example of an ectomycorrhizal fungus that appears to lack host specificity is *Cortinarius amazonicus* Singer & Araujo. This species occurs in Brazil in association with both *Glycoxylon* (Sapotaceae) and certain legumes (Singer et al. 1983). However, some potentially host-specific fungi have also been reported. Thus far, *Pseudotulostoma volvata* Mill. & Henkel appears to form ectomycorrhizas only with tree species of the genus *Dicymbe* that occur in monodominant stands in Guyana (Miller et al. 2001; Henkel et al. 2006). Based on the limited amount of information that is available, it is likely that, as in temperate forests, a mix of host-specific and broad host range ectomycorrhizal fungi occurs throughout the lowland neotropics. With recent advances in molecular approaches, it seems likely that, in the near future, significant advances will be made in the fields of phylogeography and in the accurate identification of fungal symbionts and their host plant roots within these complex ecosystems.

References

Ashford, A.E., and Allaway, W.G. 1982. A sheathing mycorrhiza on *Pisonia grandis* R. Br. (Nyctaginaceae) with development of transfer cells rather than a Hartig net. New Phytol. **90**: 511–519.

Ba, A.M., and Thoen, D. 1990. First synthesis of ectomycorrhizas between *Afzelia africana* Sm. (Caesalpinioideae) and native fungi from West Africa. New Phytol. **114**: 99–103.

Bas, C. 1978. Studies in *Amanita*—I. Some species from Amazonia. Persoonia, **10**: 1–22.

Bereau, M., Gazel, M., and Garbaye, J. 1997. Les symbioses mycorhiziennes des arbres de la forêt tropical humide de Guyane française. Can. J. Bot. **75**: 711–716.

Buyck, B., and Ovrebo, C.L. 2002. New and interesting *Russula* species from Panama. Mycologia, **94**: 888–901.

Cairney, J.W.G., Rees, B.J., Allaway, W.G., and Ashford, A.E. 1994. A basidiomycete isolated from a *Pisonia* mycorrhiza forms sheathing mycorrhizas with transfer cells on *Pisonia grandis* R. Br. New Phytol. **126**: 91–98.

Carneiro, M.A.C., Siqueira, J.O., Moreira, F.M.S., de Carvalho, D., Botelho, S.A., and Junior, O.J.S. 1998. Micorriza arbuscular em espécies arbóreas e arbustivas natives de ocorrência no sudeste do Brasil. Cerne, **4**: 129–145.

Castellano, M.A., Trappe, J.M., Maser, Z., and Maser, C. 1989. Key to spores of the genera of hypogeous fungi of north temperate forests with special reference to animal mycophagy. Mad River Press, Inc., Eureka, Calif.

Chilvers, G.A., Lapeyrie, F.F., and Horan, D.P. 1987. Ectomycorrhizal vs. endomycorrhizal fungi within the same root system. New Phytol. **107**: 441–448.

Danielson, R.M. 1984. Ectomycorrhizal associations in jack pine stands in northeastern Alberta. Can. J. Bot. **62**: 932–939.

Dennis, R.W.G. 1970. Fungus flora of Venezuela and adjacent countries. Royal Botanical Gardens, Kew, UK; Cramer, Vaduz, Liechtenstein. Kew Bull. Additional Ser. III.

Gregory, P.H. 1973. The microbiology of the atmosphere 2nd ed. John Wiley & Sons, New York.

Halling, R.E., and Mueller, G.M. 2002. Agarics and boletes of neotropical oakwoods. *In* Tropical mycology. Vol. 1. Macromycetes. *Edited by* R. Watling, J.C. Frankland, A.M. Ainsworth, S. Isaac, and C.H. Robinson. CAB International, Wallingford, UK. pp. 1–10.

Halling, R.E., and Mueller, G.M. 2005. Common mushrooms of the Talamanca Mountains, Costa Rica. New York Botanical Garden, Bronx, N.Y.

Hanson, A.M., Hodge, K.T., and Porter, L.M. 2003. Mycophagy among primates. Mycologist, **17**: 6–10.

Haug, I., Weiss, M., Homeier, J., Oberwinkler, F., and Kottke, I. 2005. Russulaceae and Thelephoraceae form ectomycorrhizas with members of the Nyctaginaceae (Caryophyllales) in the tropical mountain rainforest of southern Ecuador. New Phytol. **165**: 923–936.

Hawksworth, D.L., Kirk, P.M., Sutton, B.C., and Pegler, D.N. 1995. Aisworth and Bisby's dictionary of the fungi. 8th ed. CAB International, Wallingford, UK.

Henkel, T.W., Terborgh, J., and Vilgalys, R.J. 2002. Ectomycorrhizal fungi and their leguminous hosts in the Pakaraima Mountains of Guyana. Mycol. Res. **106**: 515–531.

Henkel, T.W., James, T.Y., Miller, S.L., Aime, M.C., and Miller, O.K., Jr. 2006. The mycorrhizal status of *Pseudotulostoma volvata* (Elaphomycetaceae, Eurotiales, Ascomycota). Mycorrhiza, **16**: 241–244.

Hoogiemstra, H., and Cleef, A.M. 1995. Pleistocene climate change and environmental and generic dynamics in the north Andean montane forests and paramo. *In* Biodiversity and conservation of neotropical montane forests. *Edited by* S.P. Churchill, H. Balslev, E. Forero, and J.L. Luteyn. New York Botanical Garden Press, Bronx, N.Y. pp. 35–49.

Horak, E. 1979. Flora Cryptogamica de Tierra del Fuego. Tomo 11. Fascículo 6. Fungi, Basidiomycetes Agaricales y Gasteromycetes secotioides. Fundación para la Educación, la Ciencia y la Cultura, Buenos Aires.

Janos, D.P. 1985. Mycorrhizal fungi: agents or symptoms of tropical community composition? *In* Proceedings of the 6th North American Conference on Mycorrhizae, 25–29

June 1984, Corvallis, Oregon. *Edited by* R. Molina. Forest Research Laboratory, Corvallis, Ore. pp. 98–103.

Kreisel, H. 1970. Ektotrophe mykorrhiza bei *Coccoloba uvifera* in Kuba. Biol. Rundsch. **9**: 97–98.

Kropp, B.R. 2001. Familiar faces in unfamiliar places: mycorrhizal fungi associated with Caribbean pine. Mycologist, **15**: 137–140.

Lapeyrie, F.F., and Chilvers, G.A. 1985. An endomycorrhiza–ectomycorrhiza succession associated with enhanced growth of *Eucalyptus dumosa* seedlings planted in a calcareous soil. New Phytol. **100**: 93–104.

Lodge, D.J. 1989. The influence of soil moisture and flooding on formation of VA-endo- and ectomycorrhizae in *Populus* and *Salix*. Plant Soil, **117**: 243–253.

Lodge, D.J. 1996. Microorganisms. *In* The food web of a tropical forest. *Edited by* D.P. Reagan and R.B. White. The University of Chicago Press, Chicago, Ill. pp. 55–108.

Lodge, D.J., and Cantrell, S. 1995. Fungal communities in wet tropical forests: variation in time and space. Can. J. Bot. **73**(Suppl. 1): 1391–1398.

Lodge, D.J., and Wentworth, T.R. 1990. Negative associations among VA-mycorrhizal fungi and some ectomycorrhizal fungi inhabiting the same root system. Oikos, **57**: 347–356.

Millar, C.L. 1998. Early evolution of pines. *In* Ecology and biogeography of *Pinus*. *Edited by* D.M. Richardson. Cambridge University Press, Cambridge, Mass. pp. 69–91.

Miller, O.K., Jr., Lodge, D.L., and Baroni, T.J. 2000. New and interesting mycorrhizal fungi from Puerto Rico, Mona, and Guana Islands. Mycologia, **92**: 558–570.

Miller, O.K., Jr., Henkel, T.W., James, T.Y., and Miller, S.L. 2001. *Pseudotulostoma*, a remarkable new volvate genus in the Elaphomycetaceae from Guyana. Mycol. Res. **105**: 1268–1272.

Miller, S.L., Aime, M.C., and Henkel, T.W. 2002. Russulaceae from the Pakaraima Mountains of Guyana. I. New species of pleurotoid *Lactarius*. Mycologia, **94**: 545–553.

Molina, R., Massicotte, H., and Trappe, J.M. 1992. Specificity phenomena in mycorrhizal symbioses: community ecological consequences and practical implications. *In* Mycorrhizal functioning and integrated plant fungal process. *Edited by* M.F. Allen. Chapman & Hall, New York. pp. 357–423.

Moser, M., and Horak, E. 1975. *Cortinarius* Fr. Und nahe verwandte gattungen in Sudamerica. J. Cramer, Vaduz, Liechtenstein. Beih. Nova Hedwigia Heft 25.

Moyersoen, B. 1993. Ectomicorrizas y micorrizas vesiculo–arbusculares en caatinga amazonia del sur de Venezuela. Scientia Guianae, **3**: 1–82.

Mueller, G.M., and Strack, B.A. 1992. Evidence for a mycorrhizal host shift during migration of *Laccaria trichodermophora* and other agarics into neotropical oak forests. Mycotaxon, **45**: 249–256.

Norris, D.O. 1969. Observations on the nodulation status of rainforest leguminous species in Amazonia and Guyana. Trop. Agric. (Trinidad), **46**: 145–151.

Pegler, D.N. 1983. Agaric flora of the Lesser Antilles. Her Majesty's Stationary Office, London. Kew Bull. Additional Ser. IX.

Simmons, C., Henkel, T., and Bas, C. 2002. The genus *Amanita* in the Pakaraima Mountains of Guyana. Persoonia, **17**: 563–582.

Singer, R., and Araujo, I. 1979. Litter decomposition and ectomycorrhiza in Amazonian forests. I. A comparison of litter decomposition and ectomycorrhizal basidiomycetes in latosal-terra-firme forest and white podzol campinarana. Acta Amazonica, **9**: 25–41.

Singer, R., Araujo, I., and Ivory, M.H. 1983. The ectotrophically mycorrhizal fungi of the neotropical lowlands especially central Amazonia. Nova Hedwigia Beih. **77**: 1–352.

Smith, S.E., and Read, D.J. 1997. Mycorrhizal symbiosis 2nd ed. Academic Press Inc., San Diego, Calif.

St. John, T.V. 1980. A survey of mycorrhizal infection in an Amazonian rainforest. Acta Amazonica, **10**: 527–533.

St. John, T.V., and Uhl, C. 1983. Mycorrhizae in the rain forest at San Carlos de Rio Negro, Venezuela. Acta Cient. Venez. **34**: 233–237.

Thoen, D. 1993. Looking for ectomycorrhizal trees and ectomycorrhizal fungi in tropical Africa. *In* Aspects of tropical mycology. *Edited by* S. Isaac, J.C. Frankland, R. Watling, and A.J.S. Whalley. Cambridge University Press, Cambridge, UK. pp. 193–205.

Thoen, D., and Ba, A.M. 1989. Ectomycorrhizas and putative ectomycorrhizal fungi of *Afzelia africana* Sm., and *Uapaca guineensis* Müll. Arg. in southern Senegal. New Phytol. **113**: 549–559.

Torti, S.D., and Coley, P.D. 1999. Tropical monodominance: a preliminary test of the ectomycorrhizal hypothesis. Biotropica, **31**: 220–228.

Torti, S.D., Coley, P.D., and Kursar, T.A. 2001. Causes and consequences of monodominance in tropical lowland forests. Am. Nat. **157**: 141–153.

Chapter 13
Ecophysiology of sporocarp development of ectomycorrhizal basidiomycetes associated with boreal forest gymnosperms

J. André Fortin and Mohammed S. Lamhamedi

Introduction

For more than 50 years, research into the biology and ecology of mycorrhizal fungi and their host plants has been attracting an ever-increasing number of scientists. As a result, the fundamental role of mycorrhizal fungi in plant evolution, plant growth, and ecosystem dynamics is now widely recognized (Smith and Read 1997; Leake et al. 2004). However, studies investigating the biology of these symbiotic associations have largely focused on the mycorrhizas, their extraradical mycelia, and the advantages for the plant host in terms of growth and fitness. By contrast, fungal fitness, measured in terms of spore production and ecophysiology, has been rarely investigated.

With regards to ectomycorrhizal (ECM) fungi, the few studies that have been conducted have focused on the reproductive structures formed, fungal systematics, ecological distribution of sporocarps, and development of techniques for industrial-scale production of edible species (Yun and Hall 2004). Very few scientists have addressed the physiological mechanisms underlying the initiation and development of ECM fungal sporocarps under natural conditions.

In this chapter, we review this subject, revisit some of our past contributions, and add to these more recent, and as yet unpublished, observations. Finally, we suggest the direction that future fundamental and applied research into the ecophysiology of sporocarp development should take. Because we have worked exclusively with Basidiomycetes, we refer to sporocarps as basidiomes.

Carbon flow in ectomycorrhizal systems

Several analytical studies linked with the investigation of carbon (C) flow in ECM systems have been published, and questions regarding the C cost of these symbiotic relationships have often been raised (e.g., Bryla and Eissenstat 2005). However, most of the studies on C flow in ECM systems have been restricted to observations and measurements conducted at the ECM root and extraradical mycelium level or at the ecosystem level (e.g., Steinmann et al. 2004). Here, we will consider C flow at the level of the ECM system as a whole, that is, all parts of the host tree and all parts of the fungus—including their basidiomes. Using information from short-term and longer term studies, we will also consider C flow over a complete growing season in the field; the aim being to conciliate the divergent information available on the subject. To conclude this section, we propose an holistic view of C flow in ECM systems.

In 1942, Björkman (see review in Hacskaylo 1973) presented a hypothesis to explain the dependency of ECM fungi on forest trees and why these fungi were not parasites. Briefly, the hypothesis suggested that a surplus of hexoses (reducing sugars) in the roots was essential for the development and functioning of ectomycorrhizas. Consequently, under these conditions, ECM fungi were considered to be the passive recipients of C that was surplus to that required by the plant. To test this hypothesis, Björkman (1942) supplied ECM seedlings with increasing amounts of

---

J.A. Fortin.[1] Centre d'étude de la forêt, Pavillon C.-E.-Marchand, Université Laval, Québec, QC G1V 0A6, Canada.
M.S. Lamhamedi. Direction de la recherche forestière, ministère des Ressources naturelles et de la Faune, 2700, rue Einstein, Québec, QC G1P 3W8, Canada.

[1]Corresponding author (e-mail: j.andre.fortin@videotron.ca).

N. At low N concentrations, ectomycorrhizas were formed; however, with increasing concentrations of N, their formation was reduced and even inhibited. In the low-N treatment, root tissues contained higher concentrations of hexoses than in high N treatments. The author proposed that a certain amount of N was necessary for the formation and normal functioning of ectomycorrhizas, allowing a certain level of free hexoses to be maintained in the root. By contrast, the further addition of N favoured protein synthesis, which used the surplus C.

In the late 1950s, Harley and his collaborators started a series of now-famous experiments using excised European beech (*Fagus sylvatica* L.) ectomycorrhizas and ^{14}C (Lewis and Harley 1965a, 1965b); a source–sink mechanism between the short roots and their mycorrhizal mantles was clearly demonstrated. Before passing into the fungal mantle, sucrose was hydrolyzed into glucose and fructose. In the mantle, the glucose was taken up by the fungal hyphae and converted into trehalose, a form that cannot diffuse back into the root. On this basis, Björkman's hypothesis explaining the movement of C from the host to the fungus was completely rejected (Harley 1969).

More recently, Wallander and Nylund (1991) submitted a new hypothesis that aimed to reconcile the views of Björkman with those of Harley (1969). This new hypothesis proposed that at low N concentrations, surplus C would be stored in the fungal mycelium, rather than in the root as had been suggested by Björkman, and that, once the fungus had exhausted the N within the substrate, this surplus C would be used to produce fungal mycelium and fruit bodies. However, there are two fundamental problems with this hypothesis. Firstly, because the development of fungal hyphae and especially of sporocarps requires large amounts of N, where does the fungus find this if the principal source has been exhausted? Secondly, Wallander and Nylund's hypothesis also supposed that C accumulated in the fungal mycelium before being used for basidiome production. However, we have demonstrated that ectomycorrhizal basidiome development depends on current photosynthesis, not on C accumulated in the mycelium (Lamhamedi et al. 1994).

When studying C flow in ECM systems, most researchers assume that the physiology of ectomycorrhizas and the associated mycelium is constant throughout the year; however, few considerations have been given to fungal reproduction. Continuing on from the ideas of Wallander and Nylund (1991), we propose to consider the ECM system from a fungal perspective rather than from that of the host tree. We will attempt to show that, at different periods of the year, the ECM fungi can take advantage of the different physiological states of their host for the benefit of the different phases of their own life cycle. We will also consider a possible saprophytic capacity of ectomycorrhizal fungi that, we hope, will allow a better understanding of C flow in ECM systems.

Phenology of the ectomycorrhizas of boreal forests gymnosperms

Active meristematic tissues, including developing buds and cambium, growing mycorrhizal and non-mycorrhizal roots, flowers, and developing seeds, are all strong C sinks that may compete with each other for resources. Generally, the more active the tissue, the stronger is its potential as a C sink. At the beginning of the growing season in the boreal forest, developing buds and emerging shoots offer a stronger C sink for stored and current photosynthates than do roots and their associated mycorrhizal fungi. As a result, root growth is temporarily reduced below its potential (Landis et al. 1999; Grossnickle 2000; Lamhamedi et al. 2001, 2003). However, once the emerging foliage begins to photosynthesize, it becomes a net C source, and this C is available for export to other parts of the tree, including roots and mycorrhizal fungi. Nevertheless, it is only during the second half of the growing season that root and basidiome development usually reach their peak.

This phenomenon is also observed in boreal forest nurseries. Studies done using container-grown black spruce (*Picea mariana* (Mill.) BSP) and white spruce (*Picea glauca* (Moench) Voss) seedlings, showed a three- to four-fold increase in root dry mass (DM) between mid-August and late October (Lamhamedi et al. 2001, 2003). Our observations also revealed that this increase in root DM was generally synchronized with the development of ECM fungal basidiomes. Similarly, phenological observations on naturally growing mature balsam fir (*Abies balsamea* L. (Mill.); Forêt Montmorency, Québec; 47°20′N, 71°07′W; Langlois and Fortin 1984) showed that the growth of the tree's aerial tissues was out of phase with root activity. On the study site, bud burst of leader shoots occurred in late June, followed by bud burst on progressively lower branches. By mid-July, the terminal bud of

Chapter 13. Ecophysiology of sporocarp development of ectomycorrhizal basidiomycetes associated with boreal forest gymnosperms

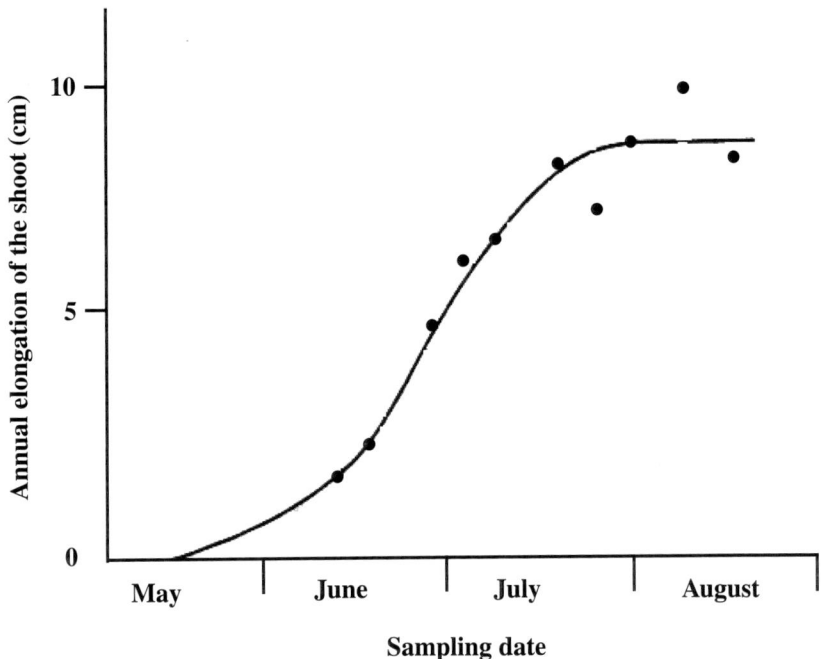

Fig. 13.1. Annual pattern of terminal internode elongation of *Abies balsamea* in the Montmorency Experimental Forest (Québec) during the 1978 growing season (adapted from Fig. 6, Langlois and Fortin 1984).

the leader shoot had already formed (Fig. 13.1). This was followed by the progressive formation of buds from the top of the tree to the bottom—a process that was completed by mid-August. Bud formation on the lowest branches was followed by an increase in ECM fungal development and the growth of a few coarse and rapidly elongating nonmycorrhizal long roots. These observations support the idea that, at the beginning of the growing season, the sink for photosynthates is located at the top of the tree and that it is progressively switched toward the bottom of the tree. This is also supported by the observations of Loach and Little (1973).

On the study site used by Langlois and Fortin (1984), soil temperatures (at 5 cm below the surface) during the second half of June were close to freezing, reaching a maximum of 15 °C by late July. Soil temperature is a major regulator of root growth (Lawrence and Oechel 1983*a*, 1983*b*). For example, growth of black spruce roots with a diameter between 0.5 and 1.5 mm, reaches an optimum at 20 °C and stops when soil temperatures drop below 5 °C (Tryon and Chapin 1983). Because root activity is minimal in June, their capacity to act as a strong C sink is also likely to be minimal. The progressive allocation of C towards the bottom of the tree during the second half of the growing season should be coupled with a greater physiologically activity (e.g., P absorption) of the roots. This is reflected by the fact that, in general, boreal tree species appear to maintain active root growth late into the fall (Lamhamedi and Bernier 1994).

After the development of new needles, new roots have high priority in terms of C allocation (Waring and Pitman 1985). It has been shown that new root growth depends on current photosynthates that are transported via the phloem (Lippu 1998; Pellicer et al. 2000). Any excess C is likely to be channelled into storage reserves (in the roots, stem, and canopy), radial growth, and the production of secondary metabolites that serve as defensive compounds (Waring and Pitman 1985). However, several factors, including tree age (Yoder et al. 1994), can affect photosynthetic rates and subsequent C allocation, root growth, and basidiome development.

Seasonal fluctuations of phosphorus absorption by balsam fir ectomycorrhizas

The absorption of P is an important function of ECM associations and requires energy. Langlois and Fortin (1984) periodically measured P uptake by short mycorrhizal and long nonmycorrhizal balsam fir roots collected from mature trees over an entire year. The use of ^{32}P clearly revealed annual patterns of P absorption. In long nonmycorrhizal

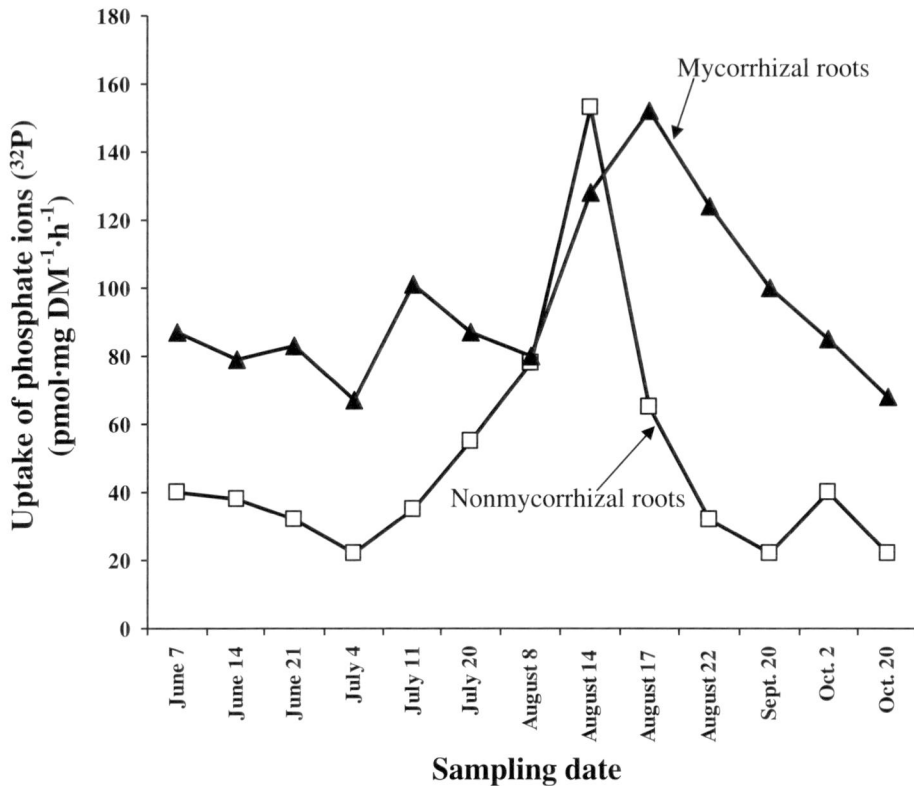

Fig. 13.2. Seasonal pattern of phosphate ion absorption by brown ectomycorrhiza (solid triangles) and nonmycorrhizal long roots (open squares) of *Abies balsamea* collected in the Montmorency experimental forest (Québec) during the 1978 growing season (adapted from Figs. 1 and 2, Langlois and Fortin 1984). DM, dry mass.

roots (Fig. 13.2), the rate of P uptake was very low at the beginning of the season; however, it slowly started to increase at the end of July, before rising very sharply and peaking in mid-August. Thereafter, it rapidly dropped back to the rate observed at the beginning of the summer. A similar but slightly different pattern (Fig. 13.2) was observed for "smooth brown" ECM, which is the most abundant morphotype found on balsam fir on the study site. Briefly, the rate of absorption was higher than that observed for long nonmycorrhizal roots, and the mid-August peak, which was as high as that of long roots, was achieved a few days later. The marked difference between mycorrhizal and nonmycorrhizal roots was that the decrease in P absorption rates after the mid-August peak was much slower in the former and only reached the low absorption rates of long roots at the end of October. That peaks in the P absorption rate for both types of roots were observed one month after closure of the host tree's terminal bud, suggests a strong relationship with the seasonal shift in C allocation from the aerial part of the tree during the first half of the growing season to the root system during the second half.

Seasonal production of ectomycorrhizal fungal basidiomes

In Canada, observations made in jack pine (*Pinus banksiana* Lamb.) stands at high latitudes (>53°N) showed profuse ectomycorrhizal fungal basidiome development. Most of the production occurred after mid-August, again suggesting that enhanced C allocation to the roots later in the season influences ECM fungi and plays an important role in the timing of basidiome production. At high latitudes, the decrease in day length at the end of the summer is much more rapid than at lower latitudes, and this probably explains why profuse basidiome production occurs over a relatively short period. However, it has been shown that photosynthesis is still active on warm sunny days during late fall (Lamhamedi and Bernier 1994; Grossnickle 2000). At high latitudes, basidiome production for many species continues until the first important snowfall, which probably reduces physiological activity in the aerial parts of the host tree.

In lower latitude coniferous forests (e.g., at 45°N

Chapter 13. Ecophysiology of sporocarp development of ectomycorrhizal basidiomycetes associated with boreal forest gymnosperms

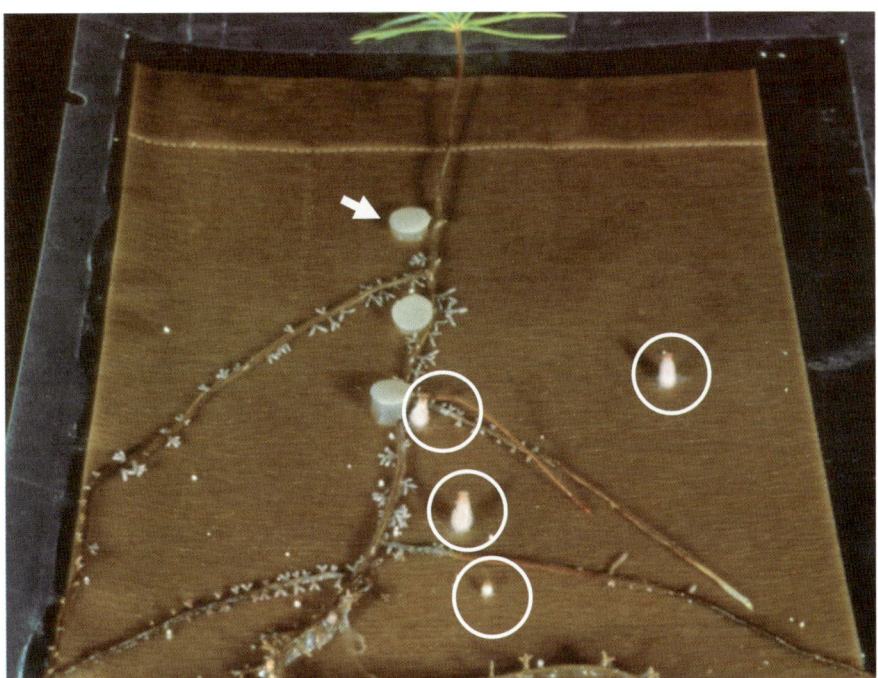

Fig. 13.3. *Pinus strobus* seedling inoculated with *Laccaria bicolor* (inoculum plug diameter: 3 mm) developing in a growth pouch. Basidiome primordia are produced exclusively using photosynthates from the host (photograph kindly provided by C. Godbout).

in the valley of the St. Lawrence River, Québec), abundant basidiome production also occurs in the second half of the summer, but it starts progressively and ends later than at higher latitudes. In forest nurseries, large numbers of ECM fungal basidiomes are also observed at the end of August and the beginning of September. Although this is a general rule in the boreal forest, the fructification of some species follows a completely different pattern. For example, the chanterelle (*Cantharellus* aff. *cibarius* Fr.) starts fruiting much earlier (i.e., during the second half of July) and grows much slower than many other species; its average lifespan is 44 days (see Chapter 14). Development of this species continues until the beginning of September. This slow development could be due to a very slow transfer of C from the host tree rather than a fast transfer from mid-August, as is observed in the case of boletes.

Symbiotism and saprophytism in ECM systems

In a recent review paper, Read et al. (2004) illustrated the capacity of ECM fungal mycelia to rapidly colonize different raw organic materials, such as leaf litter or nematode necromass, taking advantage of organic-bound nutrients before actively proliferating in the surrounding medium. This suggests that, under certain conditions, some ectomycorrhizal fungi may not depend exclusively on their host trees for C. This, in turn, suggests that different species of ECM fungi may use different balances of symbiotism and saprophytism to complete their life cycle. Based on basidiome production, we suggest that ECM fungi could probably be divided into three functional groups: (*i*) strictly symbiotic species that produce basidiomes solely from host-derived C (e.g., *Laccaria bicolor* (Maire) Orton) (Fig. 13.3)—however, this does not imply that these fungi have no saprophytic capacity at all; (*ii*) hemisaprophytic species that cannot complete their life cycle without obtaining part of their C needs from raw organic matter (i.e., plant, animal, and (or) microbial necromass) and would require symbiotic relationship for their survival but could not complete their life cycle without a certain degree of saprophytic activity—*Paxillus involutus* (Batsch:Fr.) Fr. could well be a striking example of this type of life strategy (Laiho 1970); and (*iii*) facultative ECM species able to complete their life cycle solely using a saprophytic strategy but that can form ectomycorrhizas occasionally (e.g., *Morchella* spp.; Buscot and Kottke 1990).

Experimental production of ECM basidiomes

Several researchers have obtained ECM fungal basidiomes under more or less controlled conditions. However, none of these have been able to explain why production is so abundant during the fall under boreal conditions (Debaud and Gay 1987; Danell and Camacho 1997; Guérin-Laguette et al. 2000; Yamada et al. 2001a, 2001b). The abundant fall production of ECM fungal basidiomes in natural forests and in tree nurseries, especially at high latitudes, coupled with the available data concerning balsam fir phenology and seasonal P absorption by ECM suggests that photoperiodism might be one of the primary factors explaining the phenology of ECM fungal basidiomes production.

To verify this hypothesis, Godbout and Fortin (1990) grew white pine (*Pinus strobus* L.) seedlings inoculated with *L. bicolor* under a long photoperiod (18 h) for 12 weeks. The seedlings were then submitted to a short photoperiod (9 h). After four weeks under the short photoperiod (but without a reduction in temperature), bud set was complete, numerous basidiome initials were present in the substrate, and basidiome development occurred in most of the pots (Fig. 13.4a). Basidiomes of other ECM fungal species (e.g., *Suillus neoalbidipes* Palm & Stewart) were also obtained using the same strategy (Fig. 13.4b).

A recent study that we conducted in a boreal forest nursery showed that an artificial short photoperiod treatment (8 h, achieved by blackout; Figs. 13.5a and 13.5b) during the active growth phase of *L. bicolor* inoculated *P. mariana* seedlings, halted growth of aerial tissues, induced terminal bud formation and induced basidiome formation (Fig. 13.5c) (unpublished results). These results confirm that photoperiod can play a determinant role in ECM fungal basidiome initiation and development under nursery and growth chamber (Godbout and Fortin 1990) conditions.

The growth chamber experiment conducted by Godbout and Fortin (1990) was continued for a further eight weeks following the reduction in photoperiod. During this period, seedlings remained dormant; however, basidiome formation continued. Although numerous basidiome initials were present in the substrate in each container, basidiomes only developed one at a time. It took about one week for a basidiome to fully develop from an initial. Godbout and Fortin (1992) demonstrated that critical N and P inputs were necessary to obtain and optimize fruit body production. A correlation was also observed between seedling and basidiome DM. A decrease in temperature alone did not influence basidiome development and, therefore, cannot be considered as the primary factor responsible for the initiation of basidiome development, at least not with *L. bicolor*. However, for certain species, such as *Boletus edulis* Bull.: Fries, it is likely that lower temperatures play a secondary role once the flow of C has reached the root system.

The reproducibility of the *P. strobus* – *L. bicolor* model described above, allowed a clear correlation to be drawn between net photosynthesis and basidiome biomass (Lamhamedi et al. 1994). Under higher light intensity (330 µmol·m^{-2}·s^{-1}), basidiome development was completed within 10–12 days. By contrast, under reduced light intensity (50 µmol·m^{-2}·s^{-1}), full development was reached but only after about 20 days, and the basidiomes produced were smaller (Fig. 13.6).

When nearly mature basidiomes were excised, a rapid and sharp decline in net photosynthesis was recorded, and stomata closed (Fig. 13.7). After a few days, a new basidiome started developing from a dormant initial, and net photosynthesis and stomatal conductance returned to the values observed prior to the removal of the first basidiome. In another experiment, the presence of ^{14}C was detected in *L. bicolor* basidiomes a few hours after treating the host plant with ^{14}CO$_2$ (C. Godbout, personal communication). These results have been corroborated by field studies showing that daily variations in the δ^{13}C isotopic signature of recently fixed C in trees is mirrored, after a few days, by the isotope signature of root and rhizosphere respiration, which includes ectomycorrhizal fungal mycelium and basidiomes (Ekblad and Högberg 2001; Steinmann et al. 2004). Recently, Högberg et al. (2007), working in the boreal forest, pulse-labelled Scots pine (*Pinus sylvestris* L.) saplings (4 m) with ^{13}CO$_2$ for 1.5 h and showed that tracer levels peaked after four to seven days in ECM roots. Previously, in another boreal pine forest study, Högberg et al. (2001) compared basidiome production between control and girdled plots. The removal of phloem from girdled trees disrupted the transport of photosynthate from the crown to the roots and their mycorrhizal fungi. The DM of ECM basidiomes harvested from the control plots was nearly 200 times higher than that of the basidiomes from the girdled plots. The results of these studies under field (Högberg et al. 2001, 2007) and controlled conditions (Lamhamedi

Chapter 13. Ecophysiology of sporocarp development of ectomycorrhizal basidiomycetes associated with boreal forest gymnosperms

Fig. 13.4. (*a*) Fall production of *Laccaria bicolor* basidiomes in association with inoculated container-grown *Pinus strobus* seedlings grown in Leach Cone-Tainers (3.8 × 10 cm). (*b*) *Suillus neoalbidipes* basidiome (cap diameter: 50 mm) produced in association with a group of *Pinus banksiana* seedlings that share a clone of this ectomycorrhizal fungus after exposure to reduced photoperiod.

et al. 1994) clearly demonstrate that the growth and development of ECM fungal basidiomes are dependent on the host plant's current rate of photosynthesis, rather than on C reserves accumulated in the mycorrhizas or the external mycelium, as is the case for saprophytic fungi, e.g., *Agaricus bisporus* (Lge.) Sing. (Delmas 1989).

Following on from the *L. bicolor* basidiome-development experiments, we monitored older mycorrhizal *P. strobus* plants for 12 months following the short-photoperiod treatment. It is well known that growth of dormant trees will not resume until after the shoots have been exposed to low temperatures over a prolonged period to reach the

Fig. 13.5. Midsummer production of *Laccaria bicolor* basidiomes obtained by experimentally reducing day length: (*a*) normal daylight exposure; (*b*) the use of plastic covers to shorten day length (arrow shows control seedlings); and (*c*) each treated white spruce seedling produced one *L. bicolor* basidiome.

chilling requirement (Perry 1971; Sakai and Larcher 1987). Thus, seedlings could be kept dormant for at least four months, even if the photoperiod was increased. Under such conditions and providing that the fruit bodies were collected before they reach full maturity, the production of basidiomes could be maintained over several months. However, if a basidiome was left to reach full maturity, a second initial started developing and produced a much smaller basidiome, after which a third initial started to develop but aborted (Fig. 13.8). Although no measurements were conducted, it was obvious that the total mass of the successive generations of harvested *L. bicolor* basidiomes exceeded the mass of the host seedling. In this case, the C cost of the ectomycorrhizal association was probably close to 100%. Hence it is necessary to define the physiological state of the mycorrhizal system when evaluating the energy cost of the association.

Hormonal relationships

Much has been written on the role of auxins in the development and functioning of ectomycorrhizas (e.g., Slankis 1973). However, Wallander et al. (1994) concluded that "No support was found for the hypothesis that indole acetic acid (IAA) is important in regulating mycorrhizal colonization." Nevertheless, we are of the opinion that the controversies about the role of auxins in the formation and functioning of ectomycorrhizas, and their relationships with C sinks and N nutrition are far from being solved. The problem with most, if not all, of the previous studies is that they are based on instantaneous data regarding concentrations of auxins, C, and N in the tissues rather than on the dynamics of these substances. It is quite possible to have the same concentration of C in two root samples, but the rate of transfer in one sample might be double that of the other sample.

Chapter 13. Ecophysiology of sporocarp development of ectomycorrhizal basidiomycetes associated with boreal forest gymnosperms

Fig. 13.6. Effect of different light intensity (from left to right: 20, 50, and 90 µmol·m^{-2}·s^{-1}) on size and development of *Laccaria bicolor* basidiomes (from Fig. 3, Lamhamedi et al. 1994).

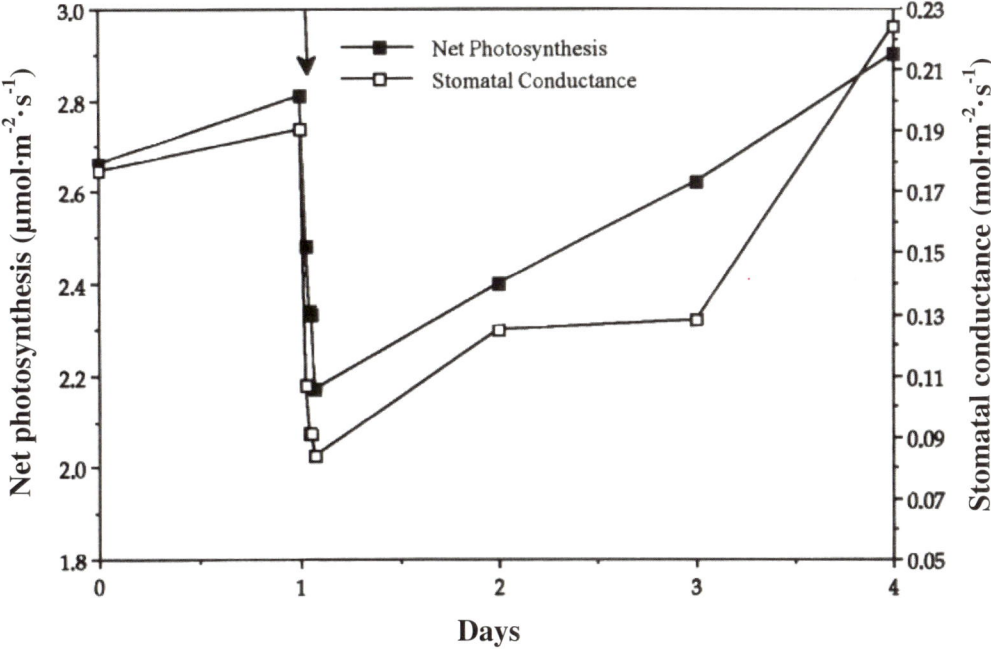

Fig. 13.7. Changes in net photosynthesis and stomatal conductance of *Pinus strobus* seedlings following removal of a mature *Laccaria bicolor* basidiome (adapted from Fig. 4c, Lamhamedi et al. 1994).

ECM fungi (Fortin 1967, 1970) as well as lichen fungi (Fortin et al. 1972) both produce IAA and both require external tryptophan to do so. We suggested that IAA produced by ECM fungi act as self-inhibitory factors that might be relieved by absorption of IAA in the root through rapid conversion to indole aspartic acid (Fortin 1970). We postulated that the root is producing the necessary tryptophan; thus, a feedback equilibrium mechanism could regulate the interaction between symbionts. We also suggested that critical concentrations of IAA might affect membrane permeability, modu-

Fig. 13.8. The dominance effect of maturing *Laccaria bicolor* basidiomes associated with *Pinus strobus* seedlings grown in Leach Cone-Tainers (3.8 × 10 cm), suppresses further development of younger basidiomes. The host plant is *Pinus strobus*.

lating exsudation of glucose from the root into the fungus. A first demonstration of such hypotheses should be easier using lichens rather than ECM symbionts. This subject should be revisited using modern techniques.

Although we do not pretend to solve the question concerning auxins, C, and N, we are of the opinion that the hormonal activity of a basidiome must also be taken in account when studying C flow in ECM systems. We have seen that removal of the basidiome directly affects stomatal conductance and photosynthesis, but how does removal of a developing basidiome affect the opening and closing of stomata so rapidly? Our unpublished observations on continuous basidiome production on dormant seedlings, as well as those showing that production rapidly slows and stops when basidiomes are not collected, strongly suggest that the fruit body can exert hormonal control on the physiology of the host tree. Our observations also suggest that the maturation of the basidia is likely to be involved in this phenomenon. Clearly, this research area requires much more attention.

Toward a comprehensive understanding of carbon flow in ECM systems

Much work still remains to be done before we can fully understand C flow in diverse ECM systems under different ecological situations and at different stages in the life cycle of the host plant and the fungus. However, the facts and personal observations that we have presented in this chapter highlight new avenues that should lead to the development of a more comprehensive theory concerning C flow in these systems as a whole and over an entire year.

From the standpoint of the ECM fungus, basidiome development seems to coincide with the availability of a large supply of C from its host, at least for most boreal species. Therefore, it is not surprising to observe reduced ECM development and physiological activity (P absorption) during the period when C resources are being channelled into growth of aerial parts of the tree (i.e., in height and diameter). Conversely, on warm sunny fall days after bud formation, the host is still actively photosynthesizing, but the C is no longer required by aerial tissues and is channelled into root growth. This transfer coincides with increased physiological activity (P absorption) of ectomycorrhizas and the production of ECM fungal fruit bodies. These observations suggest that ECM fungal fruit body growth is fuelled by C that is surplus to that required by the tree for its own growth and maintenance, as suggested by Björkman (1942). To support this, it would not be necessary to record a higher concentration of sugars in the mycorrhiza but, rather, a faster rate of C flow—through no return diffusion—from the host to the ECM system, which is in accordance with the source–sink mechanism described by Lewis and Harley (1965a, 1965b).

Our observations on direct relations between net photosynthesis and basidiome development show that C does not appear to accumulate in the mantle of the ectomycorrhizas, even under high light intensity. This might be explained by an increase in the rate of C transfer from the host to the fungus under higher light intensities. However, the mechanism by which the excision of maturing basidiomes so rapidly affects stomatal conductance and the rate of net photosynthesis (Fig. 13.7) remains to be explained.

The basidiome excision experiments suggest that the basidiome exerts a hormonal-like control on stomatal conductance as well as on the rate of net photosynthesis. Therefore, the measurement of C flow in ectomycorrhizas requires that the physiological state of the tree and the fungus be well defined. Our experiments and observations suggest that the C cost of ectomycorrhizas can be very low during the rapid growth of aerial parts of the tree and very high during the period where the aerial parts become dormant. Experimentally, it was possible to create an artificial system, where nearly 100% of the fixed C was allocated to the ECM fungus.

Once a reduction in the photoperiod had induced bud formation on the host tree, *L. bicolor* produced dozens of fruit body initials; however, these only developed one at a time. This suggests that the other basidiome initials were repressed. Even in a container with several dozen seedlings interconnected by a common mycelium, only one basidiome matured at a time (unpublished results). Continuous picking of the basidiomes prior to maturity, seemed to keep the C sink open; however, maturation of the basidiome led to its closure. The timing of this effect with maturation of the basidia, suggests a major change in the basidiome's hormonal status following the shedding of spores. A comparative approach using fruit development as a model may prove rewarding.

It is easy to suggest that hormones might play a key role in this phenomenon; however, it is difficult to identify such hormones and their mode of action. Over the last 50 years, the existence of a growth hormone in basidiomes has been evoked (e.g., Gruen 1965); however, to the best of our knowledge, this compound has never been isolated.

Most of the experiments and observations outlined above help explain the possible reasons behind fall production of ECM fungal fruit bodies in the boreal region; however, it must be borne in mind that many of our studies have been conducted using a *L. bicolor – P. strobus* model and that the mycobiont in question is a strictly symbiotic species. Nevertheless, even in the boreal region, certain species do not wait until fall before producing their fruit bodies. A typical example is *P. involutus*, and we have suggested a possible hemisaprophytic strategy for such species. It is conceivable that some ECM fungi might have two annual growth phases: one essentially symbiotic during late summer and early fall that corresponds to active C allocation from trees following bud set and a somewhat saprophytic phase that benefits from warmer soil temperatures in the middle of the summer and allows early basidiome development. If this hypothesis proves to be correct, determining the relative proportions of symbiotic and saprophytic C entering the mycorrhizal system will not be easy. However, the method proposed by Hobbie (2006), which discriminates between C fixed during the current year and that fixed several years before, allows the sporocarps of fungi to be classified as either having been produced by symbiotic or saprophytic fungi. Such an approach might be useful for discriminating between obligate and hemisaprophytic ectomycorrhizal fungi, if such a category exists.

As yet, we do not completely understand all the aspects of C flow in ECM systems. Future work in this field must take into account all phases of the annual cycle of the host plant and its mycorrhizal fungi, and should also include investigations into potential saprophytic activities of the latter.

From a practical viewpoint, control of basidiome initiation and development by manipulating the photoperiod and, in turn, the physiological state of the ECM host plant, should provide the potential for continuous greenhouse production of basidiomes of edible species. To obtain fully developed basidiomes of boletes or amanitas, which are generally large, it should be possible to use numerous seedlings interconnected through a common mycelium.

References

Björkman, E. 1942. Uber die Bedingungen der Mykorrhizabildung bei Kiefer und Fichte. Symb. Bot. Ups. **6**: 1–191.

Bryla, D.R., and Eissenstat, D.M. 2005. Respiratory costs of mycorrhizal associations. Advances in photosynthesis and respiration. *In* Plant respiration: from cell to ecosystem. *Edited by* H. Lambers and M. Ribas-Carbo. Springer, Dordrecht, the Netherlands. pp. 207–224.

Buscot, F., and Kottke, I. 1990. The association of *Morchella rotunda* Boudier with roots of *Picea abies* (L.) Karst. New Phytol. **116**: 425–430.

Danell, E., and Camacho, F.J. 1997. Successful cultivation of the golden chanterelle. Nature (London), **385**: 303.

Debaud, J.C., and Gay, G. 1987. *In vitro* fruiting under controlled conditions of the ectomycorrhizal fungus *Hebeloma cylindrosporum* associated with *Pinus pinaster*. New Phytol. **105**: 429–435.

Delmas, J. 1989. Les champignons et leur culture. Flammarion, La Maison Rustique, Paris. pp. 41–48.

Ekblad, A., and Högberg, P. 2001. Natural abundance of ^{13}C in CO_2 respired from forest soils reveals speed of link between tree photosynthesis and root respiration. Oecologia (Berl.), **127**: 305–308.

Fortin, J.A. 1967. Action inhibitrice de l'acide 3-indolyl-acétique sur la croissance de quelques basidiomycètes mycorhiziens. Physiol. Plant. **20**: 528–532.

Fortin, J.A. 1970. Interaction entre basidiomycètes mycorhizateurs et racines de pin en présence d'acide indolyl-acétique. Physiol. Plant. **23**: 365–374.

Fortin, J.A., Thibault, J.R., and Morisset, P. 1972. Présence d'auxine chez un lichen (*Cladonia alpestris*) et transformation du tryptophane en acide indolyl-acétique par trois mycobiontes. Nat.-Can. **99**: 213–218.

Godbout, C., and Fortin, J.A. 1990. Cultural control of basidiome formation in *Laccaria bicolor* with container-grown white pine seedlings. Mycol. Res. **94**: 1051–1058.

Godbout, C., and Fortin, J.A. 1992. Effects of nitrogen fertilization and photoperiod on basidiome formation of *Laccaria bicolor* associated with container-grown jack pine seedlings. Can. J. Bot. **70**: 181–185.

Grossnickle, S.C. 2000. Ecophysiology of northern spruce species: the performance of planted seedlings. NRC Research Press, Ottawa, Ont.

Gruen, H.E. 1965. Growth regulation in fruit bodies of *Agaricus bisporus*. Mushroom Sci. **6**: 103–120.

Guérin-Laguette, A., Plassard, C., and Mousain, D. 2000. Effects of experimental conditions on mycorrhizal relationships between *Pinus sylvestris* and *Lactarius deliciosus* and unprecedented fruit-body formation of the Saffron milk cap under controlled soilless conditions. Can. J. Microbiol. **46**: 790–799.

Hacskaylo, E. 1973. Carbohydrate physiology of ectomycorrhizae. *In* Ectomycorrhizae, their ecology and physiology. *Edited by* G.C. Marks and T.T. Kozlowski. Academic Press, New York. pp. 207–230.

Harley, J.L. 1969. The biology of mycorrhiza. Leonard Hill, London.

Hobbie, E.A. 2006. Carbon allocation to ectomycorrhizal fungi correlates with belowground allocation in culture studies. Ecology, **87**: 563–569.

Högberg, P., Nordgren, A., Buchmann, N., Taylor, A.F.S., Ekblad, A., Högberg, M.N., Nyberg, G., Ottosson-Löfvenius, M., and Read, D.J. 2001. Large-scale forest girdling shows that current photosynthesis drives soil respiration. Nature (London), **411**: 789–792.

Högberg, P., Högberg, M.N., Göttlicher, S.G., Betson, N.R., Keel, S.G., Metcalfe, D.B., Campbell, C., Schindlbacher, A.,

Hurry, V., Lundmark, T., Linder, S., and Näsholm, T. 2007. High temporal resolution tracing of photosynthate carbon from the tree canopy to forest soil microorganisms. New Phytol. **177**: 220–228.

Laiho, O. 1970. *Paxillus involutus* as a symbiont of forest trees. Ph.D. thesis, University of Helsinki, Helsinki, Finland.

Lamhamedi, M.S., and Bernier, P.Y. 1994. Ecophysiology and field performance of black spruce (*Picea mariana*): a review. Ann. Sci. For. **51**: 529–551.

Lamhamedi, M.S., Godbout, C., and Fortin, J.A. 1994. Dependence of *Laccaria bicolor* basidiome development on current photosynthesis of *Pinus strobus* seedlings. Can. J. For. Res. **24**: 1797–1804.

Lamhamedi, M.S., Lambany, G., Margolis, H.A., Renaud, M., Veilleux, L., and Bernier, P.Y. 2001. Growth, physiology and leachate losses in *Picea glauca* seedlings (1 + 0) grown in air-slit containers under different irrigation regimes. Can. J. For. Res. **31**: 1968–1980.

Lamhamedi, M.S., Margolis, H.A., Renaud, M., Veilleux, L., and Auger, I. 2003. Effets de différentes régies d'irrigation sur la croissance, la nutrition minérale et le lessivage des éléments nutritifs des semis d'épinette noire (1 + 0) produits en récipients à parois ajourées en pépinière forestière. Can. J. For. Res. **33**: 279–291.

Landis, T.D., Tinus, R.W., and Barnett, J.P. 1999. The container tree nursery manual. Vol. 6. Seedling propagation. US Dep. Agric. Agric. Handb. 674.

Langlois, C.-G., and Fortin, J.A. 1984. Seasonal variations in the uptake of ^{32}P phosphate ions by excised ectomycorrhizae and lateral roots of *Abies balsamea*. Can. J. For. Res. **14**: 412–415.

Lawrence, W.T., and Oechel, W.C. 1983a. Effects of soil temperature on the carbon exchange of taiga seedlings. I. Root respiration. Can. J. For. Res. **13**: 840–849.

Lawrence, W.T., and Oechel, W.C. 1983b. Effects of soil temperature on the carbon exchange of taiga seedlings. II. Photosynthesis, respiration, and conductance. Can. J. For. Res. **13**: 850–859.

Leake, J., Johnson, D., Donnelly, D., Muckle, G., Boddy, L., and Read, D.J. 2004. Networks of power and influence: the role of mycorrhizal mycelium in controlling plant communities and agroecosystem functioning. Can. J. Bot. **82**: 1016–1045.

Lewis, D.H., and Harley, J.L. 1965a. Carbohydrate physiology of mycorrhizal roots of beech. I. Identity of endogenous sugars and utilisation of exogenous sugars. New Phytol. **64**: 224–237.

Lewis, D.H., and Harley, J.L. 1965b. Carbohydrate physiology of mycorrhizal roots of beech. III. Movement of sugars between host and fungus. New Phytol. **64**: 256–269.

Lippu, J. 1998. Redistribution of ^{14}C-labelled reserve carbon in *Pinus sylvestris* seedlings during shoot elongation. Silva Fenn. **32**: 3–10.

Loach, K., and Little, C.H.A. 1973. Production, storage, and use of photosynthate during shoot elongation in balsam fir (*Abies balsamea*). Can. J. Bot. **51**: 1161–1168.

Pellicer, V., Guehl, J.M., Daudet, F.A., Cazet, M., Rivière, L.M., and Maillard, P. 2000. Carbon and nitrogen mobilization in *Larix ×eurolepsis* leafy stem cuttings assessed by dual ^{13}C

and ^{15}N labelling: relationships with rooting. Tree Physiol. **20**: 807–814.

Perry, T.O. 1971. Dormancy of trees in winter. Science (Washington, D.C.), **171**: 29–36.

Read, D.J., Leake, J.R., and Perez-Moreno, J. 2004. Mycorrhizal fungi as drivers of ecosystem processes in heathland and boreal forest biomes. Can. J. Bot. **82**: 1243–1263.

Sakai, A., and Larcher, W. 1987. Frost survival of plants. Responses and adaptation to freezing stress. Springer-Verlag, New York.

Slankis, V. 1973. Hormonal relationships in mycorrhizal development. *In* Ectomycorrhizae, their ecology and physiology. *Edited by* G.C. Marks and T.T. Kozlowski. Academic Press, New York. pp. 231–298.

Smith, S.E., and Read, D.J. 1997. Mycorrhizal symbiosis. 2nd ed. Academic Press, London.

Steinmann, K., Siegwolf, R.T.W., Saurer, M.K., and Körner, C. 2004. Carbon fluxes to the soil in a mature temperate forest assessed by ^{13}C isotope tracing. Oecologia (Berl.), **141**: 489–501.

Tryon, P.R., and Chapin, F.S., III. 1983. Temperature control over root growth and root biomass in taiga forest trees. Can. J. For. Res. **13**: 827–833.

Wallander, H., and Nylund, J.E. 1991. Effects of excess nitrogen on carbohydrate concentration and mycorrhizal development of *Pinus sylvestris* L. seedlings. New Phytol. **119**: 405–411.

Wallander, H., Nylund, J.E., and Sundberg, B. 1994. The influence of IAA, carbohydrate and mineral concentration in host tissue on ectomycorrhizal development on *Pinus sylvestris* in relation to nutrient supply. New Phytol. **127**: 521–528.

Waring, R.H., and Pitman, G.B. 1985. Modifying lodgepole pine stands to change susceptibility to mountain pine beetle attack. Ecology, **66**: 889–897.

Yamada, A., Ogura, T., and Ohmasa, M. 2001*a*. Cultivation of mushrooms of edible ectomycorrhizal fungi associated with *Pinus densiflora* by *in vitro* mycorrhizal synthesis. II. Morphology of mycorrhizas in open-pot soil. Mycorrhiza, **11**: 67–81.

Yamada, A., Ogura, T., and Ohmasa, M. 2001*b*. Cultivation of mushrooms of edible ectomycorrhizal fungi associated with *Pinus densiflora* by *in vitro* mycorrhizal synthesis. I. Primordium and basidiocarp formation in open-pot culture. Mycorrhiza, **11**: 59–66.

Yoder, B.J., Ryan, M.G., Waring, R.H., Schoettle, A.W., and Kaufmann, M.R. 1994. Evidence of reduced photosynthetic rates in old trees. For. Sci. **40**: 513–527.

Yun, W., and Hall, I.R. 2004. Edible ectomycorrhizal mushrooms: challenges and achievements. Can. J. Bot. **82**: 1063–1073.

Chapter 14
Ecology and management of edible ectomycorrhizal mushrooms in eastern Canada

Marie-France Gévry and Normand Villeneuve

Introduction

The income that can be generated from the harvest and sale of edible ectomycorrhizal (ECM) forest mushrooms can considerably increase the potential revenue obtained from exploited, and in certain cases protected, forests (Cherkasov 1988). Commercial harvesting is now well established in Europe, Asia, and western North America (Hosford et al. 1997; Boa 2004). However, for the latter, this is a relatively recent activity. In western Canada, commercial harvesting only began in the 1970s, when Japan started importing *Tricholoma magnivelare* (Peck) Redhead as a substitute for its own closely related but declining pine mushroom species, *Tricholoma matsutake* (S. Ito & S. Imai) Singer (de Geus and Berch 1997; Redhead 1997, 2000; Wiensczyk and Berch 2002). By 1997, Canada had become the second largest exporter (after China) of *T. magnivelare* to Japan (Weigand 2000).

Today, more than 40 species of edible ECM forest mushrooms are marketed in Canada. At present, the bulk of the harvest is done in British Columbia (de Geus 1995; Berch and Cocksedge 2003). The most sought-after species are *Cantharellus cibarius* Fr. and *T. magnivelare*, closely followed by *Boletus* aff. *edulis* (referred to below as *Boletus edulis* Bull.:Fr.), *Cantharellus subalbidus* A.H. Sm. & Morse, and species of *Morchella* (which are possibly ECM). Boletes other than *B. edulis* are marketed; however, the demand for these is much lower. The annual value of wild mushrooms exported from Canada now greatly exceeds $40 million (Fortin 2003).

Therefore, this is an extremely valuable resource, and a number of studies have recently been done in western Canada and the United States in an attempt to avoid the decline in productivity that has followed intensive harvesting of certain species in Asia and Europe (Molina et al. 1993; Hosford et al. 1997; Berch and Wiensczyk 2001; Kranabetter et al. 2002, 2005; Wiensczyk and Berch 2002; Lim et al. 2003; Pilz et al. 2003; Tanino et al. 2005). These studies have provided a great deal of information concerning the biology and ecology of some of the most sought-after species (e.g., *T. magnivelare*, *C. cibarius*, and species of *Morchella*). Other studies have investigated the impact of forest management on the diversity of edible species and on epigeous mushroom production, again with the aim of developing a sustainable harvesting and management strategy for this resource (Amaranthus and Pilz 1996; O'Dell et al. 1996; Hosford et al. 1997; Amaranthus et al. 1998; Pilz et al. 1999; Berch 2000; Pilz and Molina 2002; Ehlers et al. 2003; Luoma et al. 2004, 2006; Egli et al. 2006).

In central Canada, notably in the region of Nipawin in northern Saskatchewan, edible forest mushrooms have also been commercially harvested since approximately 1990. The concerns of local harvesters in this region regarding the protection of forest habitat suitable for the production of high-value edible species recently led to a three-year study on *C. cibarius* and *T. magnivelare*. This was done to provide guidelines for the development of laws to ensure locally sustainable harvesting of these and other wild mushroom species (Tanino et al. 2005).

M.-F. Gévry.[1] Chaire de recherche sur la forêt habitée, Centre d'études nordiques, Université du Québec à Rimouski, 300, allée des Ursulines, Rimouski, QC G5L 3A1, Canada.

N. Villeneuve. Ministère des Ressources naturelles et de la Faune, Direction de l'environnement et de la protection des forêts, 880, chemin Sainte-Foy, local 6.50, Québec, QC G1S 4X4, Canada.

[1]Corresponding author (e-mail: mf_gevry@hotmail.com).

By contrast, in eastern Canada, the marketing of edible forest mushrooms has remained largely underdeveloped and is generally limited to supplying local markets and restaurants. However, small quantities of *C. cibarius* and species of *Morchella* are exported. In Québec in particular, several factors have been invoked in an attempt to explain why the harvest of this resource has remained largely unexploited. These include a general lack of knowledge concerning edible forest mushrooms, the fact that the local market is dominated by a few cultivated saprophytic species, and the past weakness of the local market for wild species (Villeneuve 1995). Logistical problems linked with the storage and long-distance transport of forest mushrooms have also hindered the development of a commercially viable edible forest mushroom harvesting industry in eastern Canada (Redhead 2000).

In spite of these difficulties, the abundance of the resource and a growing international demand suggest that a commercial edible forest mushroom harvesting industry could be developed in Québec (Miron 1995; Villeneuve 1995; Fortin and Piché 2000; Deslandes and Pic 2001; Boudreau et al. 2003). Furthermore, there is currently a growing interest in edible forest mushrooms in the province. This is manifest by the recent publication of a number of identification guides (e.g., Després et al. 2002; McNeil 2006); by an increase in the number of companies harvesting, buying, and selling forest mushrooms; and by the recent formation of an association for the commercialization of forest mushrooms (Association pour la commercialisation des champignons forestiers; www.acchf.org). The latter aims to favour exchanges between harvesters and potential buyers and to provide information concerning the harvest, storage, processing, and marketing of edible forest mushrooms.

However, this current interest cannot make up for the present lack of knowledge concerning the distribution of edible ECM forest mushrooms in eastern Canada and the factors affecting productivity. The short fruiting period, the interannual variability in productivity, and the complexity of ECM fungal–plant relationships make these organisms particularly difficult to census. Unfortunately, because of differences in climate and forest composition, the information available from studies done on ECM fungi in western North America (e.g., Kinugawa and Goto 1978; Menge and Grand 1978; Ito and Ogawa 1979; Pilz and Molina 1996; Durall et al. 1999; Kranabetter et al. 2002, 2005) and Europe (e.g., Thoen 1976; Ohenoja 1978, 1988; Garbaye et al. 1979; Salo 1979; Wästerlund and Ingelög 1981; Ohenoja and Koistinen 1984; Brandrud 1987; Arnolds 1988, 1991, 1995; Egli et al. 1990; Straatsma et al. 2001; Bonet et al. 2004; Salerni and Perini 2004; Martínez de Aragón et al. 2007) cannot be used to make assumptions about these fungi in eastern Canadian forests. Therefore, the collection of data concerning the biology and ecology of ECM fungal species producing edible sporocarps in the eastern part of the country is of utmost importance. Such information will allow the development of guidelines for private and commercial harvesters that will help ensure the sustainability of the resource.

This chapter is divided into three sections and reviews recent advances that have been made concerning the distribution, ecology, and productivity of edible epigeous ECM fungi in eastern Canada, particularly in the forests of Québec. The first section examines the principal factors influencing the spatial and temporal distribution of different ECM species in the wild. The second section uses the results of different regional surveys to highlight the factors that determine the production of sporocarps of edible species. Finally, the third section briefly outlines some of the current studies being done in Québec.

Ectomycorrhizal mushrooms in Québec

Pioneer mycologists working in eastern Canada were among the first to collect and identify ECM fungal species in Québec (Mounce and Jackson 1937; Jackson 1948; Pomerleau 1951; Groves 1962; Groves and MacRae 1963; Jackson and Cazort 1979). These historical records have since served as the basis for a number of descriptive works treating ECM forest mushrooms (e.g., Pomerleau and Smith 1962; Pomerleau 1966, 1975, 1980, 1984; Redhead 1979; Stanis 1979; Kallio 1980; Rendall 1980; Lamoureux and Després 1997), and for studies investigating the niche of certain ECM fungal species. The genera *Boletus*, *Suillus*, *Cantharellus*, *Amanita*, *Russula*, and *Lactarius* are among those that have been studied in the greatest detail.

Research teams in Québec have used in vitro experiments to investigate the symbioses formed between certain host plants, notably *Alnus crispa* (Ait.) Pursh, *Alnus rugosa* (Du Roi) Spreng., *Populus tremuloides* Michx., *Pinus strobus* L., and *Larix laricina* (Du Roi) K. Koch, and certain ECM fungi (Fortin 1968; Piché and Fortin 1982; Godbout and

Fortin 1983, 1985; Samson and Fortin 1986). Many of the mycobionts investigated belong to genera within the Agaricales (e.g., *Boletus*, *Leccinum*, *Suillus*, and *Gomphidius*) that are known to include species that produce edible sporocarps.

In situ observation, notably using the hosts *Pinus banksiana* Lamb. and *Picea mariana* (Mill.) BSP, and subsequent ecological analyses (Danielson 1984; Randall and Grand 1986; McAfee and Fortin 1987, 1989; Visser 1995), have provided a greater understanding of the environmental conditions that control the formation of specific ECM relationships in the wild. These initial in vivo observations, which were rare at the time, have since been joined by a suite of studies describing the habitat associated with certain macromycetes and the ecological conditions controlling sporocarp production (Villeneuve 1985, 1993; Lessard 1986; Villeneuve et al. 1989, 1991; Nantel and Neumann 1992). A number of similar studies have been done at other locations in North America (e.g., Fogel 1976; Kinugawa and Goto 1978; Menge and Grand 1978; Bills et al. 1986; Cibula and Ovrebo 1988; Redhead 1989) and elsewhere (e.g., Darimont 1973; Lisiewska 1974; Petersen 1977; Ohenoja 1978; Kalamees 1979; Arnolds 1981; Watling 1981; Tyler 1985; Barkman 1987; Bonet et al. 2004; Martínez de Aragón et al. 2007).

Finally, the current interest in developing an industry based on the commercial harvest of edible forest mushrooms in eastern Canada has stimulated research teams in Québec to investigate the environmental conditions that favour the development and fructification of edible ECM mushrooms. This has been done through studies (Villeneuve 1995; Deslandes and Pic 2001; Fallu 2003) and surveys (Miron 1994, 1995; Guérette 2001; Boudreau et al. 2003, Gévry 2008) that aimed to show the region's potential to support a large-scale commercial harvesting industry.

Distribution, abundance, and productivity of edible forest mushrooms

To date, relatively few studies have investigated the ecological factors controlling the distribution and abundance of edible forest mushrooms in Québec. The conclusions drawn from these vary greatly depending on the geographical region investigated and the study's aims. The latter can be divided into (*i*) those investigating the influence of ecological factors on the growth and the distribution of fungal species and (*ii*) those that also integrate the effect of environmental factors on sporocarp production. Furthermore, to gain a better insight into the ecological processes involved, investigations have been done at the landscape (macrohabitat) and the forest stand (microhabitat) scale.

Distribution and abundance at the landscape scale

The studies done in southern Québec by Villeneuve et al. (1989), and Nantel and Neumann (1992) were largely based on the method developed by Bills et al. (1986), which allowed the evaluation of the spatial frequency of a species of macromycetes from repeated observations of sporocarps in adjacent microquadrats. This technique allows the analysis of the factors influencing fructification, irrespective of mushroom phenology and interspecific variation in productivity. Therefore, it facilitates the comparison of the ecology of diverse species in different habitats and in different seasons. To date, the above studies, together with those published by Villeneuve et al. (1991) and Villeneuve (1993), are the only studies that have allowed the spatial frequency of macromycetes in diverse habitats in Québec to be evaluated. These studies are briefly outlined below.

Floristic composition

Among the factors considered at the landscape scale, the floristic composition of the forest stand showed the strongest correlation with the distribution and species composition of ECM fungal assemblages. At least in part, this is due to a certain degree of host specificity among ECM fungi. For example, there is a strong relationship between *Suillus granulatus* (L.) Roussel and *P. strobus*; between *T. magnivelare* and *P. banksiana* or *Pinus resinosa* Sol. ex Ait.; between *Leccinum snellii* A.H. Sm., Thiers & Watling and *Gyroporus cyanescens* (Bull.) Quél. and the host *Betula alleghaniensis* Britt.; between *Rozites caperatus* (Pers.) P. Karst, *Tricholoma equestre* (L.) P. Kumm., and *C. cibarius* and the host *P. mariana*; and between *Suillus cavipes* (Opat.) A.H. Sm. & Thiers and *L. laricina*.

The specificity of certain ECM fungal–plant associations is a major factor explaining the presence of a given ECM fungal species in diverse habitats. In certain cases, the strength of these relations can even lead a fungal species to transgress its normal ecological limits if its host's distribution is altered following sporadic disturbance of the forest. This is particularly true during forest succession when opportunistic host trees, such as *Betula papyrifera* Marsh. and *Populus balsamifera* L., take advantage of fire disturbance and timber harvesting to colonize areas that were previously occupied by more shade-tolerant species, such as *Acer saccharum* Marsh. and *Fagus grandifolia* Ehrh. Studies indicate that a significant part of the cortege of fungi typically associated with a given host species follow their host, even on to sites where conditions (e.g., underlying parent material, soil type, and drainage) may be suboptimal for their own growth.

On disturbed sites that are colonized by light-demanding host species, spores stored in the soil spore bank could assure the rapid synthesis of host-specific fungal associations. The spores of certain ECM fungal species can remain dormant in the soil for many years and are only stimulated to germinate by the presence of a suitable host root (Watling 1981). Therefore, the composition of a given cortege of ECM fungi is influenced not only by the composition of the forest presently occupying the site, but also by its previous composition during succession. Thus, the perturbation history of a given forest may explain part of the ECM fungal species variation observed between stands with the same floristic composition.

Finally, the specificity of certain ECM fungal–plant associations may alter the distribution, frequency, and diversity of the ECM fungal species within a given bioclimatic zone. In Québec as elsewhere, the vegetation changes with increasing latitude (Fig. 14.1) or altitude, and each zone is characterized by different host tree species (e.g., typically from south to north, *F. grandifolia* is replaced by *B. alleghaniensis*, which is replaced by *Abies balsamea* (L.) Mill., which is replaced by *P. mariana* in turn) with their specific assemblages of ECM fungal species. The spatial distribution of these assemblages appears to be more strongly linked to the dominant host than to macroclimatic factors, latitudinal gradients, or altitudinal gradients.

Structure of forest cover

The structural characteristics of the forest cover, and more particularly stand age and tree density (i.e., open or closed stands), also have a strong influence on the spatial frequency and diversity of ECM fungal species (Villeneuve 2000). For example, under certain environmental conditions, *P. mariana* grows in open stands in which the forest floor is dominated by lichens. In such forests, the spatial frequency of sporocarps of a given ECM fungal species is <10% of that found for the same ECM fungal species in closed stands in which the forest floor is dominated by mosses.

Host species density also affects the diversity of ECM fungi on a given site. A lower diversity of ECM fungal species was observed in open stands (10 species/400 m^2) than in closed stands (15 species/400 m^2). However, open lichen-dominated *P. mariana* forests were richest in edible ECM fungal species (e.g., *C. cibarius*, *G. cyanescens*, *Leccinum atrostipitatum* A.H. Sm., Thiers & Watling, and *T. equestre*); whereas *R. caperatus* was the only edible ECM fungal species commonly found in closed moss-dominated stands.

Distribution and abundance of edible forest mushrooms at the forest stand scale

Although the assemblage of ECM fungi present in a given habitat is influenced principally by floristic composition and stand structure, the nature of the edaphic environment (e.g., physical and chemical characteristics of the litter, soil, and surface deposits) also has an important influence (Villeneuve et al. 1989, 1991; Villeneuve 1993). Although the specificity of certain ECM associations can cause certain fungi to occupy sites outside their normal range, their known sensitivity to other environmental factors means that they are rarely present over the complete range of edaphic conditions occupied by their hosts (Nantel and Neumann 1992; Lodge et al. 2004). Furthermore, several species of ECM fungi appear to avoid sites with certain understory types or those with specific conditions of soil fertility and drainage.

Understory vegetation

The different floral components of the understory modify microhabitat conditions in different ways and may affect the development and fructification of edible forest mushrooms (Villeneuve 1993, 2000). Forest floor vegetation influences soil humidity by creating a boundary layer, which alters airflow patterns and so reduces surface evaporation. The layer of vegetation also buffers the soil against temperature extremes. Furthermore, understory vegetation affects the physical and chemical characteristics of the soil and the quality of the litter layer. The latter is due to the concentration of available nutrients within the litter and to the release, by some plants, of allelopathic substances that may accumulate within the organic horizons.

Observations suggest that the spatial frequency and diversity of ECM fungal species are negatively

Chapter 14. Ecology and management of edible ectomycorrhizal mushrooms in eastern Canada

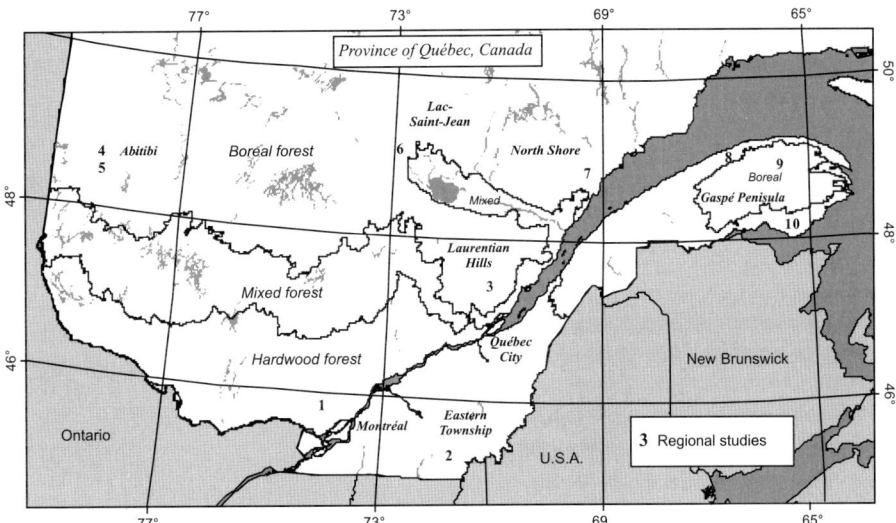

Fig. 14.1. A map of southern Québec showing the locations of former and ongoing regional studies investigating the ecology and productivity of edible forest mushrooms. The forest vegetation zones are after Saucier et al. (1998). The study areas shown are those of (1) Nantel and Neumann (1992), (2) Fallu (2003), (3) Villeneuve et al. (1989), (4) Miron (1995), (5) Maneli and collaborators (ongoing), (6) Rochon and collaborators (ongoing), (7) N. Villeneuve (unpublished data), (8) Gévry (2008), (9) Gévry and collaborators (ongoing), and (10) Guérette (2001).

affected by the presence of dense shrub- or herb-dominated understories (Villeneuve 2000; Guérette 2001). This effect might be explained by the phenomena of direct root interference or of competition for soil nutrients between the roots of shrubs or herbs (i.e., forbs and grasses) and those of the plant host or even the ECM fungal mycelium.

Lichens within the ground layer have also been suggested to negatively affect the development of ECM fungi in open *P. mariana* stands (Villeneuve et al. 1989). This effect is mainly associated with the presence of species of *Cladina*. By contrast, observations made in open *P. mariana* stands along the North Shore region of Québec suggest that the lichen cover can also play a positive role because it restricts the growth of ericaceous shrubs and grasses that are known to negatively influence a range of physical and chemical parameters within the rhizosphere (N. Villeneuve, unpublished data).

Finally, the presence of mosses in the ground layer has a positive influence on the development of certain ECM fungal species. In the Laurentian Hills, the cumulative spatial frequency of a range of ECM fungal species attained a maximum of 120% in stands where the forest floor was dominated by mosses within the Hypnaceae. Furthermore, certain edible species, such as *Catathelasma ventricosum* (Peck) Singer, *Craterellus tubaeformis* (Fr.) Quél., *Hydnum repandum* L., and *R. caperatus*, are known to exhibit a marked preference for stands with moss-dominated forest floors (Villeneuve et al. 1989; Miron 1995, 2000; Guérette 2001; Gévry 2008).

Quality of the humus and richness of the soil

The mycelium of the majority of ECM fungi proliferates close to the surface of the soil within the organic horizons (litter and humus). A given ECM fungal species is likely to have a particular suite of requirements concerning humus quality and soil nutrient composition. However, because the properties of the humus are largely linked to the composition of the forest cover, it becomes problematic to directly isolate the influence of forest soil on ECM fungi. Nevertheless, studies done in southern Québec suggest that the ECM fungal community may be influenced by certain factors affecting the quality of the humus (e.g., thickness of the litter layer, the quantity of organic matter in the humus layer, and pH) and the soil's exchangeable cation (especially Ca and Mg) and N content.

Drainage and surface deposits

In Québec as elsewhere, soil drainage and the nature of the surface deposits (e.g., texture, depth, and degree of peat development) are among the most important edaphic factors controlling the distribution of ECM fungal species at both the landscape and stand scale. The sensitivity of fungi to water stress influences their distribution along soil moisture gradients. In general, ECM fungal species obtain maximum development in sites that are mesic and subhydric (well to partially drained); however, several species show a particular tolerance for rapidly drained sites. On the latter, ECM host plants may be largely dependant on their associated fungi to enhance water uptake and drought resistance (Slankis 1974). In certain situation, this allows host species to dominate the drier parts of the soil moisture gradient. Species of several edible ECM fungal genera (e.g., *Boletus*, *Leccinum*, *Suillus*, *Sarcodon*, and *Tricholoma*) are thought to be particularly resistant to rapidly drying soils and are common on the sandy deposits of fluvioglacial, marine, or lacustrine origin (Villeneuve 2000).

By contrast, relatively few ECM fungal species occur on sites characterized by poorly drained organic soils and only a few edible ECM species (e.g., *R. caperatus* and *C. tubaeformis*) are considered to prefer such sites (Nantel and Neumann 1992; Miron 2000).

Production of epigeous sporocarps

Several species of edible mushrooms, such as *C. cibarius* and *B. edulis*, form symbioses with a large range of host species and in stands of variable age (Molina et al. 1999). Nevertheless, most species only produce large quantities of sporocarps in certain types of habitat (Molina et al. 1993). If a commercial-scale harvest is to be developed in Québec, it is important to have a thorough understanding of the biotic and abiotic conditions and processes that favour sporocarp production.

Recent studies done in Québec (Fig. 14.1) have begun to identify some of the most favourable conditions for the production of commercial quantities of edible mushrooms. Two of the main factors identified are linked to forest dynamics: host tree age and the degree of natural or anthropogenic disturbance. However, the phenology of sporocarp production and the influence of seasonal and current weather conditions on productivity have been rarely studied in Québec. Nevertheless, studies done elsewhere provide useful information that may help explain the results obtained in the boreal forest region of Québec.

Stand age

The influence of forest succession and aging on the fruiting of ECM fungi is still poorly understood (Laganà et al. 2002; Lodge et al. 2004). However, for several fungi, host tree age alone is one of the most important factors determining fructification (Miron 1995, 2000; Villeneuve 1995, 2000; Guérette 2001; Bonet et al. 2004). Sporocarp production generally reaches a peak in young even-aged stands exhibiting a maximal annual increase in volume (Villeneuve 2000). After this peak, increasing host age and changes in stand composition and litter quality generally result in a reduction in the level of productivity (Dighton et al. 1986; O'Dell et al. 1992; Egli and Ayer 1997; Guinberteau and Courtecuisse 1997; Smith et al. 2002; Jones et al. 2003; Kranabetter et al. 2005).

Studies done in stands of *P. banksiana* in the Québec North Shore region showed that the annual production of edible mushrooms reached a peak in young stands (i.e., 20–40 years old) and declined thereafter (N. Villeneuve, unpublished data). This is supported by the findings of an investigation into the production of *Sarcodon squamosus* (Schaeff.) Quél. in *P. banksiana* stands in the Abitibi region (Miron 1995, 2000). In addition, surveys conducted in plantations on the southern part of the Gaspé Peninsula (Québec) revealed that *C. cibarius* shows a marked preference for young (i.e., 15–30 years old) *Picea* plantations (Guérette 2001). Several other species, including many members of the genera *Suillus* and *Leccinum*, show a similar preference for young productive pine stands (Villeneuve 2000). The above-mentioned ECM fungal species are also negatively affected by the thickening of the humus (mor) layer that occurs with increasing stand age.

Other published data show that older forests (i.e., ≥60 years old) exhibit an annual production of fungi that is between three and six times lower than that reported for younger forests (<60 years old) (Villeneuve 1995). For example, in forests dominated by species of *Pinus*, the mean annual dry edible fungal biomass dropped from 7 kg/ha in young stands to <2 kg/ha in the oldest stands. The early fruiting ECM fungi that are responsible for this change include several species of *Suillus*, certain species of *Boletus*, and to a lesser extent, *C. cibarius* and *C. tubaeformis*. A similar trend was also observed in old (i.e., ≥80 years old) moss-dominated *P. mariana* stands on the Gaspé Peninsula (Guérette 2001).

By contrast, the production of sporocarps by certain species of edible ECM fungi shows a positive correlation with host age. For example, in the Abitibi region, *T. magnivelare* only begins fruiting in 40-year-old *P. banksiana* stands, and it reaches its maximum annual biomass production in stands that are >80 years old (Miron 2000).

Natural and anthropogenic disturbance

During forest succession, natural or anthropogenic

disturbance may affect fungal productivity. For example, forest fire and blowdown stimulate the production of sporocarps of species of *Morchella* (Duchesne and Weber 1993), and certain other species are favoured by a limited degree of anthropogenic disturbance, such as the presence of cut stumps in managed forests. *Hypomyces lactifluorum* (Schwein.) Tul. & C. Tul., which parasitizes members of the Russulaceae (most often *Russula brevipes* Peck) and produces an edible orange-coloured fungal complex (referred to below as the lobster mushroom), appears to be favoured by soil compaction along forest trails and the presence of fillings or cuttings along forest roads (Villeneuve 2000). Such conditions also seem to be favorable to *C. cibarius* in the northern part of the Gaspé Peninsula in Québec (Gévry 2008).

The favourable reaction of some fungi to habitat disturbance suggests that certain forest management practices could be used to increase productivity, especially of highly sought-after species. The most effective practices are likely to be those that affect the density, vigour, and growth of host trees (e.g., plantation thinning and pruning and additions of N and K) (Villeneuve 2000). Although no such studies have yet been done in Québec, other research groups have investigated a number of methods; however, the results vary greatly with the species and the region studied (Menge and Grand 1978; Ohenoja 1978, 1988; Garbaye et al. 1979; Ito and Ogawa 1979; Wästerlund and Ingelög 1981; Garbaye and Le Tacon 1982; Wilklund et al. 1995; Kropp and Albee 1996; Egli and Ayer 1997; Dahlberg et al. 2001; Kranabetter and Kroeger 2001; Salerni and Perini 2004; Pilz et al. 2006).

Phenology
An investigation of the factors controlling the fruiting of edible forest mushrooms would be incomplete without considering the effect of climatic factors on seasonal variations and interannual variations in productivity. Nevertheless, the data from eastern Canada are rather limited.

The capacity of an ECM fungus to form sporocarps in a given region is largely influenced by the altitude and latitude of the site, which affects site temperature and precipitation levels in turn (Ohenoja 1993). The latter two are the main factors controlling sporocarp growth within forests (Lamoureux 1993). At the local scale, annual precipitation is likely to be an excellent indicator of the diversity and the structure of the fungal community (O'Dell et al. 1999) because it is particularly important for the initiation of fructification (Ohenoja and Metsänheimo 1982).

In Québec, each season (with the exception perhaps of winter) stimulates the development of different species of fungi. However, the vast majority of species form sporocarps during the autumn (Lamoureux and Sicard 2001). Nevertheless, under temperate climates where summer rains are abundant (or at higher latitudes and altitudes if the temperatures are high enough (Ohenoja and Metsänheimo 1982), sporocarp production can be equally abundant during the summer (Lodge et al. 2004). In Europe, recent findings show that the growth period of certain fall-fruiting species has been significantly extended since the 1950s because of the effect of climate change (Gange et al. 2007), which include elevated temperatures in August coupled with increased precipitation in October. In a number of extreme cases, sporocarp production has been extended through to the spring.

In Québec, the climate is generally humid throughout the year, and it is the lower temperatures at the end of autumn that limit sporocarp production. Studies have also shown that the length of the fruiting season decreases northwards and with increasing altitude, and from the coast inland. In southern Québec, 95% of the annual production of edible ECM mushrooms occurs over a five to eight week period (i.e., late summer and autumn) (Villeneuve 2000).

Regional surveys

Recent surveys done in Québec (Fig. 14.1) have confirmed the potential for the development of a commercially viable edible ECM mushroom harvesting industry based on several highly sought-after species, including *B. edulis*, *C. cibarius*, *C. tubaeformis*, species of *Morchella*, and *T. magnivelare* (Table 14.1). *Sarcodon squamosus*, lobster mushrooms, and several less well-known species of *Boletus*, *Leccinum*, *Suillus*, and *Tricholoma* also seem to have commercial potential (Miron 1995, 2000; Villeneuve 2000; Deslandes and Pic 2001).

Abitibi

To ascertain the possibility of developing a commercial forest mushroom harvesting industry, the annual biomass of sporocarps was quantified along transects in diverse forest sites in the Abitibi region in western Québec (Miron 1994, 1995). In 1995,

Table 14.1. The commercial importance, distribution, and abundance of the principal edible ectomycorrhizal fungi recorded in regional studies.

Species*	Commercial index[†]	Miron 1994, 1995	Villeneuve et al. 1989	Nantel and Neumann 1992	Boudreau et al. 2003[‡]	Guérette 2001	Fallu 2003	Gévry 2008	N. Villeneuve (unpublished data)	Family
Cantharellus cibarius Fr.	44	S	S	S	—	A	—	A	S	Cantharellaceae
Morchella spp.	42	S	—	—	—	P	—	—	—	Morchellaceae
Boletus aff. *edulis* Bull.: Fr.	41	S	—	—	—	A	S	P	S	Boletaceae
Hypomyces lactifluorum (Schwein.) Tul. & C. Tul.	41	A	—	—	—	—	S	P	A	Hypocreaceae
Craterellus tubaeformis (Fr.) Quél.	40	A	—	—	—	—	—	S	—	Cantharellaceae
Gyroporus cyanescens (Bull.) Quél.	39	S	S	—	—	—	—	—	—	Gyroporaceae
Hydnum repandum L.	39	P	—	—	—	S	—	S	—	Hydnaceae
Tricholoma magnivelare (Peck) Redhead	39	A	—	—	P	—	—	—	A	Tricholomataceae
Leccinum aurantiacum (Bull.) Gray – *Leccinum piceinum* Pilát & Dermek[§]	38	A	S	—	P	A	A	A	A	Boletaceae
Boletus subglabripes Peck	35	—	S	—	P	—	S	A	S	Boletaceae
Leccinum atrostipitatum A.H. Sm., Thiers, & Watling	34	—	A	—	—	—	—	S	A	Boletaceae
Sarcodon squamosus (Schaeff.) Quél.	34	A	—	—	P	—	—	—	S	Bankeraceae
Tricholoma caligatum (Viv.) Ricken	29	A	—	—	P	—	—	—	S	Tricholomataceae
Rozites caperatus (Pers.) P. Karst	24	A	S	A	P	—	—	S	S	Cortinariaceae
Catathelasma ventricosum (Peck) Singer	23	—	—	—	—	A	—	A	S	Tricholomataceae
Lactarius thyinos A.H. Sm.	23	S	—	S	—	—	A	S	—	Russulaceae
Lactarius lignyotus Fr.	19	—	—	A	—	—	—	—	—	Russulaceae
Russula cyanoxantha (Schaeff.) Fr.	18	—	A	S	—	—	—	—	—	Russulaceae
Tricholoma equestre (L.) P. Kumm.	18	A	S	—	P	—	—	—	A	Tricholomataceae
Sarcodon imbricatus (L.) P. Karst.	17	S	—	—	P	—	—	—	—	Bankeraceae
Boletus subtomentosus L.	—	—	S	S	—	—	—	—	S	Boletaceae
Cantharellula umbonata (J.F. Gmel.) Singer	—	S	—	—	P	—	—	—	S	Cantharellaceae
Chalciporus piperatus (Bull.) Bataille	—	—	—	S	—	—	S	A	—	Boletaceae

Table 14.1 (*concluded*).

Species*	Commercial index†	Miron 1994, Villeneuve et al. 1989, 1995	Nantel and Neumann 1992	Boudreau et al. 2003‡	Guérette 2001	Fallu 2003	Gévry 2008	N. Villeneuve (unpublished data)	Family
Gomphidius glutinosus (Schaeff.) Fr.	—	S	—	P	—	—	—	—	Gomphidiaceae
Gomphus floccosus (Schwein.) Singer	—	—	—	—	A	—	P	—	Gomphaceae
Lactarius deterrimus Gröger	—	—	—	P	—	S	A	—	Russulaceae
Leccinum holopus (Rostk.) Watling	—	—	S	—	—	—	—	S	Boletaceae
Leccinum insigne A.H. Sm., Thiers & Watling	S	—	A	—	—	—	—	S	Boletaceae
Leccinum scabrum (Bull.) Gray	—	—	S	—	—	—	P	S	Boletaceae
Leccinum snellii A.H. Sm., Thiers & Watling	—	S	—	—	—	—	P	S	Boletaceae
Russula paludosa Britzelm.	—	—	A	—	—	—	—	—	Russulaceae
Suillus americanus (Peck) Snell	—	—	—	P	—	—	—	—	Suillaceae
Suillus brevipes (Peck) Kuntze	—	—	—	P	—	—	—	A	Suillaceae
Suillus cavipes (Opat.) A.H. Sm. & Thiers	—	S	—	P	—	—	—	S	Suillaceae
Suillus granulatus Roussel	—	S	S	P	—	—	—	S	Suillaceae
Suillus grevillei (Klotzsch) Singer	—	S	—	P	—	—	—	—	Suillaceae
Suillus luteus L. ex Fr.	—	—	—	—	—	—	—	S	Suillaceae
Suillus neoalbidipes M.E. Palm & E.L. Stewart	—	—	—	—	—	—	—	A	Suillaceae
Suillus pictus (Peck) A.H. Sm. & Thiers	—	—	A	—	—	S	—	S	Suillaceae
Suillus serotinus (Frost) Kretzer & T.D. Bruns	—	—	—	P	—	—	—	—	Suillaceae
Suillus tomentosus (Kauffman) Singer	—	P	—	—	—	—	—	A	Suillaceae

Note: Abundances were defined as follows: P, present but abundance not determined; A, abundant; S, sporadic.
*Scientific nomenclature follows *Index Fungorum* (Commonwealth Agricultural Bureaux International, Centraal Bureau voor Schimmelcultures, and Manaaki Whenua – Landcare Research; www.indexfungorum.org).
†According to Deslandes and Pic (2001); not available for all species.
‡Abundances not specified.
§*Leccinum aurantiacum* sensu lato (both species combined).

Table 14.2. Annual productivity and dry biomass of the principal edible epigeous forest mushrooms in 1997 in *Pinus banksiana* stand on the North Shore of Québec.

Edible species	Stand age and soil drainage class			
	<30 years, mesic	<30 years, subhydric	>30 years, mesic	>30 years, subhydric
Dry biomass (kg)				
Leccinum aurantiacum s.l.	10.805	8.883	8.126	3.189
Suillus brevipes	8.403	9.221	0.888	1.367
Hypomyces lactifluorum	4.359	11.823	0.202	2.362
Suillus tomentosus	2.694	2.779	2.263	3.337
Tricholoma equestre	2.155	2.722	4.932	0.233
Suillus neoalbidipes	2.746	4.212	1.197	0.859
Tricholoma magnivelare	—	—	2.794	—
Leccinum atrostipitatum	—	—	0.009	1.743
Leccinum holopus	0.344	0.422	0.266	0.269
Boletus subtomentosus	0.020	0.032	0.344	0.581
Russula brevipes	0.185	0.258	0.078	0.195
Leccinum scabrum	0.130	0.310	0.107	0.147
Suillus luteus	—	—	0.178	0.308
Suillus cavipes	—	—	—	0.477
Suillus granulatus	0.457	0.003	0.003	0.008
Other edible species	0.391	0.130	0.570	0.419
Total	32.689	40.795	21.957	15.493
No. of sites studied	30	30	30	30
Total area of study sites (ha)	6	6	6	6
Annual productivity (kg/ha)	5.448	6.799	3.659	2.582

39 forest sites with different dominant tree species, surface deposits and drainage were examined.

The *P. banksiana* stands on well-drained sandy deposits were the most productive in terms of edible species, with important quantities of *B. edulis*, *C. cibarius*, *Leccinum aurantiacum* (Bull.) Gray sensu lato (including *Leccinum piceinum* Pilát & Dermek), *S. squamosus*, *Tricholoma caligatum* (Viv.) Ricken, *T. equestre*, *T. magnivelare*, several species of *Suillus*, and lobster mushrooms being produced (Miron 1994, 1995, 2000) (Table 14.1). Of these, *S. squamosus* showed the highest potential for commercial harvesting with an annual dry biomass exceeding 50 kg/ha on certain sites (Miron 1995).

By contrast, the *P. mariana* stands, particularly those situated on poorly drained clay soils, produced few edible fungi. However, they did produce good quantities of *R. caperatus*, especially on sites where the moss *Pleurozium schreberi* (Brid.) Mitt. dominated the ground layer (Miron 1994, 2000). Nevertheless, in general, poorly drained sites were better suited to saprophytic wood specialists, such as *Armillaria ostoyae* (Romagn.) Herink, that are less affected by high soil moisture levels (Miron 1995). Although *C. tubaeformis* occurred on organic-rich forest soils, these sites, together with peat bogs, generally produced very limited quantities of edible mushrooms (Miron 1995, 2000).

North Shore of Québec

In 1997, the number and biomass of sporocarps of epigeous edible mushrooms was calculated in 120 *P. banksiana* stands along the Québec North Shore (i.e., the region along the northern shore of the Saint Lawrence River; N. Villeneuve, unpublished data). The sites were divided into four categories according to stand age (<30 years old or >30 years old) and soil drainage (mesic or subhydric). Measurements were done weekly from the end of July until mid-October inside permanent 1000 m × 2 m corridors (0.2 ha).

During the study, 39 species of edible ECM fungi were recorded, including eight species of *Suillus*, six species of *Leccinum*, six species of *Boletus*, and four species of *Tricholoma*. The highly sought-after species *L. aurantiacum*, *Suillus luteus* L. ex Fr., *Suillus tomentosus* (Kauffman) Singer, *T. magnivelare*, *T. equestre*, and the lobster mushroom, were found in important numbers (Table 14.2). Other species, such as *B. edulis*, *Cantharellula umbonata* (J.F. Gmel.) Singer, *C. cibarius*, *C. ventricosum*, *R. caperatus*, *S. squamosus*, and *T. caligatum*, were also found but in smaller quantities (Table 14.1). Although sporocarp production started at the end of July, the main production was concentrated over a period of approximately 30 days between August 25 and September 25.

Young stands (<30 years old) on subhydric sites showed the highest annual productivity of edible species with a dry biomass of 6.8 kg/ha followed by the young stands on mesic sites with a dry biomass of 5.4 kg/ha. The oldest stands (>30 years old) only produced 3.7 kg/ha (dry biomass) of sporocarps on mesic sites and 2.6 kg/ha (dry biomass) on subhydric sites (Table 14.2).

Some fungi, such as *Suillus brevipes* (Peck) Kuntze, *Suillus neoalbidipes* M.E. Palm & E.L. Stewart, *S. granulatus*, and the lobster mushroom, showed a higher productivity in young stands. By contrast, *T. magnivelare*, *L. atrostipitatum*, *Boletus subtomentosus* L., *S. luteus*, and *S. cavipes* preferred older stands (Table 14.2).

The drainage of the surface deposits also proved to have an influence on the productivity of edible mushrooms. The mesic stations, although being less productive overall, produced more sporocarps of *L. aurantiacum*, *T. magnivelare*, and *S. granulatus*. However, the subhydric sites were more favourable for the production of sporocarps of *L. atrostipitatum*, *S. cavipes*, and lobster mushrooms (Table 14.2).

Gaspé Peninsula

Over two consecutive years, the potential of a commercial harvest of forest mushrooms in *Picea* spp. plantations and in a few natural stands on the Gaspé Peninsula were evaluated (Guérette 2001). Among the seven sites surveyed in 2000, the plantations were clearly the most productive, with fresh weights of 18–96 kg/ha; the highest annual productivity was observed in a 20-year-old *Picea glauca* (Moench) Voss plantation. In general, the study highlighted the potential of exploring this region, particularly for *C. cibarius* but also for other edible mushrooms such as *B. edulis*, *L. auranticum*, and *C. ventricosum*.

In a recent study, Gévry (2008) investigated the edible mushroom potential of eight forest stands types (mostly situated in private coastal woodlands) in the northern part of the Gaspé Peninsula. In 2007, *Picea abies* (L.) Karst. plantations, *P. glauca* natural forests, and mixed coniferous forests (*P. glauca* and *A. balsamea*) offered the best potentials with mean mushroom yields of 35, 23, and 19 kg/ha, respectively. The species *Lactarius deterrimus* Gröger occurred in great abundance in *P. abies* plantations, reaching a peak of 165 kg/ha in an unmanaged 30-year-old stand. *Cantharellus cibarius* was frequent, and yield reached 78 kg/ha in mixed coniferous stands and 34 kg/ha in *P. glauca* stands. The results of this study also show a good potential for the harvest of *C. ventricosum*, *Boletus subglabripes* Peck, and *Chalciporus piperatus* (Bull.) Bataille. Further surveys are planned for the area to confirm these results.

James Bay

Several species of gastronomic interest occur in open stands of *P. mariana*, *P. banksiana*, and *L. laricina* on the territory of the James Bay in northwestern Québec (Table 14.1) (Boudreau et al. 2003). Furthermore, the relatively open nature of these forests makes it easy to locate the sporocarps. Moreover, the mixed *P. mariana* and *P. banksiana* forests and the pure *P. banksiana* stands that cover much of the territory are not commercially exploited for timber and are accessible, at least in part, by a network of roads.

In September 2003 (a particularly dry year), large quantities of *S. brevipes*, *Suillus americanus* (Peck) Snell, and *S. granulatus* were frequently observed in open lichen-dominated *P. banksiana* stands on well-drained deposits of sandy gravel. *B. subglabripes* was also frequent in well-drained *P. banksiana* stands. By contrast, *Suillus grevillei* (Klotzsch) Singer, *Suillus serotinus* (Frost) Kretzer & T.D. Bruns, and *Gomphidius subroseus* Kauffman were mainly found in the presence of *L. laricina* in open *P. mariana* forests on stations that were imperfectly or poorly drained (Boudreau et al. 2003).

Eastern Townships

Finally, in southern Québec, a weekly sampling of six sites done during a single dry year indicated that mixed successional forests were the most suited for the harvest of edible forest mushrooms, even though numbers were relatively low compared with other regions (Fallu 2003). Among the species of ECM fungi found, *Lactarius thyinos* A.H. Sm. and *L. aurantiacum* sensu lato showed the greatest potential for commercial harvesting (Table 14.1). However, because of the low sampling effort and the dry conditions during the single summer of observations, this study does not allow us to draw any conclusions concerning the real potential of this region.

Current research

The recent rapid expansion of the edible forest mushroom industry in North America shows to what extent it is important to accurately establish the links between environmental factors and the distribution of diverse species of edible mushrooms. This is crucial if teams of harvesters in the field are to be effective. It will also determine guidelines to assure sustainable management of the resource in the face of commercial-scale harvesting.

In Eastern Canada and in Québec particularly, the ecology of edible forest mushrooms and the processes influencing their productivity in the wild are not yet fully understood, despite the obvious contribution of the studies reviewed in this chapter. Furthermore, numerous fundamental questions remain concerning the productivity of the boreal forest and the factors that influence the distribution and the abundance of each species (e.g., climate, topography, soil, surface deposits, floral composition of forest stands, stand age, and disturbances). With this in mind, several research projects investigating the biology, ecology, and management of edible mushrooms were recently started in Québec. These projects aim to quantify mushroom productivity in natural forests dominated by *P. banksiana*, *A. balsamea*, *P. mariana*, and *P. glauca* in the boreal forest zone of the Gaspé Peninsula and in the Lac Saint-Jean and Abitibi regions.

In a three-year study investigating the phenology of fruiting and the role of forest cover and abiotic factors in determining the abundance of selected edible species on the Gaspé Peninsula, 895 permanent quadrats were established along transects in 14 representative forest types during summer 2005 (M.-F. Gévry, unpublished data). These quadrats were designed to be surveyed every seven days from mid-July to late September over a three-year period (i.e., 2005, 2006, and 2007). Preliminary results indicate that *L. piceinum*, *C. ventricosum*, *B. edulis*, *L. deterrimus*, *R. caperatus*, and *H. repandum* are the most abundant species. To date, the greatest abundance of edible forest mushrooms has occurred in *P. glauca* and *P. abies* plantations. However, these plantations were characterized by a low species richness compared with that observed in natural forest stands. In the latter, the 30- to 50-year-old conifer stands showed the greatest abundances. Biotic and abiotic factors measured at each station will be used to establish (*i*) fungus–habitat relationships and (*ii*) biomass variations between forest types. Meteorological data will also be correlated with species productivity. Furthermore, species diversity of the Boletes and the genus *Lactarius* is being investigated to improve our knowledge of these taxa on the Gaspé Peninsula. *Cantharellus cibarius* and lobster mushrooms were rarely recorded in the study plots, which were located mainly in inland sites; however both these mushrooms are apparently abundant in sites closer to the coast. This suggested that further studies should be done to evaluate the potential of those sought-after species.

A second study investigating the ecology of *C. cibarius* and the lobster mushroom in *P. banksiana* stands in the boreal forest of eastern Canada is being conducted in the Lac Saint-Jean area (C. Rochon, personal communication). This study aims to (*i*) establish the environmental factors at the forest and fungal colony scale affecting the fructification of *C. cibarius*; (*ii*) identify plant species and stand characteristics that could be used as indicators for finding *C. cibarius* in the boreal forest of eastern Canada; (*iii*) define the links between host tree growth and sporocarp production of *C. cibarius*; and (*iv*) determine the links between soil and stand characteristics that influence the distribution and fructification of the lobster mushroom. During the first year of the project, numerous colonies of *C. cibarius* and lobster mushrooms were identified and the surrounding stand age, composition, structure, associated vegetation, soil characteristics, and meteorological data were recorded. These data permitted the selection of a number of ecological characteristics linked with the presence of sporocarps. In addition, fungal interactions with site disturbance were tested. During the second and third years of the project, the physiology of the ECM fungal – host tree system will be investigated using tree growth, and root and sporocarp respiration.

A third study aims to identify inter- and intra-stand factors responsible for the distribution and productivity of six species of edible forest mushrooms (i.e., *C. cibarius*, *L. aurantiacum* sensu lato, *S. squamosus*, *T. equestre*, *T. magnivelare*, and the lobster mushroom) associated with *P. banksiana* in the Abitibi region (D. Maneli, personal communication). Permanent quadrats are being used to assess the abundance of sporocarps. The assumed spatial distribution of the fungal colonies is being plotted using a system of markers to identify the positions of each sporocarp of each species. This distribution will be analysed as a function of the biotic and abiotic factors characterizing the habitat (e.g., origin of the stand, age, disturbances, presence of clearings, density of the different vegetation layers of the forest floor, the amount of sunlight incident on the forest floor, and the physical and chemical properties of the humus and soil).

Finally, a series of studies was initiated in 2006 to gain a better understanding of the impact of different forest management practices on the productivity of epigeous edible ECM mushrooms. The influence of thinning, pruning, and fertilizer applications will be tested in young plantations and natural forest stands. The mastering of such techniques would allow the development of a more multifaceted approach to forest management permitting the exploitation of a diversity of forest resources while assuring long-term sustainability of biological diversity.

In conclusion, previous and current studies have confirmed the annual production of a large quantity of several highly sought-after species of edible forest mushrooms in Québec. Furthermore, a number of habitat-associated factors have been identified that should allow the prediction of the abundance of certain species in certain parts of the territory. Therefore, Québec seems to have the potential for the establishment of a commercial edible forest mushroom harvesting industry. This should encourage and consolidate the harvesting and marketing activities that are slowly being set up in the province. In this perspective, it will also be vitally important that the information gathered concerning the diversity and the productivity of ECM fungi be used for the long-term conservation and sustainable exploitation of this natural resource.

Acknowledgements

The authors thank Andrew P. Coughlan for his aid during the revision and translation of this chapter. We also thank Caroline Rochon and David Maneli for their contribution to the text and Bruno Lévesque for the production of the map showing the location of the different regional studies. Finally, we also thank Shannon M. Berch for kindly accepting to help us during our review of the scientific literature describing the harvest of forest mushrooms in western Canada.

References

Amaranthus, M., and Pilz, D. 1996. Productivity and sustainable harvest of wild mushrooms. *In* Managing forest ecosystems to conserve fungus diversity and sustain wild mushroom harvests. *Edited by* D. Pilz and R. Molina. USDA For. Serv. Gen. Tech. Rep. PNW-GTR-371. pp. 42–61.

Amaranthus, M.P., Weigand, J.F., and Abbott, R. 1998. Managing high-elevation forests to produce American matsutake (*Tricholoma magnivelare*), high quality timber and nontimber forest products. West. J. Appl. For. **13**: 120–128.

Arnolds, E. 1981. Ecology and coenology of macrofungi in grasslands and moist heatlands in Drenthe, the Netherlands. Part 1. Introduction and synecology. J. Cramer, Vaduz, Liechtenstein.

Arnolds, E. 1988. The changing macromycete flora in the Netherlands. Trans. Br. Mycol. Soc. **90**: 391–406.

Arnolds, E. 1991. Decline of ectomycorrhizal fungi in Europe. Agric. Ecosyst. Environ. **35**: 209–244.

Arnolds, E. 1995. Conservation and management of natural populations of edible fungi. Can. J. Bot. **73**(Suppl. 1): 987–998.

Barkmann, J.J. 1987. Methods and results of mycocoenological research in the Netherlands. *In* Studies on fungal communities. *Edited by* G. Pacioni. Aquila Publishing, Market Rasen, UK. pp. 7–38.

Berch, S.M. 2000. Integrating wild mushroom harvesting into forest management. *In* Les champignons forestiers : récolte, commercialisation et conservation de la ressource, Québec. *Edited by* J.A. Fortin and Y. Piché. Centre de recherche en biologie forestière, Université Laval, Québec, Que. pp. 13–16.

Berch, S.M., and Cocksedge, W. 2003. Commercially important wild mushrooms and fungi of British Columbia: what the buyers are buying. British Columbia Ministry of Forests, Research Branch, Victoria, B.C. Tech. Rep. 006.

Berch, S.M., and Wiensczyk, A.M. 2001. Ecological description and classification of some pine mushroom (*Tricholoma magnivelare*) habitat in British Columbia. British Columbia Ministry of Forests and Southern Interior Forest Extension and Research Partnership, Victoria, B.C. Res. Rep. 19.

Bills, G.F., Holtzmann, G.I., and Miller, O.K., Jr. 1986. Comparison of ectomycorrhizal basidiomycete communities in red spruce versus northern hardwood forests of West Virginia. Can. J. Bot. **64**: 760–768.

Boa, E. 2004. Wild edible fungi: a global overview of their use and importance to people. Food and Agriculture Organization of the United Nations, Rome.

Bonet, J.A., Fischer, C.R., and Colinas, C. 2004. The relationship between forest age and aspect on the production of sporocarps of ectomycorrhizal fungi in *Pinus sylvestris* forests of the central Pyrenees. For. Ecol. Manage. **203**: 157–175.

Boudreau, F., Boulet, B., Gauthier, R., and Paradis, J.-F. 2003. Observations sur les macromycètes de la région de Radisson, municipalité de la Baie-James, Québec. Herbier Louis-Marie, Université Laval, Québec, Que.

Brandrud, E. 1987. Mycorrhizal fungi in 30 years old, oligotrophic spruce (*Picea abies*) plantation in SE Norway: a one year permanent plot study. Agarica, **8**: 48–58.

Cherkasov, A.P. 1988. Classification of nontimber resources in the USSR. *In* Proceedings of the Finnish–Soviet Symposium on Nontimber Forest Resources, 25–29 Aug. 1986, Jyväkylä, Finland. *Edited by* I. Bannine and M. Raatikainen. Acta Bot. Fenn. **139**: 3–5.

Cibula, W.G., and Ovrebo, C.L. 1988. Mycosociological studies of mycorrhizal fungi in two loblolly pine plots in Mississippi and some relationships with remote sensing. *In* Remote Sensing for Resource Inventory: Planning and Monitoring. Proceedings of the 2nd Forest Service Remote Sensing Application Conference, 11–15 Apr. 1988, Slidell, La. *Edited by* J.D. Greer. American Society for Photogrammetry and Remote Sensing, Falls Church, Va. pp. 268–307.

Dahlberg, A., Schimmel, J., Taylor, A.F.S., and Johanneson, H. 2001. Post-fire legacy of ectomycorrhizal fungal communities in the Swedish boreal forest in relation to fire severity and logging intensity. Biol. Conserv. **100**: 151–161.

Danielson, R.M. 1984. Ectomycorrhizal associations in jack pine stands in northeastern Alberta. Can. J. Bot. **62**: 932–939.

Darimont, F. 1973. Recherches mycosociologiques dans les forêts de Haute-Belgique. Essai sur les fondements de la sociologie des champignons. Inst. R. Sci. Nat. Belg. Mem. **170(1)**: 1–220.

de Geus, N. 1995. Botanical forest products in British Columbia: an overview. British Columbia Ministry of Forests, Integrated Resources Policy Branch, Victoria, B.C.

de Geus, P.M.J., and Berch, S.M. 1997. The pine mushroom industry in British Columbia. *In* Mycology in sustainable development: expanding concepts, vanishing borders. *Edited by* I.H. Chapela and M.E. Palm. Parkway Publishers, Boone, N.C. pp. 55–67.

Deslandes, J., and Pic, C. 2001. Mise en valeur alimentaire et médicinale des plantes et champignons de sous-bois de la forêt feuillue de l'Outaouais, phase 1, rapport préliminaire. Institut québécois d'aménagement de la forêt feuillue, Québec, Que.

Després, J., Lamoureux, Y., Boyer, R., Archambault, R., and Jean, A. 2002. Mille et un champignons du Québec [CD-ROM]. Cercle des mycologues de Montréal, Montréal, Que.

Dighton, J., Poskitt, J.M., and Howard, D.M. 1986. Changes in occurrence of basidiomycete fruit bodies during forest stand development: with specific reference to mycorrhizal species. Trans. Br. Mycol. Soc. **87**: 163–171.

Duchesne, L.C., and Weber, M.G. 1993. High incidence of the edible morel *Morchella conica* in a jack pine, *Pinus banksiana*, forest following prescribed burning. Can. Field-Nat. **107**: 114–116.

Durall, D.M., Jones, M.D., Wright, E.F., Kroeger, P., and Coates, K.D. 1999. Species richness of ectomycorrhizal fungi in cutblocks of different sizes in the interior cedar–hemlock forest of northwestern British Columbia: sporocarps and ectomycorrhizae. Can. J. For. Res. **29**: 1322–1332.

Egli, S., and Ayer, F. 1997. Est-il possible d'améliorer la production de champignons comestibles en forêt? L'exemple de la réserve mycologique de la Chanéaz en Suisse. Rev. For. Fr. **49**(No. Spec.): 235–243.

Egli, S., Ayer, F., and Chatelain, F. 1990. Der Einfluss des Pilzsammelns auf die Pilzflora. Mycol. Helv. **3**: 417–428.

Egli, S., Peter, M., Buser, C., Stahel, W., and Ayer, F. 2006. Mushroom picking does not impair future harvests—results of a long-term study in Switzerland. Biol. Conserv. **129**: 271–276.

Ehlers, T., Berch, S.M., and MacKinnon, A. 2003. Inventory of non-timber forest product plant and fungal species in the Robson Valley. B.C. J. Ecosyst. Manage. **4**: 38–52.

Fallu, J. 2003. Évaluation du potentiel de récolte des champignons forestiers comestibles dans les boisés de l'Estrie, Québec. M.Sc. thesis, Département de géographie et télédétection, Université de Sherbrooke, Sherbrooke, Que.

Fogel, R. 1976. Ecological studies of hypogeous fungi. II. Sporocarp phenology in a western Oregon Douglas fir stand. Can. J. Bot. **54**: 1152–1162.

Fortin, J.A. 1968. Les mycorhizes ectotrophes des arbres forestiers. Nat. Can. **95**: 287–305.

Fortin, J.A. 2003. Potentiel commercial et biodiversité des champignons nordiques. *In* Mycologie nordique : recueil de textes, atelier de Radisson, Québec. Centre d'études nordiques, Université Laval, Québec, Que.

Fortin, J.A., and Piché, Y. 2000. Les champignons forestiers : récolte, commercialisation et conservation de la ressource. Centre de recherche en biologie forestière, Université Laval, Québec, Que.

Gange, A.C., Gange, E.C., Sparks, T.H., and Boddy, L. 2007. Rapid and recent changes in fungal fruiting patterns. Science (Washington, D.C.), **316**: 71.

Garbaye, J., and Le Tacon, F. 1982. Influence of mineral fertilization and thinning intensity on the fruit body production of epigeous fungi in an artificial spruce stand (*Picea abies* Link.) in north-eastern France. Acta Oecol. Oecol. Plant. **3**: 153–160.

Garbaye, J., Kabre, A., Le Tacon, F., Mousain, D., and Piou, D. 1979. Fertilisation minérale et fructification des champignons supérieurs en hêtraie. Ann. Sci. For. **36**: 151–164.

Gévry, M.-F. 2008. Projet d'intégration de la récolte des champignons forestiers comestibles dans la communauté—Secteur de Mont-Louis : description du projet, résultats des inventaires et perspectives d'avenir locales. Comité de bassin de la rivière Mont-Louis, Mont-Louis, Que.

Godbout, C., and Fortin, J.A. 1983. Morphological features of synthesized ectomycorrhizae of *Alnus crispa* and *A. rugosa*. New Phytol. **94**: 249–262.

Godbout, C., and Fortin, J.A. 1985. Synthesized ectomycorrhizae of aspen: fungal genus level of structural characterization. Can. J. Bot. **63**: 252–262.

Groves, J.W. 1962. Edible and poisonous mushrooms of Canada. Canada Department of Agriculture, Ottawa, Ont. Publ. 1112.

Groves, J.W., and MacRae, R. 1963. The fungus records of Mr. H.A.C. Jackson from L'Islet Co., Québec. Can. Field-Nat. **77**: 179–202.

Guérette, M. 2001. Évaluation du potentiel multiressource en Gaspésie. Groupement forestier Baie-des-Chaleurs, Bonaventure, Que.

Guinberteau, J., and Courtecuisse, R. 1997. Diversité des champignons (surtout mycorhiziens) dans les écosystèmes forestiers actuels. Rev. For. Fr. **49**: 25–39.

Hosford, D., Pilz, D., Molina, R., and Amaranthus, M. 1997. Ecology and management of the commercial harvested American matsutake mushroom. USDA For. Serv. Gen. Tech. Rep. PNW-GTR-412.

Ito, T., and Ogawa, M. 1979. Cultivating method of the mycorrhizal fungus, *Tricholoma matsutake* (Ito et Iman) Sing. II: Increasing number of shiros (fungal colonies) of *T. matsutake* by thinning the understory vegetation. J. Jpn. For. Soc. **61**: 163–173.

Jackson, H.A.C. 1948. Notes on the higher fungi collected in Lasalle, Qué., 1930–1940. Can. Field-Nat. **62**: 127–133.

Jackson, H.A.C., and Cazort, M. 1979. Les champignons de M. Jackson. Musées Nationaux du Canada, Ottawa, Ont.

Jones, M.D., Durall, D.M., and Cairney, J.W.G. 2003. Ectomycorrhizal fungal communities in young forest stands regenerating after clearcut logging. New Phytol. **157**: 399–422.

Kalamees, K. 1979. The role of fungal groupings in the structure of ecosystems. Eesti Nsv Tead. Akad. TOIM Biol. **28**: 206–213.

Kallio, P. 1980. Some observations on the fungi of the central Quebec–Labrador peninsula. Environmental studies in the central Quebec–Labrador peninsula: Finnish contributions. McGill University, Montréal, Que. McGill Subarct. Res. Pap. 30.

Kinugawa, K., and Goto, T. 1978. Preliminary survey on the 'matsutake' (*Armillaria ponderosa*) of North America. Trans. Mycol. Soc. Jpn. **19**: 91–101.

Kranabetter, J.M., and Kroeger, P. 2001. Ectomycorrhizal mushroom response to partial cutting in a western hemlock/ western redcedar forest. Can. J. For. Res. **31**: 978–987.

Kranabetter, J.M., Trowbridge, R., Macadam, A., McLennan, D., and Friesen, J. 2002. Ecological descriptions of pine mushroom (*Tricholoma magnivelare*) habitat and estimates of its extent in northwestern British Columbia. For. Ecol. Manage. **158**: 249–261.

Kranabetter, J.M., Friesen, J., Gamiet, S., and Kroeger, P. 2005. Ectomycorrhizal mushroom distribution by stand age in western hemlock – lodgepole pine forests of northwestern British Columbia. Can. J. For. Res. **35**: 1527–1539.

Kropp, B.R., and Albee, S. 1996. The effects of silvicultural treatments on occurrence of mycorrhizal sporocarps in a *Pinus contorta* forest: a preliminary survey. Biol. Conserv. **78**: 313–318.

Laganà, A., Angiolini, C., Salerni, E., Perini, C., Barluzzi, C., and De Dominicis, V. 2002. Periodicity, fluctuations and successions of macrofungi in forests (*Albies alba* Miller) in Tuscany, Italy. For. Ecol. Manage. **169**: 187–202.

Lamoureux, Y. 1993. Le monde méconnu des champignons. Quatre-Temps, **17**(3): 25–26.

Lamoureux, Y., and Després, J. 1997. Champignons du Québec. Tome 1. Les bolets. Cercle des mycologues de Montréal. Montréal, Que.

Lamoureux, Y., and Sicard, M. 2001. Connaître, cueillir et cuisiner : les champignons sauvages du Québec, Édition Fides, Montréal, Que.

Lessard, E. 1986. Deuxième année de recherches sur la sociologie des champignons supérieurs dans les érablières du comté de L'Islet, Québec. B.Sc. project, Faculté de Foresterie et Géodésie, Université Laval, Québec, Que.

Lim, S.R., Fischer, A., Berbee, M., and Berch, S.M. 2003. Is the booted tricholoma in British Columbia really Japanese matsutake? B.C. J. Ecosyst. Manage. **3**: www.forrex.org/JEM/ISS15/vol3_no1_art7.pdf.

Lisiewska, M. 1974. Macromycetes of beech forests within the eastern part of the *Fagus* area in Europe. Acta Mycol. **10**: 3–72.

Lodge, D.J., Ammirati, J.F., O'Dell, T.E., Lodge, G.M., Huhndorf, S.M., Wang, C.-H., Stokland, J.N., Schmit, J.P., Ryvarden, L., Lealock, P.R., Mata, M., Umana, L., and Wu, Q., and Czederpiltz, D.L. 2004. Terrestrial and lignicolous macrofungi. *In* Biodiversity of fungi: inventory and monitoring methods. *Edited by* G.M. Lodge, G.F. Bills, and M.S. Foster. Elsevier, Amsterdam. pp. 127–172.

Luoma, D.L., Eberhart, J.L., Molina, R., and Amaranthus, M.P. 2004. Response of ectomycorrhizal fungus sporocarp production to varying levels and patterns of green-tree retention. For. Ecol. Manage. **202**: 337–354.

Luoma, D.L., Eberhart, J.L., Abbott, R., Moore, A., Amaranthus, M.P., and Pilz, D. 2006. Effects of mushroom harvest on subsequent American matsutake production. For. Ecol. Manage. **236**: 65–75.

Martínez de Aragón, J., Bonet, J.A., Fischer, C.R., and Colinas, C. 2007. Productivity of ectomycorrhizal and selected edible saprotrophic fungi in pine forests of the pre-Pyrenees mountains, Spain: predictive equations for forest management of mycological resources. For. Ecol. Manage. **252**: 239–256.

McAfee, B.J., and Fortin, J.A. 1987. The influence of pH on the competitive interactions of ectomycorrhizal mycobionts under field conditions. Can. J. For. Res. **17**: 859–864.

McAfee, B.J., and Fortin, J.A. 1989. Ectomycorrhizal colonization on black spruce and jack pine seedlings outplanted in reforestation sites. Plant Soil, **116**: 9–17.

McNeil, R. 2006. Le grand livre des champignons de l'est du Québec, Éditions Michel Quintin, Waterloo, Ont.

Menge, J.A., and Grand, L.F. 1978. Effect of fertilization on production of epigeous basidiocarps by mycorrhizal fungi in loblolly pine plantations. Can. J. Bot. **56**: 2357–2362.

Miron, F. 1994. Champignons forestiers sauvages : potentiel de cueillette et de mise en marché, phase 1. Rapport du programme Essais, expérimentations et transfert technologique en foresterie par Champignons Laurentiens inc. Available from Ressources naturelles Canada, Service canadien des forêts, Québec, Que. Rep. 4050.

Miron, F. 1995. Champignons forestiers sauvages : potentiel de cueillette et de mise en marché, phase 2. Rapport du programme Essais, expérimentations et transfert technologique en foresterie par Champignons Laurentiens inc. Available from Ressources naturelles Canada, Service canadien des forêts, Québec, Que. Rep. 4054.

Miron, F. 2000. Récolte et commercialisation des champignons forestiers : six ans d'expérience. *In* Les champignons forestiers : récolte, commercialisation et conservation de la ressource. *Edited by* J.A. Fortin and Y. Piché. Centre de recherche en biologie forestière, Université Laval, Québec, Que. pp. 53–57.

Molina, R., O'Dell, T., Luoma, D., Amaranthus, M., Castellano,

M., and Russell, K. 1993. Biology, ecology, and social aspects of wild edible mushrooms in the forests of the Pacific Northwest: a preface to managing commercial harvest. USDA For. Serv. Gen. Tech. Rep. PNW-GTR-309.

Molina, R., O'Dell, T.E., Dunham, S., and Pilz, D. 1999. Biological diversity and ecosystem functions of forest soil fungi: management implications. *In* Proceedings: Pacific Northwest Forest and Rangeland Soil Organism Symposium, 17–19 Mar. 1998, Corvallis, Ore. *Technical Editors*: R.T. Meurisse, W.G. Ypsilantis, and C. Seybold. USDA For. Serv. Gen. Tech. Rep. PNW-GTR-461. pp. 45–58.

Mounce, I., and Jackson, H.A.C. 1937. Two Canadian collections of *Cantharellus multiplex*. Mycologia, **29**: 286–288.

Nantel, P., and Neumann, P. 1992. Ecology of ectomycorrhizal-basidiomycete communities on a local vegetation gradient. Ecology, **73**: 99–117.

O'Dell, T.E., Luoma, D.L., and Molina, R.J. 1992. Ectomycorrhizal fungal communities in young, managed, and old-growth Douglas-fir stands. Northwest Environ. J. **8**: 166–168.

O'Dell, T.E., Smith, J.E., Castellano, M., and Luoma, D. 1996. Diversity and conservation of forest fungi. *In* Managing forest ecosystems to conserve fungus diversity and sustain wild mushroom harvests. *Edited by* D. Pilz and R. Molina. USDA For. Serv. Gen. Tech. Rep. PNW-GTR-371. pp. 5–18.

O'Dell, T.E., Ammirati, J.F., and Schreiner, E.G. 1999. Species richness and abundance of ectomycorrhizal basidiomycete sporocarps on a moisture gradient in the *Tsuga heterophylla* zone. Can. J. Bot. **77**: 1699–1711.

Ohenoja, E. 1978. Mushrooms and mushroom yields in fertilized forests. Ann. Bot. Fenn. **15**: 38–46.

Ohenoja, E. 1988. Effect of forest management procedures on fungal fruit body production in Finland. Acta Bot. Fenn. **136**: 81–84.

Ohenoja, E. 1993. Effects of weather conditions on the larger fungi in different forest sites in northern Finland, 1976–1988. Ph.D. thesis, University of Oulu, Oulu, Finland. Sci. Rerum Nat. 243.

Ohenoja, E., and Koistinen, R. 1984. Fruit body production of larger fungi in Finland. 2: Edible fungi in northern Finland 1976–1978. Ann. Bot. Fenn. **21**: 357–366.

Ohenoja, E., and Metsänheimo, K. 1982. Phenology and fruit body production of macrofungi in subarctic Finnish Lapland. *In* Arctic and alpine mycology. *Edited by* G.A. Laursen and J.F. Ammirati. University of Washington Press, Seattle, Wash. pp. 390–404.

Petersen, P.M. 1977. Investigations on the ecology and phenology of the macromycetes in the Arctic. Medd. Gronl. **199**: 1–72.

Piché, Y., and Fortin, J.A. 1982. Development of mycorrhizae, extramatrical mycelium and sclerotia on *Pinus strobus* seedlings. New Phytol. **91**: 211–220.

Pilz, D., and Molina, R. (*Editors*). 1996. Managing forest ecosystems to conserve fungus diversity and sustain wild mushroom harvests. USDA For. Serv. Gen. Tech. Rep. PNW-GTR-371.

Pilz, D., and Molina, R. 2002. Commercial harvests of edible mushrooms from the forests of the Pacific Northwest United States: issues, management, and monitoring for sustainability. For. Ecol. Manage. **155**: 3–16.

Pilz, D., Smith, J., Amaranthus, M.P., Alexander, S., Molina, R., and Luoma, D. 1999. Mushrooms and timber: managing commercial harvesting in the Oregon Cascades. J. For. **97**: 4–11.

Pilz, D., Norvell, L., Danell, E., and Molina, R. 2003. Ecology and management of commercially harvested chanterelle mushrooms. USDA For. Serv. Gen. Tech. Rep. PNW-GTR-576.

Pilz, D., Molina, R., and Mayo, J.J. 2006. Effects of thinning young forests on chanterelle mushroom production. J. For. **104**: 9–14.

Pomerleau, R. 1951. Champignons de l'est du Canada et des États-Unis. Les Éditions Chantecler, Montréal, Que.

Pomerleau, R. 1966. Les Amanites du Québec. Nat. Can. **93**: 861–887.

Pomerleau, R. 1975. Identity of two uncommon amanitas in Québec. *In* Studies on higher fungi. *Edited by* H.E. Bigelow and W.D. Thiers. J. Cramer, Vaduz, Liechtenstein. pp. 191–198.

Pomerleau, R. 1980. Flore des champignons au Québec. Les Éditions La Presse, Montréal, Que.

Pomerleau, R. 1984. Supplément à la flore des champignons au Québec. Les Éditions La Presse, Montréal, Que.

Pomerleau, R., and Smith, A.H. 1962. *Fuscoboletinus*, a new genus of the Boletales. Brittonia, **14**: 156–172.

Randall, B.L., and Grand, L.F. 1986. Morphology and possible mycobiont (*Suillus pictus*) of a tuberculate ectomycorrhiza on *Pinus strobus*. Can. J. Bot. **64**: 2182–2191.

Redhead, S.A. 1979. A study of the sphagnicolous fleshy basidiomycetes in the eastern sections of the Canadian boreal forest. Ph.D. thesis, University of Toronto, Toronto, Ont.

Redhead, S.A. 1989. A biogeographical overview of the Canadian mushroom flora. Can. J. Bot. **67**: 3003–3062.

Redhead, S.A. 1997. The pine mushroom industry in Canada and the United States: why it exists and where it is going. *In* Mycology in sustainable development: expanding concepts, vanishing borders. *Edited by* M.E. Palm and I.H. Chapela. Parkway Publishers, Boone, N.C. pp. 15–54.

Redhead, S.A. 2000. Forest mushroom harvesting in Canada: past, present and future. *In* Les champignons forestiers : récolte, commercialisation et conservation de la ressource. *Edited by* J.A. Fortin and Y. Piché. Centre de recherche en biologie forestière, Université Laval, Québec, Que. pp. 1–5.

Rendall, D.L. 1980. The genus *Lactarius* occurring in the southern boreal forest region of Ontario and Québec. M.Sc. thesis, University of Toronto, Toronto, Ont.

Salerni, E., and Perini, C. 2004. Experimental study for increasing productivity of *Boletus edulis* s.l. in Italy. For. Ecol. Manage. **201**: 161–170.

Salo, K. 1979. Mushrooms and mushroom yield on transitional peatlands in central Finland. Ann. Bot. Fenn. **16**: 181–192.

Samson, J., and Fortin, J.A. 1986. Ectomycorrhizal fungi of *Larix laricina* and the interspecific and intraspecific variation in response to temperature. Can. J. Bot. **64**: 3020–3028.

Saucier, J.-P., Bergeron, J.-F., Grondin, P., and Robitaille, A. 1998. Les régions écologiques du Québec méridional (3e version) : un des éléments du système hiérarchique de classification écologique du territoire mis au point par le ministère des Ressources naturelles du Québec. L'Aubelle, S1–S12.

Slankis, V. 1974. Soil factors influencing formation of mycorrhizae. Annu. Rev. Phytopathol. **12**: 437–457.

Smith, J.E., Molina, R., Huso, M.M.P., Luoma, D.L., McKay, D., Castellano, M.A., Lebel, T., and Valachovic, Y. 2002. Species richness, abundance, and composition of hypogeous and epigeous ectomycorrhizal fungal sporocarps in young, rotation-age, and old-growth stands of Douglas-fir (*Pseudotsuga menziesii*) in the Cascade Range of Oregon, U.S.A. Can. J. Bot. **80**: 186–204.

Stanis, V.F. 1979. *Russula* species occurring in the boreal region of Ontario and Québec, Canada. M.Sc. thesis, University of Toronto, Toronto, Ontario.

Straatsma, G., Ayer, A., and Egli, S. 2001. Species richness, abundance and phenology of fungal fruit bodies over 21 years in Swiss forest plot. Mycol. Res. **105**: 515–523.

Tanino, K.K., Ivanochko, G., Jessup, C., Nelson, J., and Hrycan, W. 2005. Stage I: Sustainable harvest of wild mushrooms in northern Saskatchewan. Final report. Saskatchewan Government, Agriculture Department Fund, Saskatoon, Sask. Project ADF 20000235.

Thoen, D. 1976. Facteurs physiques et fructifications des champignons supérieurs dans quelques pessières d'Ardenne Méridionale (Belgique). Bulletin mensuel de la société Linnéenne de Lyon, **45**: 269–284.

Tyler, G. 1985. Macrofungal flora of Swedish beech forest related to soil organic matter and acidity characteristics. For. Ecol. Manage. **10**: 13–29.

Villeneuve, N. 1985. Étude mycosociologique de quelques érablières sucrières du comté de L'Islet, Québec. B.Sc. project, Faculté de Foresterie et Géodésie, Université Laval, Québec, Que.

Villeneuve, N. 1993. Organisation cénologique et écologie des macromycètes terrestres dans les forêts des laurentides québécoises. Ph.D. thesis, Université Laval, Québec, Que.

Villeneuve, N. 1995. Estimation de la productivité naturelle des champignons comestibles dans les forêts de l'Est québécois. Dessau Environnement et Aménagement Inc., Saint-Romuald, Que.

Villeneuve, N. 2000. Diversité et productivité des champignons forestiers : les apports de la recherche et de l'inventaire. *In* Les champignons forestiers : récolte, commercialisation et conservation de la ressource. *Edited by* J.A. Fortin and Y. Piché. Centre de recherche en biologie forestière, Université Laval, Québec, Que. pp. 91–100.

Villeneuve, N., Grandtner, M.M., and Fortin, J.A. 1989. Frequency and diversity of ectomycorrhizal and saprophytic macrofungi in the Laurentide mountains of Quebec. Can. J. Bot. **67**: 2616–2629.

Villeneuve, N., Grandtner, M.M., and Fortin, J.A. 1991. The coenological organization of ectomycorrhizal macrofungi in the Laurentide mountains of Quebec. Can. J. Bot. **69**: 2215–2224.

Visser, S. 1995. Ectomycorrhizal fungal succession in jack pine stands following wildfire. New Phytol. **129**: 389–401.

Wästerlund, I., and Ingelög, T. 1981. Fruit body production of larger fungi in some young Swedish forests with special reference to logging waste. For. Ecol. Manage. **3**: 269–294.

Watling, R. 1981. Relationships between macromycetes and the development of higher plant communities. *In* The fungal community: its organization and role in the ecosystem. *Edited by* D.T. Wicklow and G.C. Carroll. Marcel Dekker, New York. pp. 427–458.

Weigand, J.F. 2000. Wild edible mushroom harvests in North America: market econometric analyses. *In* Les champignons forestiers : récolte, commercialisation et conservation de la ressource. *Edited by* J.A. Fortin and Y. Piché. Centre de recherche en biologie forestière, Université Laval, Québec, Que. pp. 35–43.

Wiensczyk, A.M., and Berch, S.M. 2002. Ecological description and classification of some pine mushroom habitat in British Columbia. BC J. Ecosyst. Manage. **1**: 119–125.

Wiklund, K., Nilsson, L.-O., and Jacobsson, S. 1995. Effect of irrigation, fertilization, and artificial drought on basidioma production in a Norway spruce stand. Can. J. Bot. **73**: 200–208.

Index

Abies balsamea, 162–164, 178, 185, 186
Absorptive hyphal networks, 43, 46, 47
Acacetin, 59
Acaulospora cavernata, 3
Acaulospora laevis, 44
Acer saccharum, 2, 178
Agriculture, 1, 51, 61, 71, 77, 78, 80, 88–90, 93, 105–112
AHN, 43, 46, 47
Alfalfa, 56
Alkaline soils, 2, 108, 123, 131, 142, 145
Allelopathy, 178
Allium ampeloprasum var. *porrum*, 89, 90
Alnus crispa, 176
Alnus rugosa, 176
ALS, 43, 46, 47
ALS spore, 46
Amanitaceae, 157, 158
Amphinema byssoides, 117, 118, 120, 122
Anastamoses, 75
Andira inermis, 156
Appressorium, 1, 15, 21, 23, 44, 45
Arbuscular mycorrhiza, 1–3, 6, 15, 25, 26, 52, 54, 56, 71, 72, 77, 110, 116, 153, 154, 156, 157
Arbuscule-like structures, 43, 46, 47
Arbutoid mycorrhiza, 6, 7
Armillaria ostoyae, 184
Ascomycetes, 6, 116, 117, 122, 131, 132, 135
Aspen, 118, 119, 123, 176
Autoregulation, 15, 27–29
Auxiliary cells, 41
Auxins, 20, 21, 26, 168, 170
Azotobacter chroococcum, 58
Azoxystrobin, 79
Bacillus coagulans, 54
Bacillus subtilis, 111
Balsam poplar, 118, 119, 178
Barley, 2, 27, 89, 129
BAS, 46, 47, 157, 158
BAS spore, 46
Basidiomycetes, 4, 6, 117, 122, 161
Bean (Broad), 90
Benomyl, 79
Beta vulgaris, 89
Betula alleghaniensis, 178
Betula papyrifera, 177, 178
Biocontrol, 55, 57–60
Bioremediation, 130
Björkman's hypothesis, 162

Black spruce, 106, 118–121, 123, 124, 162, 163, 166, 177–180, 184–186
Blueberry, 4
Blumenin, 59
Boletus edulis, 166, 175, 180, 181, 184–186
Boletus subglabripes, 182, 185
Boletus subtomentosus, 182, 184, 185
Boreal forest, 6, 115, 119, 120, 123, 129–131, 136, 161, 162, 165, 166, 180, 186
Branched absorbing structures, 46, 47, 157, 158
Branching factor, 17, 18, 29, 44
Brassica juncea, 89
Brassica napus, 2, 89
Burkholderia cepacia, 60
Canadian Prairie Provinces, 119
Cantharellula umbonata, 182, 185
Cantharellus cibarius, 175–178, 180–182, 184–186
Cantharellus subalbidus, 175
Capsicum annuum, 90
Captan, 79
Carbon cost of ectomycorrhizal association, 168
Carbon credits, 109
Carbon flow in ectomycorrhizal systems, 161
Carbon sequestration, 107, 116
Carrot, 21, 43, 44, 52, 87, 89, 90
Catathelasma ventricosum, 179, 182, 185, 186
Cedrus atlantica, 142–144
Cephalanthera austinae, 4
Ceratobasidium cornigerum, 5
Ceratorhiza goodyerae-repentis, 5
Chenopodiaceae, 87, 89
Chloridium paucisporum, 7, 116
Chlorosis, 145–148
Chlorsulfuron, 79
Clavibacter michiganensis, 53
Climate change, 71, 107, 181
Coenocytic hyphae, 39, 46, 47
Commercial harvesting of ectomycorrhizal fungi, 175, 177, 184, 186
Commercial inocula, 78, 96, 107, 118
Commercial nurseries, 118, 134
Common mycelial network, 75
Compatible solutes, 130, 132
Compost, 117, 146
Container-grown tree seedlings, 116
Corallorhiza maculata, 5
Corallorhiza trifida, 5
Cortinarius amazonicus, 159
Craterellus tubaeformis, 179–182, 184

Crop breeding and arbuscular mycorrhizal fungi, 78
Crop rotations and arbuscular mycorrhizal fungi, 77, 78
Crushed carpophore inoculum, 118
Cryopreservation, 98, 99
Cryptococcus laurentii, 54
Cupressus sempervirens, 145, 146
Daucus carota, 21, 43, 44, 52, 87, 89, 90
Defence regulation, 22
Depletion zone, 72, 88
Diclofop, 79
Dicymbe corymbosa, 154
Dikaryotic culture, 144
Disease-suppressive soils, 56
Douglas-fir, 108, 120, 122
Drought stress, 109, 122, 129, 130
Drought tolerance, 74, 78, 120, 141, 144
Ecophysiology of sporocarp development, 161–171
Ecosystem restoration, 129–136
Ectendomycorrhiza, 6, 7, 116, 155, 156
Ectomycorrhiza, 5–7, 26, 116–118, 129, 142–144, 153–157, 159, 161–165, 168, 170, 171
Ectomycorrhizal basidiomycetes, 161–171
Edaphic factors controlling fungal distribution, 154, 178, 179
Edible ectomycorrhizal mushrooms, 175–187
Egress of hyphae from roots, 45
Elkhorn-like structures, 45
Engelmann spruce, 122
Eperua falcata, 156
Eperua grandiflora, 156
Eperua leucantha, 156
Eperua purpurea, 156
Ergosterol, 79
Ericoid mycorrhizal fungi, 4, 117, 136
E-strain, 7, 116, 120, 122
Eucalyptus, 140, 143, 157
Eucalyptus camaldulensis, 143
Extraction of arbuscular mycorrhizal propagules from soil, 93–95
Extraradical hyphae, 1, 3, 39, 45, 46, 52, 72–74
Fagus grandifolia, 178
Fan-like structure, 44, 45
FB, 40–43, 45–47
Fenpropimorph, 79
Fertilizer application, 88, 89, 110, 116–119, 148
Field trials, 97, 115, 119, 121, 123, 130, 134–135
Fine branching hyphae, 40–43, 45–47
Flavonoids, 16, 17, 19–21, 29, 42, 59, 75
Forest nurseries, 7, 115–117, 139–141, 143, 145, 148, 162, 165–166
Forestry, 1, 2, 105–111, 117, 123
Fosetyl-Al, 79
Fungal community richness, 75

Fusarium oxysporum, 18, 57, 58, 60
Gaultheria shallon, 4
Germinative hyphae, 40, 41, 44–47, 98
Germ-tube, 40–43, 47, 54
Gigaspora gigantea, 18, 40, 41, 43, 44
Gigaspora margarita, 40, 42, 44, 58
Gigaspora rosea, 17, 19, 40, 42–44, 79
Glomalin, 73, 74
Glomeromycetes *in vitro* collection (GINCO), 2, 97
Glomeromycota, 2, 21, 87
Glomus aggregatum, 3
Glomus borealis, 3
Glomus caledonium, 41
Glomus clarum, 3, 41
Glomus constrictum, 3
Glomus coronatum, 60
Glomus etunicatum, 54, 58, 59, 79
Glomus fasciculatum, 56, 58
Glomus geosporum, 3
Glomus hoi, 3
Glomus intraradices, 3, 19, 26, 40–44, 46, 52, 54, 55, 58, 60, 79, 98, 106, 107, 110, 121
Glomus macrocarpum, 3
Glomus melanosporum, 3
Glomus microaggregatum, 3
Glomus monosporum, 58
Glomus mosseae, 3, 18, 19, 41–44, 54, 56, 59, 60, 79, 99
Glomus radiatum, 3
Glomus versiforme, 42, 46
Glomus vesiculifer, 3
Glucose, 3, 118, 132, 162, 170
Glucose-Yeast-Malt-Extract medium, 118
Glycerol, 95, 132
Gomphidius subroseus, 185
Goodyera repens, 4, 5
Greenhouse gas emissions, 71
Greenhouse trials, 130
Gyroporus cyanescens, 177, 178, 182
Hebeloma crustuliniforme, 118, 131–135
Hebeloma cylindrosporum, 106, 120, 121
Hebeloma longicaudum, 118, 120–124
Hebeloma mesophaeum, 142, 144
Helianthus annuus, 2, 89
Hordeum vulgare, 2, 27, 89, 129
Horse-tailed structures, 46
Host specificity, 4, 75, 93, 116, 117, 158, 159, 177
Hydnum repandum, 179, 182, 186
Hyphal branching, 15–18, 20, 26, 44
Hyphal network, 44, 47, 73, 74, 108
Hypomyces lactifluorum, 181, 182, 184–187
In situ trapping of arbuscular mycorrhizal fungi, 94
In vitro propagation of arbuscular mycorrhizal fungi, 97

In vitro screening of mycorrhizal fungi, 130
Inoculation, 7, 19, 54–57, 60, 72, 75, 89, 96, 98, 106–108, 110, 111, 115, 118–124, 129, 133–135, 139–144, 146–148
Inoculum, 2, 3, 46, 47, 57, 61, 77–80, 93–99, 105–112, 118, 119, 123, 124, 129, 135, 143, 144, 147, 148, 165
Inoculum potential, 2, 3, 78, 94, 98, 105, 111, 119, 123, 129
Inoculum production in the field, 96
Inoculum registration, 61, 105, 110–112
Inoculum storage, 98
Inocybe neotropicalis, 159
Intraradical propagules, 40, 94, 95
Iron deficiency, 145–147
Jack pine, 7, 118–122, 129, 132–136, 164, 167, 177, 180, 184–187
Kresoxim-methyl, 79
Kyoto Protocol, 109
Laccaria bicolour, 106, 118–124, 131–135, 165–171
Laccaria proxima, 120–122
Laccaria trichodermophora, 153
Lactarius deterrimus, 183, 185, 186
Lactarius thyinos, 182, 186
Larix laricina, 176, 177, 185
Larix sibirica, 121, 123, 124
Leccinum atrostipitatum, 178, 182, 184, 185
Leccinum aurantiacum, 182, 184–186
Leccinum piceinum, 182, 184, 186
Leccinum snellii, 177, 183
Leek, 79, 89, 90
Legumes, 15, 16, 19, 20, 23, 25, 26, 28, 29, 56, 73, 75, 77, 89, 106, 155, 159
Lichens, 169, 170, 178, 179
Liquid inocula, 107–110, 118, 144
Lobster mushroom, 181, 182, 184–187
Lodgepole pine, 5, 120–122, 124
Maize, 2, 17, 18, 23, 26, 56, 78, 79, 89, 90
Mannitol, 132
Medicago sativa, 19, 20, 24, 26, 28, 56
Melin-Norkran medium, 118, 131
Mikro-Tek, 107, 109, 123
Molecular signalling, 7, 8
Monokaryotic culture, 144
Monotropa hypopitys, 7, 8
Monotropa uniflora, 7
Monotropoid mycorrhiza, 7
Morocco, 139–143, 148
Mustard, 89
Mycophagy, 158
Mycorrhizal dependency, 87–90
Mycorrhizal inoculum potential, 98, 105, 111, 128
Mycorrhizosphere, 45, 51, 52, 56, 58, 60, 61, 140, 146
NaCl, 129–134
NaOH, 129

Native Plant Industries, 106
Natural mycorrhization, 141, 142
Neea obovata, 155
Negative geotropism of hyphae, 40, 41
Neotropical lowland forests, 153–159
Neotropics, 153, 155, 157–159
N-feruloyltyramine, 42
Nitrogen uptake, 55
Nothofagaceae, 158
Nutrient uptake by arbuscular mycorrhizal fungi, 72
Nyctaginaceae, 155, 157
Oil sands, 47, 107, 122, 123, 129
Oil seed rape, 89
Orchid mycorrhiza, 4, 5
Pathogenic fungi, 15, 18, 78
Paxillus involutus, 118, 121, 123, 124, 165, 171
Pea, 23–25, 60, 79, 89, 90
Pepper (Green/Sweet), 90
Pesticides and arbuscular mycorrhizal fungi, 2, 79, 80, 110
pH, 2, 6, 40, 57, 72, 75, 118, 119, 131, 140, 142, 145–148, 179
Phaseolus vulgaris, 23, 90
Phenology of ectomycorrhiza, 162, 163
Phenology of ectomycorrhizal host plants, 162, 163
Phialophora finlandia, 7, 116
Phosphorus, 3, 55, 56, 87, 88, 90, 148, 163
Phytobial-remediation, 130
Phytohormone, 26–28, 74
Phytophthora fragariae, 57, 58
Phytophthora nicotianae, 57–60
Phytoremediation, 130
Picea abies, 117, 185, 186
Picea glauca, 117, 118, 120–124, 129, 132–136, 162, 168, 185, 186
Picea mariana, 106, 118–121, 123, 124, 162, 163, 166, 177–180, 184–186
Pinaceae, 5, 116, 153
Pinus banksiana, 7, 118–122, 129, 132–136, 164, 167, 177, 180, 184–187
Pinus canariensis, 143
Pinus contorta, 5, 122, 123
Pinus halepensis, 142–146, 148
Pinus pinaster, 142–144, 148
Pinus pinea, 145–148
Pinus strobus, 165–167, 169–171, 176, 177
Pinus sylvestris, 7, 117, 166
Pisolithus tinctorius, 118–121, 123, 124, 144, 148
Pisonia grandis, 155
Pisum sativum, 23, 60, 79, 89, 90
Plant defence, 15, 21, 23, 26, 57, 60
Plant growth promoting rhizomicroorganisms, 53, 54, 60
Plant protection, 93

Platanthera hyperborea, 5
Plot volume index, 121, 122, 124
Polyol, 133, 134
Populus alba, 23, 123
Populus balsamifera, 118, 119, 178
Populus tremuloides, 118, 119, 123, 176
Pot cultures, 45, 46, 56, 96–98, 105, 106
Potato, 2, 59, 89, 90
Pot-culture inoculum, 95
Premier horticulture, 106, 110
Production of ectomycorrhizal basidiomes, 161–171
Proline, 132–134
Propagation of arbuscular mycorrhizal fungi, 94, 95
Propiconazole, 79
Pseudomonas aeruginosa, 55
Pseudomonas chlororaphis, 53
Pseudomonas fluorescens, 58, 60
Pseudomonas putida, 55
Pseudotsuga menziesii, 108, 122
Pterospora andromedea, 8
Pure culture inoculum, 95–97, 118, 144
Pythium ultimum, 57
Quercus suber, 142–144
Reclamation, 107, 115, 119, 122–124, 130, 133–135
Relative field mycorrhizal dependency, 87–90
Resistance to pathogens, 116
Restoration, 105, 115, 124, 129
RH, 3, 40–47
Rhamnetin, 59
Rhizobium meliloti, 56
Rhizoctonia cerealis, 5
Rhizoctonia solani, 5, 59
Rhizoid-like branches, 40, 43
Rhizopogon rubescens, 119–122
Rhizopogon vinicolor, 108, 120–124
Rhizopogon vulgaris, 143, 144, 148
Rhizoscyphus ericae, 3, 4
Rhizosphere, 1, 3, 17, 52, 53, 56, 58, 60, 73, 88, 94–96, 98, 144, 145, 166, 179
Rhizotec Laboratories Inc., 106
Rhodotorula mucilaginosa, 54
Root depletion zone, 72, 98
Root exudates, 16–19, 28, 29, 41, 42, 44, 47, 57, 59, 73
Root-organ cultures, 2, 43, 45, 52, 97, 98, 106, 107, 110
Rozites caperatus, 177–180, 182, 184–186
Runner hyphae, 3, 40–47
Russulaceae, 5, 7, 157, 158, 181–183
Saccharomyces cerevisiae, 54
Saccharomyces kunashirensis, 54
Salal, 4
Salt-stress tolerance, 130, 131
Sarcodes sanguinea, 8
Sarcodon squamosus, 180–182, 184–186

Sclerocystis rubiformis, 3
Scutellospora calospora, 3
Secondary spores, 41
Siberian larch, 121, 123, 124
Single spore cultures, 96
Sinorhizobium meliloti, 59
Soil aggregation, 73, 75
Soil characteristics, 186
Soil disturbance and ectomycorrhizal fungal productivity, 181
Soil microbial biomass, 51, 73, 74
Soil microorganisms, 40, 46, 51–55, 57–61, 72, 73, 98
Soil organic matter, 51, 73
Soil temperature, 163
Solanum lycopersicum, 17, 19, 24, 27, 28, 56, 58–60, 79, 90
Solanum tuberosum, 2, 59, 89, 90
Solid inoculum, 144
Sphaerosporella brunnea, 7, 116, 120
Spore dispersal, 158
Spore dormancy, 40, 98
Spore germination, 15–17, 39, 40, 44, 54, 79
Sporocarp production, 177, 180, 181, 185, 186
Stand age and fruiting of ectomycorrhizal fungi, 180
Strigolactones, 17–20, 44, 45
Submontane forests, 153
Sugar beet, 89
Suillus americanus, 183, 185
Suillus brevipes, 183–185
Suillus cavipes, 177, 183–185
Suillus granulatus, 177, 183–185
Suillus grevillei, 183, 185
Suillus luteus, 117, 183–185
Suillus neoalbidipes, 166, 167, 183–185
Suillus serotinus, 183, 185
Suillus tomentosus, 118, 121, 123, 124, 131–135, 183–185
Sunflower, 2, 89
SymbioTech Research Inc., 115, 123
Thelephora americana, 117, 121, 122, 134
Thelephora terrestris, 117, 120
Thin-walled hyphae, 43, 46
Tillage and arbuscular mycorrhizal fungal communities, 2, 77, 78
Tomato, 17, 19, 24, 27, 28, 56, 58–60, 79, 90
Trehalose, 3, 74, 132, 162
Trichoderma harzianum, 54, 58
Trichoderma pseudokoningii, 54
Tricholoma caligatum, 182, 184, 185
Tricholoma equestre, 177, 178, 182, 184–186
Tricholoma magnivelare, 175, 177, 180–182, 184–186
Tricholoma matsutake, 175
Tricholoma tridentinum, 142
Triticum aestivum, 2, 78, 79, 87, 89, 90

Tsuga heterophylla, 3
Tunisia, 71, 140, 141, 145–149
Understory vegetation, 178
Vaccinium angustifolium, 4
Vesicles, 1, 15, 41, 46, 93, 94, 99
Vicia faba, 23, 90
Water absorption, 116
Water stress, 179
Western hemlock, 3
Wheat, 2, 78, 79, 87, 89, 90
White aspen, 123
White spruce, 117, 118, 120–124, 129, 132–136, 162, 168, 185, 186
Wilcoxina mikolae, 7, 116
Wilcoxina rehmii, 116
Yarrowia lipolytica, 54
Zea mays, 2, 17, 18, 23, 26, 56, 78, 79, 89, 90